Analytic Tomography

This book is about tomography, which is a way to see what is inside an object without opening it up. The unifying idea of tomography is the Radon transform, which is introduced in an informal and graphic way in Chapter 1. The remaining chapters deal with the basic and advanced properties of the Radon transform and related operators.

The book was written to appeal to the broadest possible group of readers. The first chapter, which introduces computerized tomography, x-ray imaging and the Radon transform, requires almost no mathematical background. The second chapter, which is devoted to a rigorous and detailed study of the basic properties of the Radon transform should be accessible to readers with a good undergraduate background in mathematics. The last three chapters are devoted to the more advanced areas of mathematical tomography and the Radon transform. These chapters require a more sophisticated background in mathematics. There are numerous figures and more than 600 references to literature in the field.

Andrew Markoe is Professor of Mathematics at Rider University. He is the author of 19 publications in the areas of several complex variables, Radon transforms, and mathematical tomography.

Analytic Tomography

ANDREW MARKOE

Rider University
Lawrenceville, New Jersey

CAMBRIDGE
UNIVERSITY PRESS

CAMBRIDGE
UNIVERSITY PRESS

32 Avenue of the Americas, New York NY 10013-2473, USA

Cambridge University Press is part of the University of Cambridge.

It furthers the University's mission by disseminating knowledge in the pursuit of education, learning and research at the highest international levels of excellence.

www.cambridge.org
Information on this title: www.cambridge.org/9781107438620

First published 2006
First paperback edition 2014

A catalogue record for this publication is available from the British Library

Library of Congress Cataloguing in Publication data
Markoe, Andrew, 1943–
Analytic tomography / Andrew Markoe.
p.   cm. – (Encyclopedia of mathematics and its applications)
Includes bibliographical references and index.
ISBN-13:   978-0-521-79347-6 (hardback)
ISBN-10:   0-521-79347-5 (hardback)
1. Radon transforms.   2. tomography.
I. Title.   II. Series.
QA672.M37   2005
515'.723 – dc22   2005022524

ISBN 978-0-521-79347-6 Hardback
ISBN 978-1-107-43862-0 Paperback

Contents

Introduction

This book is about tomography, which is a way to see what is inside an object without opening it up. If you are intrigued with this idea, then, no matter what your background, you will find that at least some portion of this book will provide interesting reading. If this idea is not intriguing, then I would recommend some other publication for your reading pleasure.

The unifying idea of tomography is the Radon transform, which is introduced in an informal and graphic way in chapter 1. Reading chapter 1 will give you a good idea of the precise meaning of tomography. Reading chapter 2 will give you a very good idea of the meaning of tomography and if you read the last few chapters you will have a really good understanding of this idea. However, some of the later chapters will only be accessible to specialists.

I tried to write this book with two main ideas in mind. I wanted it to appeal to the broadest possible group of readers and I wanted it to be as comprehensive as possible. Therefore, chapter 1 has almost no mathematics in it – at least it does not require the reader to have any background beyond a good course in secondary school mathematics. CT (computerized tomography) scanners are used for medical diagnosis and produce detailed pictures of the human anatomy without opening up the patient. The dedicated reader will learn, in a very graphic way, how a CT scanner works. I hope this chapter will also be interesting to specialists who will see how the Radon transform and integral convolutions correspond to some familiar everyday processes.

Chapter 2 presupposes some knowledge of calculus. I tried to write it so that a second-year undergraduate student in mathematics or the sciences would have enough background to read the chapter. However, most readers of this chapter should have a few more mathematics courses beyond the elementary calculus sequence. This chapter rigorously introduces the Radon transform which is the basis of the rest of the book. I was somewhat surprised, myself, to see that almost all the basic theory of the Radon transform could be developed with not much more than the change of variables formula for integrals and Fubini's theorem on multiple integration. Students in the basic calculus sequence should know these formulas, at least in dimension two. However, and here is where it requires some dedication, I develop the theory in n dimensions, explaining the necessary generalizations as the chapter proceeds.

Chapter 3 requires at least some graduate-level mathematics. This is where we generalize the Radon transform to the k-plane transform. The resulting analysis is much deeper than that of the first two chapters. The minimal requirements are a knowledge of real analysis (at least through the general theory of Lebesgue measure and integration and some elementary Fourier analysis) and a minimal familiarity with group theory. Readers having this sort of background should include graduate students in mathematics, even just after the first year, and many scientists and engineers. There are many more ideas necessary to understand this chapter, but I have taken care to at least explain the notation and basic concepts and to provide references for the interested reader whose background does not reach this far. There is no reason to expect that even a mathematician who is not a full-time analyst would have a good enough working knowledge of Grassmann manifolds, Haar measure, and distribution theory to really understand this chapter. Therefore, I have tried to err on the side of providing more detail, even if some ideas and arguments could be made more succinct when aimed at a specialist in the field. This will probably annoy most of my colleagues, and for this I hastily offer an apology. However, I think other readers will be thankful for the amount of detail that I have provided.

The material in chapter 3 is essentially self-contained beyond the prerequisites that I just mentioned. Any more advanced ideas are described carefully enough that a mathematically literate reader should be able to follow the arguments, although considerable effort may be required. Except for a set of measure zero, all proofs in this chapter are self-contained.

However, in the remaining chapters I do not always provide full proofs. In general, in these chapters I present some basic ideas with full details and then I expand on these ideas. But I do not always give full, or sometimes even any, proofs. When proofs are omitted I always provide appropriate references to the literature. Therefore, in Chapters 2 and 3 when "Theorem..." occurs, you can almost always be sure to find "Proof... ■" immediately following.[1] However, in subsequent chapters you may find some theorems without proofs. Sometimes I mention this, but, in general, the lack of a proof indicates that the demonstration may be found in the associated reference.

I have provided a brief summary of prerequisites in the introduction to each chapter.

Here are the exceptions to the basic policy that I have just outlined. In chapter 1 there is a technical note that requires knowledge of some elementary calculus. In chapter 3, section 3.10, there are some very interesting results on how the k-plane transform acts on L^p functions. These results require a much more extensive development of the Riesz potential than I was able to provide. In fact, I probably would have required another volume just to provide these prerequisites. Therefore, I took the liberty of stating the main ideas and results about Riesz potentials without proof but with references.

I have tried to make this book a comprehensive treatment of the subject of analytic methods in tomography. However, it was impossible to go into full detail concerning

---

[1] The symbol ■ denotes the end of a proof.

every aspect of this field. Therefore, each chapter has a section titled "Additional References and Results" which can be thought of as a guide to the literature for the reader who wants to delve more deeply into the ideas of the chapter. Most of the material in these sections consists of a reference to the literature with a brief description of the author's contribution. You will occasionally find more detail. Also, in these sections, you will occasionally find historical comments. I am not a historian, but I have tried to make these comments as accurate as possible.

Some topics could, maybe even should, be in this volume, but because of space and time restrictions they were not included. One example that comes immediately to mind is the area of impedance tomography. This area is extremely interesting, valuable, and analytic in nature. However, the background necessary to understand this exceeds both what I expect of the "generic" reader and also my ability to fit this theory into the number of pages and amount of time available to me. Similarly I could only give a brief introduction to the relation between tomography and partial differential equations, twistors, several complex variables and $\mathcal{D}$ modules. I hope that my readers will see that any field not included or only briefly treated is of the nature that would probably require a volume all by itself. However, I have tried to make at least passing mention of any area of tomography that is at all related to analysis.

Tomography, which may be justifiably defined as the study of the Radon transform, is itself part of the field of integral geometry. Tomography is divided into roughly two fields: geometric tomography and analytic tomography. Although geometric tomography is mostly concerned with probing the interior structure of geometric objects by using the techniques of geometry, analytic tomography has the same aim but uses techniques that are intimately related to both classical and modern real analysis, and sometimes also to complex analysis. These techniques include Lebesgue integration, the theory of distributions, and Fourier analysis.

Geometric tomography is treated in the excellent book by Gardner [185], but not at all by me. There are also several texts that deal with the analytic aspects of tomography (see section 2.9.2). There is some overlap between this volume and these other texts, but I believe that this volume has a unique emphasis, choice of topics, and point of view.

Boris Rubin has proposed writing a book with the tentative title "Introduction to Radon Transforms: Real Variable Methods, Integral Geometry and Harmonic Analysis." This work will treat in much more detail some of the more advanced topics of this volume, in particular, those dealing with singular and fractional integration.

A search of the mathematical literature since 1917[2] that has a relation to tomography or the Radon transform will yield well over 2,000 publications. This does not take into account publications in other fields such as medicine, physics, and engineering. I have no idea what the total number of papers on tomography is, but I would not be surprised if it is greater than 10,000. I found about 1,500 papers that might even be remotely

---

[2] 1917 is the year of publication of Johann Radon's ground-breaking paper [508]. Although one can trace the origins of tomography further back, the year 1917 is generally believed to signal the beginning of tomography.

related to this book. Because of the time and page restraints, I had to narrow this down even more. The result is the set of papers that you see in my bibliography. This is a substantial set of references, but clearly it is not exhaustive. Therefore, I apologize to any colleagues who were not mentioned or who received only a passing mention. This set includes many colleagues who work on the more applied area of the subject. Their work is interesting and valuable, but unfortunately I could not include every possible reference in this book.

The reader should be aware that this is a book on pure mathematics. This is inevitable because mathematical analysis depends on infinite processes, whereas any applied mathematical problem eventually has to deal with a finite process. Not many CT scanners exist that need to handle objects defined by general L^p functions. So, if you are planning to build a CT scanner in your garage, this book is not going to tell you how to do it. However, I believe that many engineers, physicists, and applied mathematicians will benefit from the theory that is presented here. For those of you who do want to build a CT scanner in your garage, after you read this book, you should look into Herman [296], Kak and Slaney [328], Natterer and Wübbeling [446], and Epstein [150].

Tomographers tend to be split into two, probably disjoint, classes. Members of the first class believe that nothing practical can come out of a theorem depending on, say, infinitely many x-ray projections. The other class, of which I am a member, disagrees. However, because I respect the opinions of the first class, I therefore abandon all pretense of presenting practical applications, although in a few places I make some remarks heading in that direction. This frees me to present the general theory, which happens to be a beautiful mathematical gem.

Dedication and Acknowledgments

I dedicate this book to my wife Ruth and to my children Ariana, Abigail, and Emily. It is written in honor of my mother Hyacinth Markoe and in loving memory of my father Ralph Markoe, my wife's parents Charles and Rachel Kalisky and my brother-in-law Henry Jones.

I had an enormous amount of help while writing this book. I thank my wife Ruth Markoe for her support and love and for putting up with me during this project. I also thank my children for the same reasons.

I am grateful for the support provided by Rider University in the form of research leaves and financial grants for this project. I am particularly grateful for the help I received from the interlibrary loan department of the Moore Library of Rider University.

The Institute for Advanced Study was gracious in appointing me Director's Visitor for the Spring of 2002. This gave me the opportunity to pursue this research in a very pleasant and productive environment. I thank Phillip Griffiths, the director of the Institute at that time and also Momota Ganguli and Judith Wilson-Smith, of the Institute's library, and Kate Monohan and Linda Geraci of the Director's office.

I thank the Institute of Physics Publishing, in particular, Elaine Longden-Chapman and Lara Finan, for arranging access to back issues of their journal *Inverse Problems*.

I thank the Siemens Corporation for providing images of CT scans and CT scanners.

I thank the Staff at Cambridge University Press and at TechBooks, Inc. for their efforts on behalf of producing this book. I am especially grateful to Jessica Farris of Cambridge University Press and the anonymous copy editor at TechBooks who had to deal with many fatuous errors on my behalf.

My gratitude and a salute go to my students Harry Doctor and Sharon Kobrin who proved that the first two chapters of this book could be read by undergraduates. Also my thanks go to them for help in proofreading those chapters.

Finally, I thank my many colleagues who helped me while I was writing this volume and from whom I learned so much. In particular, I thank Mark Agranovsky, Anthony Bahri, Jan Boman, Rolf ClackDoyle, Richard Gardner, Fulton Gonzalez, Eric Grinberg, Gabor Herman, Sigurdur Helgason, Alexander Katsevich, Fritz Keinert, Peter Kuchment, Rob Lewitt, Erwin Lutwak, Eric Todd Quinto, Boris Rubin, and Elias Stein.

1

Computerized Tomography, X-rays, and the Radon Transform

1.1 Introduction

The purpose of this chapter is to give an informal introduction to the subject of tomography. There are very few mathematical requirements for this chapter, so readers who are not specialists in the field, indeed who are not mathematicians or scientists, should find this material accessible and interesting. Specialists will find a graphic and intuitive presentation of the Radon transform and its approximate inversion.

Tomography is concerned with solving problems such as the following. Suppose that we are given an object but can only see its surface. Could we determine the nature of the object without cutting it open? In 1917 an Austrian mathematician named Johann Radon showed that this could be done provided the total density of every line through the object were known.[1] We can think of the density of an object at a specific point as the amount of material comprising the object at that point. The total density along a line is the sum of the individual densities or amounts of material.

In 1895 Wilhelm Roengten discovered x-rays, a property of which is their determining of the total density of an object along their line of travel. For this reason, mathematicians call the total density an *x-ray projection*. It is immaterial whether the x-ray projection was obtained via x-rays or by some other method; we still call the resulting total density an x-ray projection.

Combining Roengten's x-rays with Radon's idea gives a way of determining an unknown object without cutting it open. We call this process *tomography*.

Tomography can be applied to any object for which we can determine the x-ray projections either by actual x-rays or some other method. Tomography is used to investigate the interior structure of the following objects: the human body, rocket motors, rocks, the sun (microwaves were used here rather than x-rays), snow packs on the Alps, and violins and other bowed instruments. This list could be expanded to hundreds of objects. In this chapter we will see how tomography can be used to obtain detailed information about the human brain from its x-ray projections.

---

[1] Johann Radon (1887–1956) published the first discussion and solution of a tomographic problem (see reference [508] in the bibliography).

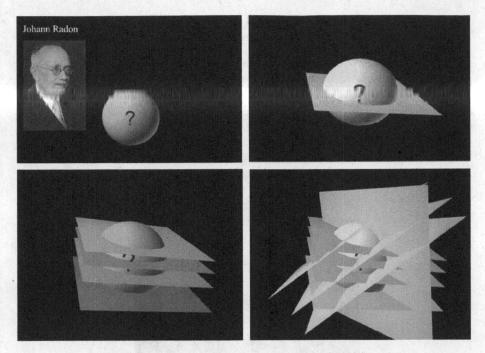

Figure 1.1. Johann Radon tries to figure out what is inside the sphere.

Allan M. Cormack and Godfrey N. Hounsfield shared the 1979 Nobel Prize in Physiology and Medicine for their contributions to the medical applications of tomography. The reference to snow packs comes from Cormack's Nobel prize lecture in Stockholm in 1979 (compare [102]).[2] Cormack remarked that the publication of his ideas on tomography took place in 1963 and 1964 and that *"There was virtually no response. The most interesting request for a reprint came from the Swiss Centre for Avalanche Research. The method would work for deposits of snow on mountains if one could get either the detector or the source into the mountain under the snow!"*

Radon not only showed how to determine a plane object from lines, but he also showed how to determine a solid object by using planes. We can visualize the discussion up to this point. In figure 1.1 Johann Radon is pondering what is inside the spherical object. In the next scene he decides to compute the total density on a single plane through the sphere. He knows that this is not enough information to determine the object, so he successively intersects with more and more planes. When he has collected the densities on all planes, then he is able to determine the object. How this may be done by using lines through a two-dimensional object is the subject of the remainder of this chapter. You do not need much background in mathematics to read this chapter – some knowledge about triangles and the ability to read a graph is really all that is required.

[2] Numbers in square brackets correspond to the list of references at the end of the book. For example, [102] refers to the article by Cormack that is listed in the references section.

1.2 Computerized Tomography (CT) and Mathematical Tomography – "Now, suddenly, the fog had cleared"

The Greek word $\tau o \mu o \sigma$ (tomos), meaning slice, is the source of the term *tomography*. This term was first used in diagnostic medicine. Since the discovery of x-rays by Roentgen, diagnosticians have attempted to produce images of human organs without the blurring and overlap of tissue that occurs in traditional x-ray pictures, such as the x-ray of the skull in the accompanying figure.

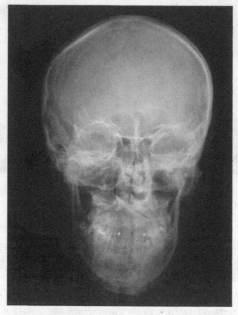

Courtesy of Ass. Prof. Dr. Mircea-Constantin Sora, MD, Ph. D., Medical University of Vienna.

We will see that tomography can produce much more detailed pictures from x-ray data. The reference to the fog clearing in the title of this section is from the presentation speech for the 1979 Nobel prize for Physiology or Medicine which was awarded, jointly to A. M. Cormack and G. N. Hounsfield in 1979. The presentation speech containing the preceding quotation was delivered by Professor Torgny Greitz of the Karolinska Medico-Chirurgical Institute and it is interesting to read the excerpts from this speech in Section 1.10.1.

Computerized Tomography, also known as *CT*, refers to the actual process of producing a detailed picture of the interior of an organism by using x-rays. *Mathematical tomography* refers to the mathematical process by which the picture is obtained. Computerized tomography is accomplished by designing a machine consisting of x-ray sources and x-ray detectors combined with a computer. The computer uses an algorithm adapted from the field of mathematical tomography to combine the data obtained from the x-rays into a detailed picture of the organs and tissue in a specific slice of a

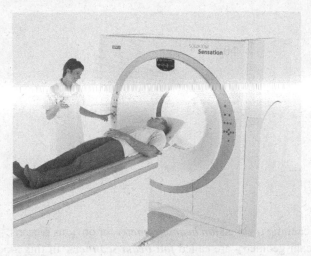

Figure 1.2. A typical CT scanner. This one is manufactured by the Siemens Corporation. Courtesy of the Siemens corporation.

patient's body. This type of machine is called a **CT scanner**.[3] A CT scanner can produce a clear and detailed image, called a **tomogram**, of the interior of a human body. This is done without cutting open the body, merely by sending x-rays through the tissue in question. How this is done is explained later in this chapter.

Some forms of tomography were used in diagnostic medicine long before computers were invented (see Section 1.10). A typical method attempted to visualize a section (slice) of a body by blurring out all the x-rays except those in the focal plane of the desired slice. Early CT scanners also concentrated on a single slice of the body. This attention on a slice (from $\tau o\mu o\sigma$) explains the origin of the word tomography in medicine. The desire was to visualize a sliced human body without actually slicing it. In mathematical tomography the slicing refers to the lines or planes that slice through the object of interest.

Figure 1.2 is a picture of a typical CT scanner.

The circular ring in the CT scanner emits x-rays from a source on one side. These x-rays are detected at the opposite side. The ring rotates so that x-rays can be beamed, in any direction, through a specific slice of the patient's body. Here is a diagram of how this operates.

---

[3] A CT scanner is also referred to as a CAT scanner, which is derived from "computer-assisted tomography," whereas CT derives from "computerized tomography." The preferred term is CT scanner, although CAT scanner is informally and ubiquitously used.

There is the story of the man who brought his sick dog to the veterinarian. Upon examination, the veterinarian pronounced the dog dead. The distraught owner replied: "That is impossible, I know my dog is listless, but certainly not dead. Is there not a more definitive test that you can do?" "Very well," replied the veterinarian, who immediately summoned a black-and-white cat. The cat proceeded to examine the dog. First, the cat only sniffed around the dog who exhibited no reaction. Then the cat hissed at the dog and finally clawed it, all without reaction from the dog. The owner finally said, "I suppose you are right, my dog is dead. How much do I owe you?" The veterinarian replied, "That will be $300." The owner retorted. "Three hundred dollars to tell me my dog is dead! That is outrageous! Why is it so much?" "The veterinarian replied, "It is $100 for the examination and $200 for the CAT scan."

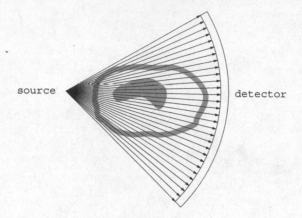

source   detector

This mode of scanning is called *fan beam geometry* for obvious reasons.[4] CT scanners that use fan beam geometry are called *fan beam scanners*. In this mode x-rays are generated at the source. They form a beam in the shape of a fan and are observed at the detector after passing through the body. In this way the total density along every line emanating from the source can be computed. By rotating the apparatus, the source and detectors move to new positions. In this way the total density along every line intersecting the body can be determined. These total densities are the input data to an algorithm that reconstructs a picture of the organs and tissue in this slice. Later we will show how these total densities can be used to reconstruct an image of the tissue. Meanwhile, see figure 1.3 for a comparison of a traditional x-ray of the head and a tomogram of a section of a human brain. Note the lack of detail of the

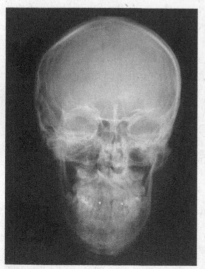
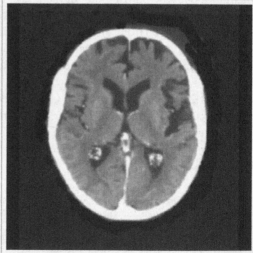

Figure 1.3. (*Left*) traditional x-ray. (*Right*) Tomographic reconstruction of a brain section. Image on the left courtesy of Ass. Prof. Dr. Mircea-Constantin Sora, MD, Ph. D., Medical University of Vienna.

---

[4] The analogous situation in three dimensions is called *cone beam geometry*.

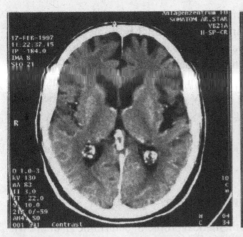
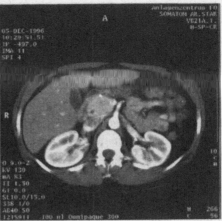

Figure 1.4. CT images. The left image is a cross section of a human brain. The right image is a cross section of a human abdomen. Courtesy of the Siemens corporation.

brain in the traditional x-ray compared with the fine detail in the tomogram. Another set of tomograms is in figure 1.4.

When an x-ray beam is sent through tissue, it experiences more attenuation by heavier tissue than by lighter tissue. For example, the skull is about twice as dense as the gray matter of the brain. Therefore, x-rays are more likely to be absorbed or scattered when passing through the skull than when passing through gray matter. Although there are some subtleties with this idea, we can make a working assumption that sending an x-ray beam through an object determines the total density of the object on the line intersected by the x-ray.

Before continuing we should mention that older CT machines used a parallel beam geometry. Their mode of operation is illustrated by the following figure.

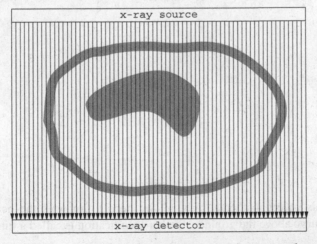

It is much more efficient to use fan beam geometry and most modern CT scanners use this method. In either method we can obtain information about the density of the object along any line, provided that the scanner is free to rotate through 180°. It is simpler to describe the mathematics for parallel beam scanners and from now on we will do so. The method of reconstructing images via parallel beam geometry can be applied to fan beam scanners because, as we already noted, information about any line intersecting the object can be obtained in either geometry. However, the computational effort in reorganizing the fan beam data into parallel beam data is substantial. There are algorithms that use the fan beam data directly and these will be described in a later chapter. The efficiency of fan beam algorithms together with the efficiency of the fan beam scanning geometry make current CT scanners much faster than older ones. The latest generation of CT scanners emit x-rays along a helical path and reconstruct three-dimensional pictures.

1.3 Objects and Functions

Tomography is an example of a classical mathematical problem: determine an unknown quantity when some given information is provided. The unknown quantity might be a real number x, which is in some relation to some known real numbers, for example, $5x + 1 = 7$. In other situations the unknown quantity might be a function with some given information about its behavior. For example, determine the unknown position of a particle given its acceleration, its initial position, and its initial velocity. In a tomographic problem the given information is a set of x-ray projections of an unknown object. The solution is an exact or approximate representation of the unknown object obtained by mathematical manipulation of the known x-ray data.

At this point we need a precise definition of the term "object." Let us take a simple object, say a two-dimensional image of the profile of a mountain:

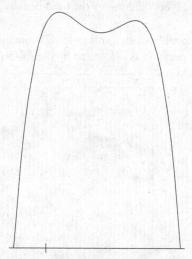

To specify this object mathematically, all we need to know is the height of each point of the curve above the ground.[5] Such a specification is called a function. Many

---

[5] In this example the exact height is given by $-\frac{1}{4}x^4 - \frac{3}{8}x^3 + \frac{1}{2}x^2 + \frac{3}{8}x + \frac{27}{4}$, where the unit for x is denoted by the mark on the ground line in the picture.

two-dimensional shapes can be represented by functions in this way. In general, a *func-tion* is a rule that uniquely assigns a value to each element of a given set. The given set is called the *domain* of the function. In this chapter we assume that the value assigned to an element of the domain is a real number.

An example of a function is given by the rule f which assigns $\frac{1}{x}$ to every nonzero real number x. Here the domain is the set of non-zero real numbers. We can define this rule by writing the equation

$$f(x) = \frac{1}{x}$$

Note that we use a letter, in this case f, to represent the rule or function. Then, for each x in the domain, $f(x)$ represents the quantity obtained by applying this rule to x. The symbol $f(x)$ is read as "the value of f at x" or, in brief, "f of x." It is important to conceive of a function as a single object and not to confuse f with $f(x)$. However, sometimes we are sloppy and use $f(x)$ to denote the function f, even though $f(x)$ actually is a real number representing the *value* of the function f at x.

It is not necessary for the domain of a function to consist of numbers. For example, we can consider the function h, which assigns to every horse its weight in kilograms. In this case the domain is the set of all horses.

In the profile of the mountain, the domain was one-dimensional (the ground line). The function f contained all the information needed to describe the mountain's profile. Although the profile of the mountain is two-dimensional, the amount of information needed to determine the profile is only one-dimensional. This is because the function describing the elevation is of the form $f(x)$, where x is a single variable that moves along a straight line.

In general, a two-dimensional object will require two variables to be completely determined. A general point in the plane is uniquely determined by the coordinates (x, y). Therefore, we can treat more general two-dimensional objects by specifying a function of two variables: $f = f(x, y)$.

Figure 1.5 is an image of an abdominal section of a human patient. The density of the tissue at each point is depicted by the amount of gray at that point. The black points are the most "gray" and represent zero density. The actual tissue has varying density. The highest-density points are the least "gray" (white) and denote bone. The other tissue is less dense and is represented by various shades of gray. To describe this picture, all we need to know is the relative amount of gray to put at each point. This amount can be specified by a real number. Therefore, for all practical purposes, this image can be represented by a function f of two variables: $f(x, y)$ represents the amount of gray to place at the point (x, y) to create this picture.

An object can thus be viewed as a function of two real variables x and y, because we are uniquely assigning a real number (a gray value) to every point (x, y) of the plane. Conversely, if we are given a function of two variables, then we can view the associated object by using the value of the function at every point as a gray value. So from the mathematical point of view, there is no difference between a function of two variables and a two-dimensional object. For this reason we use the term "object" and

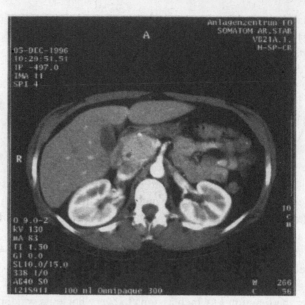

Figure 1.5. Tomogram of a human abdominal section. Courtesy of the Siemens Corporation.

the term "function" interchangeably. Also, for this reason we use letters like f and g to represent objects.

There are two main ways to exhibit the value of a function at a point. The first way is called a **density plot** and it attaches a gray level to each point in the domain. An example of a density plot is the tomogram in figure 1.5. Each gray level represents a specific real number, the lowest values of the function shown in black and the highest values shown in white.

The gray scale presented in the next figure shows how any real number between 0 and 1 can be represented as a gray level. It is not a function, it only serves to establish the correspondence of gray levels to a range of real numbers. The gray scale plays the same role as the x axis in a graph – it shows how we represent real numbers. One purpose of the gray scale is to establish the range of numbers used in the graph of the object. The range does not have to be from 0 to 1. It could be from any real number a to any larger real number b. However, the smallest value will always be represented as black and the largest as white. After this example, we will not be fussy about the actual range of values, so we will present objects without the accompanying gray scale.

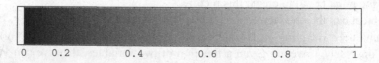

The other way of graphing an object is to place a point of height $f(x, y)$ above the location (x, y) in the plane. This type of view is called the **graph** (of the object). An example of a graph of an object may be found in figure 1.6.

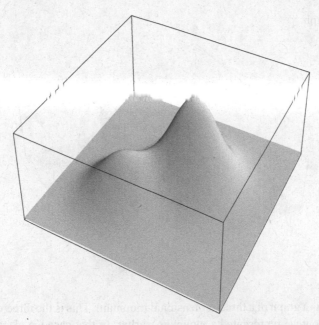

Figure 1.6. Graph of a function representing a mountain.

Simple objects may be represented by functions that take the value 1, represented by white, at all points that lie on the object, and that take the value zero, represented by black, elsewhere. Therefore, their density graphs will be exactly the shape of the object. To avoid becoming overly wordy let us agree that when we use a term such as triangular object we really mean the function that is 1 on the triangle and zero elsewhere.

Here is the density plot of a square object.

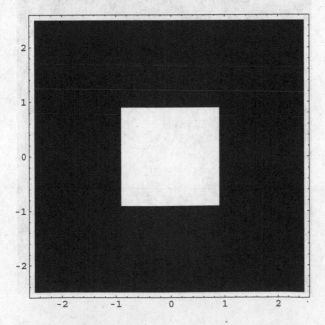

Here is its graph.

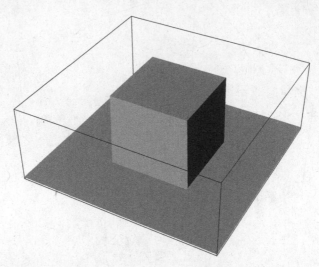

Figure 1.6 is a graph of a three-dimensional mountain. This is the three-dimensional analogue of how we represented a mountain earlier. In that case we placed the value $f(x)$ above the location x on a line. Here we place the height $f(x, y)$ above each point (x, y) in the plane.

Note that the graph of the mountain appears three-dimensional (as it should). But it actually represents a function whose domain is two-dimensional. Next we show the density plot of the mountain. It is similar to contour plots of mountainous areas and is plotted on a two-dimensional domain.

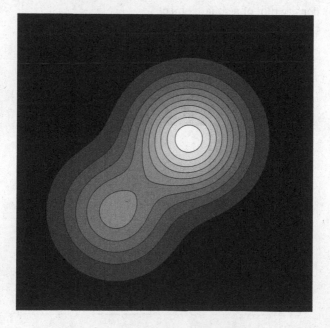

Next we show a density plot representing a transverse slice of a human cranium.

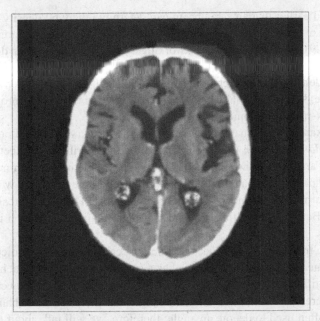

The next figure is a graph of the same cranial section. It appears three-dimensional, but because it represents a transverse section of a cranium, it actually corresponds to a two-dimensional object.

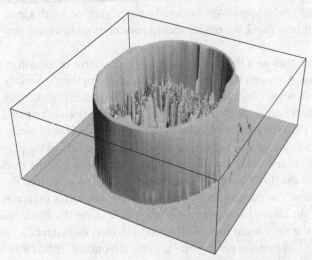

It is clear from these images that for some objects the density plot is a better representation than the graph. This is true for the square object and the brain image. On the other hand other objects may be better represented as graphs, as for the mountain. Most of the objects we need to consider will be represented as density plots. However, the graphical representation is also useful and will be used if needed.

1.4 Tomography and the Radon Transform

We mentioned earlier that an x-ray beam can determine the total density of an object along the line of travel of the x-ray. We will investigate this process in more detail later (see section 1.9). For now we make the working assumption that one can determine the total density, or amount of material, along any line intersecting a specific object. Recall that the total density of an object along a line is called an x-ray projection of the object along that line. It is immaterial whether the knowledge of these x-ray projections was obtained via x-rays or by some other method.

We now examine the situation when the densities are known for all possible lines. The term **Radon transform** (of an object) is used to describe the collection of x-ray projections of an object along all possible lines.[6] The problem of determining an object from its x-ray projections can therefore be formulated in the following way. Given the Radon transform of an unknown object, find the object. This is a typical problem in tomography. Before solving this problem, let us investigate the Radon transform in more detail.

The Radon transform of an object or function[7] consists of x-ray projections along all possible lines in the plane. Each line has a specific direction and each direction is uniquely identified by an angle. Conversely, given an angle, we can specify a unique direction. Therefore we can use the terms angle and direction interchangeably. The symbol θ is often used to denote an angle; therefore, we will talk about directions θ. We specify directions in radians that we usually express in multiples of π. Recall that a full circle can be measured as an angle of 2π radians, which is the same as 360 deg. Therefore, $\frac{\pi}{4}$ represents a 45° angle because $\frac{\pi}{4}$ is exactly one eighth of a full circle measured in radians, whereas 45° is exactly one eighth of a full circle measured in degrees. With this in mind, the reader should have no trouble visualizing any angle or direction.

We now show that we can think of the Radon transform of an object as a function of two variables. Although there are many ways of doing this, we choose the method based on the following conceptual diagram of a CT scanner (fig. 1.7): In this diagram θ denotes the direction made by the motion of the source-detector array. This direction is indicated by the arrow perpendicular to the x-ray beam. If we specify a direction θ and a distance s from the origin, then there is exactly one line in the plane that is both perpendicular to θ and at the specified distance s. In figure 1.8, the thick solid line is perpendicular to the direction $\theta = \frac{\pi}{4}$ and at the distance $s = \frac{1}{2}$ from the origin of the coordinate system. The coordinate axes are dashed and an arrow indicates the direction of the angle θ. We intend to take the x-ray projection along this thick line.

If we move the thick solid line parallel to itself, then its distance s from the origin changes, but the line remains perpendicular to the direction θ. In this way we can obtain

---

[6] In mathematics, when a function is defined with a domain consisting of other functions, then the new function is called a **transform**. Because the Radon transform assigns a set of x-ray projections to an arbitrary object (i.e., function), it is a function operating on other functions, so it is considered to be a transform.

[7] Recall that, mathematically speaking, "object" = "function."

Direction of motion

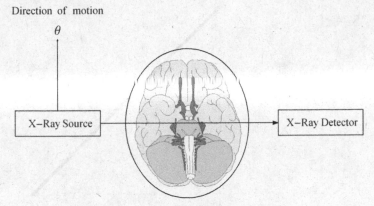

Figure 1.7.

the x-ray projections on all lines perpendicular to θ. Then by letting the direction θ vary over a range of $180°$ we can obtain the x-ray projections of an object over all possible lines in the plane.[8] Hence, we will have the Radon transform of the object.

It may seem unusual to let the angle θ represent the direction perpendicular to the direction of the x-rays, but we see that this is exactly the mechanism in figure 1.7, which is typical of early CT scanners.

This shows that we can view the Radon transform of an object f as a function Rf of two variables (θ, s): the symbol $Rf(\theta, s)$ denotes the x-ray projection of the object f along the line that is perpendicular to θ and which is located s units from the origin.

More informally, if we know the Radon transform of an object f, then we have a way of knowing the density of f along any line. Later on we will provide a graphic way of describing the Radon transform.

The notation $Rf(\theta, s)$ is most useful when we want to consider all the possible x-rays. It is useful to introduce a variation of this notation when we are dealing only with the x-rays in a single direction. For a given, fixed, direction θ, define

$$R_\theta f(s) = Rf(\theta, s)$$

This is the function of one variable which takes s into $Rf(\theta, s)$ which is precisely the **x-ray projection of f in the direction orthogonal** [9] **to** θ. Therefore, $R_\theta f(s)$ represents the total density of f along the line which is perpendicular to θ and which is located s units from the origin.

For example, consider the line that is perpendicular to $\theta = \frac{\pi}{4}$ and located a distance of $s = \frac{1}{2}$ units from the origin, as in figure 1.8. Supposing that the total density of an

---

[8] We also allow s to be negative. In this case we interpret the distance s from the origin to be in the opposite direction of θ. In this way varying θ over $180°$ gives all the lines in the plane.

[9] It is common to use the term "orthogonal" as a synonym for "perpendicular."

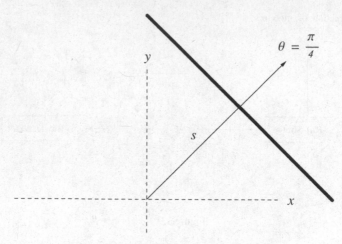

Figure 1.8. Geometry of x-ray projections.

object f along this line were 3.5, then we would write

$$R_{\frac{\pi}{4}} f \left(\frac{1}{2} \right) = 3.5$$

This is exactly the same quantity as $Rf(\frac{\pi}{4}, \frac{1}{2})$ but in a different notation.

Now we take a simple object, a square, and see how to compute its Radon transform. We begin by computing a single x-ray projection. Let f be the function representing the object that is a square of side 2 centered at the origin. Recall the earlier discussion in which we represented simple objects by functions that take the value 1, represented by white, at all points that lie on the object and that take the value 0, represented by black, elsewhere. Figure 1.9 shows a density plot of this square object. Also shown is the line perpendicular to $\theta = \frac{\pi}{4}$ at a distance $s = \frac{1}{2}$ from the origin. The value of the x-ray projection of f can then be easily calculated. Because black areas represent 0 and white areas represent 1, we can compute $Rf(\theta, s)$ by measuring the length of the intersection of the line with the white square. This is because the total density at each black point is 0 and at each white point is 1.

The equation of the given line perpendicular to θ and distance $s = \frac{1}{2}$ from the origin is $y = \frac{\sqrt{2}}{2} - x$. This line enters the square at the point $(1, \frac{\sqrt{2}}{2} - 1)$ and exits the square at the point $(\frac{\sqrt{2}}{2} - 1, 1)$. The x-ray projection of the square along this line is thus the distance between these points,[10] so $Rf(\frac{\pi}{4}, \frac{1}{2}) = 2\sqrt{2} - 1 \approx 1.8$. It would be more difficult to compute the x-ray projection of objects with varying density, but

---

[10] Recall the distance formula between points $P = (x_1, y_1)$ and $Q = (x_2, y_2)$:

$$d(P, Q) = \sqrt{(x_1 - x_2)^2 + (y_1 - y_2)^2}$$

Applying this formula to $P = (1, \frac{\sqrt{2}}{2} - 1)$ and $Q = (\frac{\sqrt{2}}{2} - 1, 1)$ gives the resulting distance. $\sqrt{9 - 4\sqrt{2}}$ which simplifies to $2\sqrt{2} - 1$.

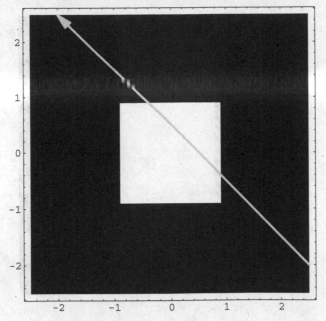

Figure 1.9. A single x-ray projection.

the basic principle is the same. In CT this computation is automatically derived from the x-ray data.

Using the x-ray projection notation, $R_\theta f(s)$, for $\theta = \frac{\pi}{4}$ and $s = \frac{1}{2}$, we have

$$R_\theta f(s) = R_{\frac{\pi}{4}} f\left(\frac{1}{2}\right) = 2\sqrt{2} - 1$$

Other x-ray projections can be computed in a similar way. The easiest example is for $\theta = 0$, because all lines perpendicular to $\theta = 0$ are vertical. The x-rays are therefore all vertical and they either intersect the square with the identical length 2 or else they completely miss the square. Therefore,

$$R_0(s) = 2$$

for any s between -1 and 1, whereas $R_0(s) = 0$ for any other real number s.

Figure 1.10 illustrates a sample of the full x-ray projection, this time in the direction orthogonal to $\frac{3\pi}{4}$. Using the notation for projections we can then say that this figure represents the x-ray projection $R_{\frac{3\pi}{4}} f$ where f is the square object.

All the x-rays in figure 1.10 are in the 45° direction, because they are perpendicular to $\theta = \frac{3\pi}{4} = 135°$. When we look at this figure we see that we have $R_{\frac{3\pi}{4}} f(s) = 0$ for very negative values of s (those less than $-\sqrt{2}$, represented by the lines in the black, upper left area of fig. 1.10, recall that s represents the signed distance of the x-ray from the origin). This is because in this area of the figure, f has the value 0 and hence the

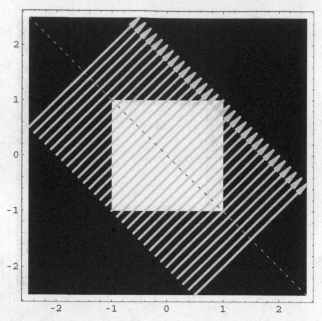

Figure 1.10. A complete projection, $R_\theta f$ orthogonal to $\theta = \frac{3\pi}{4}$.

total density of f along any of the x-rays in this area is also 0. Once $s = -\sqrt{2}$, the x-rays begin to intersect the white square. At first there is not much contribution to the x-ray projections, because for values of s greater than but close to $-\sqrt{2}$ the x-rays do not traverse very much of the white square. But once $s > -\sqrt{2}$, the values of $R_{\frac{3\pi}{4}} f(s)$ start to increase until they reach a maximum, when $s = 0$ (this will give a diagonal of the square). Then the values decrease in a symmetric way until $R_{\frac{3\pi}{4}} f(s)$ again becomes zero at $s = \sqrt{2}$. At this point the x-rays again do not intersect the square, so that the x-ray projections remain 0 for $s \geq \sqrt{2}$.

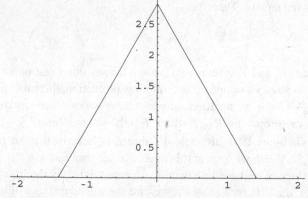

Figure 1.11. Graph of the x-ray projection $R_\theta f$ for $\theta = \frac{3\pi}{4}$.

Figure 1.11 shows the graph of $R_{\frac{3\pi}{4}} f$. This graph is plotted by computing the x-ray projection $R_{\frac{3\pi}{4}} f(s)$ for each real number s and then placing a point on the graph at a height of $R_{\frac{3\pi}{4}} f(s)$ units above the point s on the horizontal axis. You can see the increase in the values of $R_{\frac{3\pi}{4}} f(s)$ as s goes from $-\sqrt{2}$ to 0 as predicted. Also note the decrease as s goes from 0 to $\sqrt{2}$.

The next step in the study of the Radon transform is to devise a way of visualizing the entire Radon transform, instead of a single projection. The term *sinogram* is used for the density plot of the Radon transform.[11] Therefore we can create a sinogram by creating a density plot of the function of two variables $Rf(\theta, s)$.

Here is a figure showing the sinogram for the square object under consideration. The gray scale is set up so that black represents 0 and white represents $2\sqrt{2}$. The θ axis, which represents the directions, is horizontal and the s axis, which represents the distance from the origin, is vertical. To create this sinogram we merely have to place an appropriate gray value at the point located at the coordinates (θ, s). This gray value corresponds numerically to $Rf(\theta, s)$.

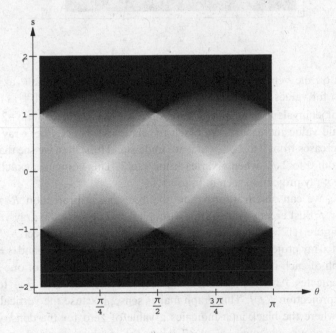

Conversely, we reverse this process to read a sinogram. If we want to know the numerical value of $Rf(\theta, s)$, then we locate the point (θ, s) in the sinogram and we convert its gray value to a real number that is therefore equal to $Rf(\theta, s)$. This is illustrated in the next figure.

---

[11] Perhaps this unusual terminology has to do with the resemblance of a typical sinogram to sine waves. Maybe a better term would have been "Radonogram."

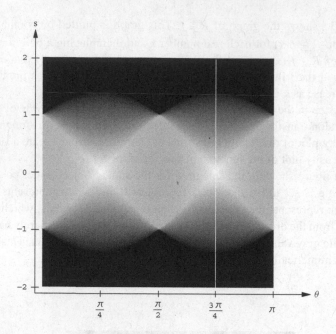

Every point on the vertical white line through the point $\frac{3\pi}{4}$ on the θ axis is of the form $(\frac{3\pi}{4}, s)$ for values of s between -2 and 2. The gray value at $(\frac{3\pi}{4}, s)$ represents $Rf(\frac{3\pi}{4}, s)$, or, equivalently, the x-ray projection $R_{\frac{3\pi}{4}} f(s)$. Between $s = -2$ and $s = 0$, the gray scale value increases from black to white and, hence, the x-ray projection $R_{\frac{3\pi}{4}} f(s)$ increases from 0 to $2\sqrt{2}$. Once we understand this, then we see that $R_{\frac{3\pi}{4}} f(s)$ decreases from 0 to $2\sqrt{2}$ when s varies from 0 to 2. This coincides exactly with the graph of the x-ray projection seen in figure 1.11.

Therefore, we can obtain information about any x-ray projection $R_\theta f$ from the sinogram for f. Just erect a vertical line through θ and observe the gray values along this line. In figure 1.12 we again have the sinogram for the square object f. Also shown are graphs of x-ray projections in the three directions $\theta = \frac{\pi}{6}, \theta = \frac{\pi}{4}$, and $\theta = \frac{2\pi}{3}$. Note that the graph of each individual x-ray projection forms a function of one variable as the theory indicates. An arrow leads from the vertical line through $\theta = \frac{\pi}{6}$ to the graph of the x-ray projection $R_{\frac{\pi}{6}} f$. This graph makes sense, because the vertical line starts at $s = -2$, where the black area indicates a value of zero for the density. The first nonzero value occurs at about $s = -1.37$, where the graph starts getting progressively less black. Therefore, the values of $R_{\frac{\pi}{6}} f$ increase until a maximum is reached at about $s = -0.37$. Then the densities are constant until about $s = 0.37$ at which point they decrease to zero at about $s = 1.37$. This gives rise to the trapezoidal graph for $R_{\frac{\pi}{6}} f$. You can check the plausibility of the other two x-ray projection graphs in the same manner.

Figure 1.13 contains examples of two human sinograms. Both plots represent the Radon transform for a section of a human body. One of these gives x-rays for a human abdominal section and the other for a human brain section. But which is which? And

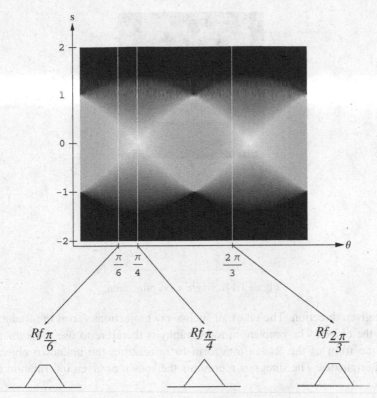

Figure 1.12. Sinogram generated by the Radon transform of the square object, with graphs of three x-ray projections.

what does each organ look like? It is hard to tell directly from the given information. We will determine the answer later.

So far we have seen that an x-ray projection, $R_\theta f$ orthogonal to a given, fixed direction θ, is a function of one variable representing the total densities of the object

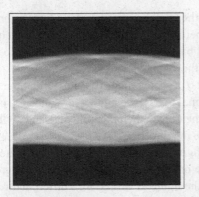
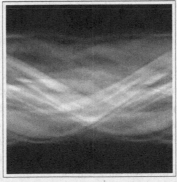

Figure 1.13. One of these graphs represents the x-rays through a human abdomen, the other represents the x-rays through a human brain. Which is which?

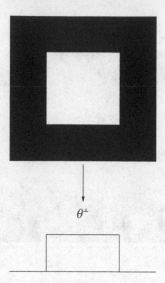

Figure 1.14. Single x-ray projection.

f in the given direction. The set of all such x-ray projections forms the Radon transform of the object. The problem in tomography is therefore to use the known x-ray data in the form of the Radon transform to reconstruct the unknown object from which it originates. The sinogram represents the known or given information for this problem.

Soon we will give an explanation of how such a reconstruction can be accomplished. The explanation is based on the method used by CT scanners in medical applications. However, we first introduce an intuitively plausible method, which appears, at first glance, to yield a reasonable reconstruction from x-ray projections in a very simple manner.

1.5 Backprojection

The x-ray projection $R_\theta f$ of an object f is the total density of f along lines orthogonal to the direction θ. We now create a dual operation, called backprojection, which takes a certain amount of material and smears it backward along lines orthogonal to θ. Let us denote the direction orthogonal to θ by the symbol $\theta^\perp$.

Let us think of the material in the object f as being made of sand. Orient the object so that gravity acts in the direction orthogonal to θ. If we allow the "sand" to spill out of the object and collect on a floor below the object, then this pile of "sand" will look like the x-ray projection $R_\theta f$ orthogonal to the direction θ.

Figure 1.14 illustrates this idea. In this figure we have taken $\theta = 0$. The material in the square object has been projected in the orthogonal direction $\theta^\perp$, thereby creating the pile of material shown at the bottom of the figure. If we choose a different direction, say $\theta = \frac{\pi}{4}$, then the situation looks like this (again we have oriented the object so that gravity acts in the direction $\theta^\perp$):

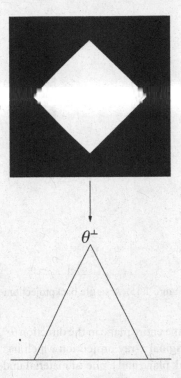

Here we get a triangular pile of "sand." This can be verified by imagining the material in the object to pour "downward" (actually in the direction $\theta^{\perp} = -\frac{\pi}{4}$).

The next figure illustrates this idea for several directions. Each "pile" has been formed by projecting material in the white square parallel to the direction $\theta^{\perp}$ between the center of the square and the center of the "pile." Each pile therefore represents the graph of an x-ray projection of the form $R_\theta f$.

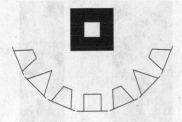

We now describe the process of backprojection. The process is dual to the operation of taking projections. Instead of starting with an object in the plane and creating a pile of material underneath, we start with a pile of material and create an object in the plane. The object is created by smearing the material back into the plane and for this reason the process is called backprojection.

Let g be the function representing the pile of material. Figure 1.15 displays the result of backprojecting this function. The pile of material has been smeared, or

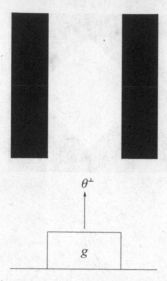

Figure 1.15. A single backprojection.

backprojected, throughout the entire plane in the direction $\theta^\perp$. The white stripe consists of the material from the original x-ray projection which has now been backprojected. Here we started with a blank plane and a pile of material underneath and we created an object, the white stripe, by the process of backprojection, in one direction. This is called **backprojection in one direction**. If the function representing the material available to be backprojected is denoted by the symbol g, then $R_\theta^\# g$ denotes its backprojection in the direction orthogonal to θ. The part of the symbol denoted by $R^\#$ is meant to remind us of a connection to the Radon transform, although the connection is not yet apparent.

Here are some more examples of this idea.

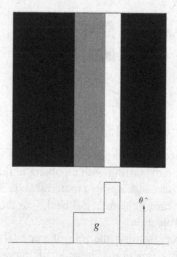

Note that the function g takes on three values: it is 1 between $-\frac{1}{2}$ and $\frac{1}{2}$, 2 between $\frac{1}{2}$, and 1, and 0 elsewhere. After backprojecting we therefore obtain four strips: two are black, corresponding to the zero values; one is gray, representing the value 1 for all points with t — coordinate between $-\frac{1}{2}$ and $\frac{1}{2}$; and one is white, corresponding to the value 2 for points with t — coordinate between $\frac{1}{t}$ and 1.

Here is one more example of a backprojection in one direction. In this case the function g takes on all values from 0 to 1:

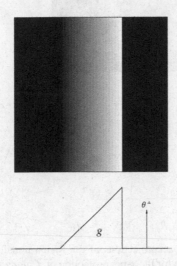

In the backprojection you can see the variation from black to various shades of gray to white and then back to black corresponding to the change in the function g from 0 to 1 and back to 0.

We would now like to perform backprojection in several directions. It makes sense to average the contributions from all these backprojections, otherwise the reconstructed object would grow to be too large. The process of averaging several backprojections from different directions is simply called **backprojection**. We use the notation $g = g(\theta, t)$ for the functions involved in the backprojection process. If we fix a particular θ, then $g(\theta, t)$ is a function of one variable that plays the role of g in backprojection in a single direction. This device allows us to backproject in any direction.

Let us now backproject the Radon transform of a simple object: the square f of side 2 centered at the origin. We saw how to compute the Radon transform in Section 1.4 and we visualized the backprojection in a single direction in figure 1.15. Let us now start backprojecting in more directions.

The following diagram shows the effect of averaging the backprojections from three directions. Note that the functions being backprojected are now x-ray projections

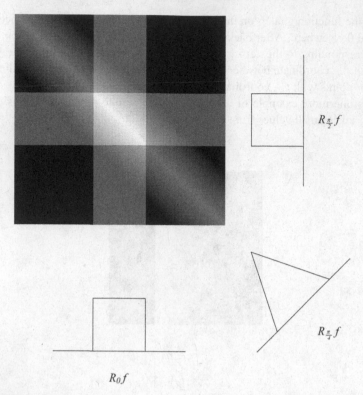

$$R_{\frac{\pi}{2}}f$$

$$R_{\frac{\pi}{4}}f$$

$$R_0 f$$

Figure 1.16. Three averaged backprojections of the Radon transform of the square object f. Recall that $R_\theta f$ is the notation for the x-ray projection of f along the line perpendicular to the direction θ. Here we have $\theta = 0, \frac{\pi}{4}, \frac{\pi}{2}$.

orthogonal to angles 0, $\frac{\pi}{4}$, and $\frac{\pi}{2}$, namely the projections $R_0 f$, $R_{\frac{\pi}{4}} f$, and $R_{\frac{\pi}{2}} f$, where f is the square object introduced before.

Here are the results of backprojecting in 10 and then in 17 directions.

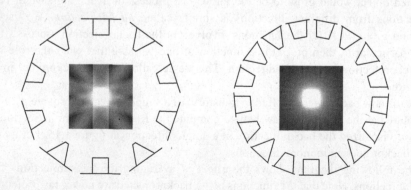

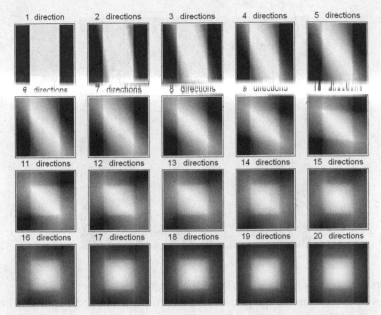

Figure 1.17. Successive backprojections of the Radon transform of a square.

It seems that the more directions used to back project, the more the result resembles the original square.

Figure 1.17 shows a backprojection formed by averaging 20 individual backprojections. The first picture in the upper left of the figure is precisely the single backprojection discussed previously (fig. 1.15). The second picture illustrates the average of this backprojection with one from a neighboring direction. The complete sequence illustrates successive averages of neighboring backprojections until a full range of directions has been achieved. Read the diagram from top left to bottom right, that is, in increasing order of the number of directions. The lower right image is the reconstruction of the square by backprojecting its Radon transform.

The image constructed by backprojecting and averaging through all directions is denoted by $R^{\#}g$, where $g = g(\theta, t)$ is the function described earlier. Recall that $g(\theta, t)$ represents the material piled up at the point t in a direction orthogonal to θ. Because all directions θ have been used to construct $R^{\#}g$ from g, a dependence on θ no longer exists. The result of backprojection, the function $R^{\#}g$, represents an object in the plane and hence is a function of the two variables x and y. It is amazing to see the reconstruction take shape from a very different initial picture through some formless blobs to finally more and more square shapes. It seems very encouraging that we have found a simple method of reconstructing a function from its Radon transform. Let us look at the results of backprojecting the Radon transform through 500 directions.

It appears that we may have reconstructed the square.

However, if we examine the graphs of the reconstruction and the original object, then we see that the reconstruction is not all that good. In Figure 1.18 we see the result of backprojecting the Radon transform of the square object f from 500 directions: We chose 500 directions because later on we will describe a modified backprojection process that gives very good reconstructions for simple objects using about this number of directions. However, the process described here gives very poor results even for simple objects. Even if we used more than 500 directions, there is no hope of doing much better from the backprojection process described here. This is true even if we use infinitely many directions. The process of taking the backprojections of the Radon transform of an object yields what is called a Riesz potential of the object. The Riesz potential is almost never very close to the object itself, so this method cannot give very good reconstructions. The more advanced reader can check this out in the discussion following the proof of theorem 2.75 in chapter 2.

The general shape of the reconstructed object is that of a square when viewed from above. But viewed as a three-dimensional graph the original object is a rectangular prism, whereas the reconstruction is far from that as figure 1.18 shows. There are ways

Figure 1.18. Reconstruction of a square by backprojection. (*Left*) Original object. (*Center*) Reconstruction by backprojection. (*Right*) The difference between the original object and its reconstruction is larger than the original object itself!

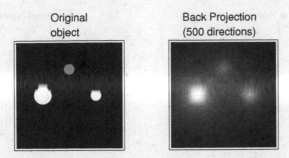

Figure 1.19.

of correcting for the poor reconstruction. One way is to look at the difference between the original object and its reconstruction. This difference can be considered to be a form of noise. Then noise reduction procedures may give a better reconstruction. We will describe this in the next section, but let us first look at another example.

This example (fig. 1.19) involves an object consisting of three disks (on the left) and its reconstruction by backprojection (on the right). Again, the reconstruction shows some of the features of the original object. But there are other features that are not part of the original picture. This is more apparent by examining the graphs:

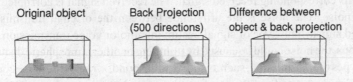

These examples show the limitations of the backprojection process. Although the reconstructed images have some of the essential features of the original object, they also have numerous features that are in error. These other features are called *noise* or *artifacts*.

Therefore, it is necessary to try to obtain better reconstructions than can be obtained by the backprojection method.

1.6 Noise Reduction and Error Correction

To a mathematician or engineer, noise is simply the difference between an object and its reconstruction. An engineer deals with objects called signals. A transmitted signal is received (reconstructed) but perhaps contaminated with noise. A mathematician deals with objects represented by functions. The mathematician may try to reconstruct a given, but unknown object by backprojection. We saw that the process of backprojection created noise: a difference between the original object and its reconstruction. The idea now is to eliminate all or at least most of the noise.

In the transmission of signals, a way of eliminating noise is to apply an averaging process called *filtering*. We will investigate this process and try to adapt it to reconstructing functions by backprojection.

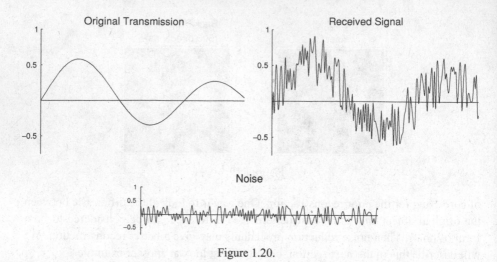

Figure 1.20.

To introduce the filtering process, we consider a signal represented by a graph of a single real variable. This signal is then transmitted over a noisy medium and received as a corrupted version of the original transmission. Figure 1.20 shows a transmitted signal and its corresponding received signal. The received signal is corrupted by high-frequency noise. The noise is the difference between the original transmission and the received signal. This type of signal occurs in radio or voice transmission. In radio transmissions, the noise could be caused by lightning or other atmospheric disturbances. For other types of transmission, such as sound, ultrasound, x-rays, etc., noise could result from any disturbance to the process of transmission or to the medium through which the signal is transmitted.

Notice that the noise in the transmission is also shown. As mentioned, this noise is merely the difference between the original signal and the received signal. Observe that the noise has almost identical positive and negative contributions. Thus if one averaged the noise, then the noise would mostly disappear. There are various ways of averaging and, in mathematics, these are performed by a process called **convolution**. A convolution takes a signal (i.e., a function of one variable) and averages it with some fixed function called a *filter*. The filter characterizes the type of averaging desired. You can understand the idea of filtering and convolution by drawing a noisy graph. Then take an eraser to try to blur the graph into a smoother shape. The graph represents the signal, the eraser represents the filter, and the process of applying the eraser to the graph is what we mean by convolution. In more precise terms, convolution is a process of moving averages: each portion of the graph is smoothed out by averaging the values in a neighborhood of each point. This process is also called *filtering* the signal. A different choice of eraser, say an eraser with a different shape or thickness, would correspond to a different convolution. The mathematical notation for a convolution is $f * k$, where f is the function being processed (signal) and k is the filter (eraser). The notation $f * k$ is read "the convolution of f and k." In chapter 2, section 3 we will see that the filter can be represented by a function and that the convolution $f * k$ is an integral of some sort involving these functions. If you have not studied the calculus, it

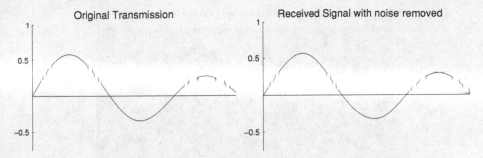

Figure 1.21.

is enough to understand that the process of forming an integral is a summation process that can be adapted to computing averages.

Figure 1.21 shows both the original transmission and the received transmission, after the convolution process has removed most of the noise.

Here are some more comparisons between the original signal, the noisy received signal, and the reconstructed signal with noise removed:

Original signal (solid line)
Received signal with noise removed
(dotted line)

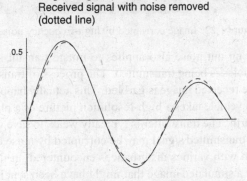

You can see that the reconstructed signal is close to but not identical with the original signal. It is, however, a much better approximation to the original transmission, as you can see in the next figure in which the noisy received signal is superimposed on the original signal.

Original signal superimposed
on received signal

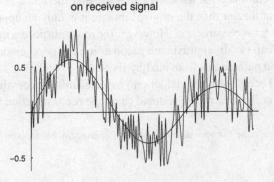

Image contaminated with noise

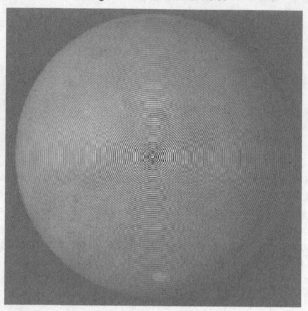

Figure 1.22. Image corrupted by high-frequency noise.

The idea of filtering out noise also applies to image transmission. In this case a function of two variables is being transmitted. The process of transmission frequently collects noise and the received image is garbled. This actually happens in space exploration where a space vehicle takes a high-resolution picture of a planet then transmits the image back to Earth. The transmitter is typically weak, to save weight on the space craft. Therefore, the transmitted signal may be corrupted by noise caused by the weak signal in conjunction with various disturbances encountered on the journey through space. Figure 1.22 is a simulated image that might have been sent by a space craft.[12] After filtering the image looks like fig 1.23. You may recognize it as a view of the planet Mars. The original image appears in fig. 1.24. Comparing this to the filtered image, we see that the filtering process has eliminated most of the noise.

In the two examples of noisy transmissions already considered, the noise had a random, high-frequency nature. We have seen that applying a certain type of filter to the noisy signals can eliminate the noise. We also see that there is a penalty: the filtered image is somewhat fuzzier than the original image; it is only an approximation to the original not an exact reconstruction. However, for most purposes, including diagnostic medicine, we can obtain approximate reconstructions good enough to distinguish diseased tissue and pathologies from healthy tissues.

The filtering process for noise reduction is so elegantly successful that we should investigate whether the idea can be adapted to image reconstruction from projections.

---

[12] For this example, the noise has been added by a computer program, but the idea is the same as in real transmissions.

Contaminated image after noise removal

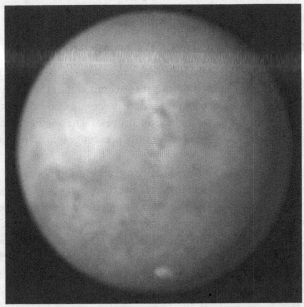

Figure 1.23. Noisy image after filtering. Notice the reduction in noise.

Original Image

Figure 1.24. Original image (courtesy of NASA.)

Indeed this can be done, but as with the filtration of high-frequency noise, the best we can expect is a good approximation to the original object.

1.7 Filtered Backprojection: How a CT Scanner Reconstructs a Picture

Filtered backprojection is the same backprojection process described before except that each x-ray projection is filtered before it is averaged into the backprojection. So you should think of the filtered backprojection process as an amalgamation of the backprojection idea with the noise reduction process. The mathematical details are presented in chapters 2 and 3, so for now we only present graphical images illustrating the technique.

The main difference between filtered backprojection in tomography and noise reduction in image processing is that in tomography there is no noise in the known data, the x-ray projections,[13] whereas, in image processing there is noise in the known data, the received image. The noise in tomography comes from the backprojection process itself. However, just as in image processing, a filter is applied to the known data, the x-ray projections. Then, when these are backprojected, the noise is mostly eliminated. The resulting picture will be a good reconstruction of the original object, if the filtering is done in the correct manner.

The reason why this process works will be covered in chapter 2, section 7. For now we present two examples that graphically illustrate the process. We show the progression of filtered backprojections that reconstructs the unknown function, but we do not describe the filter of the convolutions that produce these results.

It is important to realize that any process involving filtering can only give an approximate solution to the problem. The examples given here, however, yield very good approximations.

The first example is a simple object. We do not know what it looks like at the start, but we do know it's Radon transform. At the end of the reconstruction process we will see the original object to which we will compare the reconstruction.

The second example is a reconstruction of a human brain section from its x-ray projections.

1.7.1 Filtered Backprojection Reconstruction of a Simple Object

We are presented with an unknown object but are given its Radon transform. Let us call the unknown object f. The known data, the sinogram, is in fig. 1.25. It represents the information collected by sending x-rays through the object from all directions. As we know, this information is the Radon transform of f, and this graph is called the sinogram.

The unknown in this problem is the original object and the aim is to solve the problem by giving a good approximate reconstruction of the unknown f via filtered backprojection. Before doing this let us see what predictions we can make about the shape of the object directly from the x-ray data.

---

[13] We are considering the ideal situation in which the actual x-ray projections are the known data. In practice noise also occurs in the x-ray projections and this issue must also be dealt with.

Figure 1.25. Sinogram of $Rf(\theta, s)$, the graphical representation of the Radon transform (x-ray data) of the unknown object f.

Recall that the horizontal axis represents the directions of the x-rays. If you erect a vertical line at any particular direction, the gray levels along that line determine the x-ray data in that direction. To the unaided eye it appears that these data are the same for every direction, so the object probably has circular symmetry. Also you can see that

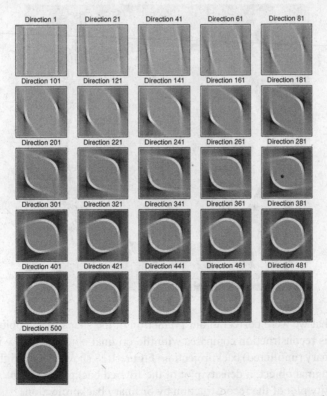

Figure 1.26. Snapshots of filtered backprojection applied to the x-ray data of the unknown object f.

as we enter the object the density is very high but quickly drops off to an intermediate value. Therefore we can guess that the object is some sort of thin ring of high density surrounding a disk of moderate density. However, it is impossible to predict the detailed structure of f merely by looking at the sinogram of the x-ray data. This will appear in a dramatic way later.

Now let us perform filtered backprojection to simulate how a CT scanner would determine the shape and density of the unknown object.

We do filtered backprojection with 500 equally spaced directions. As in ordinary backprojections, each new filtered backprojection is averaged into the preceding ones. The initial filtered backprojection is shown in figure 1.26. After each group of 20 successive backprojections we show a snapshot of the process (it would take too much room to show every step). For example, the fourth picture is the average of 61 backprojections for angles $0, \frac{\pi}{500}, 2\frac{\pi}{500}, \ldots, 60\frac{\pi}{500}$. The fifth picture averages the next ten filtered backprojections into the projections already obtained. Thus, the fifth picture is the average of 81 filtered backprojections. The final picture is the average of 500 filtered backprojections and represents our reconstruction of the unknown object from its Radon transform. We denote the final picture by the symbol f even though it is only an approximation to the original object.

Here is a larger version of the reconstruction.

It seems that we were correct in our prediction of the shape of the object. We now show how this reconstruction compares with the original object and how it compares with the ordinary (unfiltered) backprojection. Figure 1.27 shows, left to right, a density plot of the original object, a density plot of the filtered backprojection reconstruction f, and a density plot of the reconstruction by ordinary backprojection.

Figure 1.27. From left to right, original object, its reconstruction by filtered backprojection from 500 directions, and its reconstruction by ordinary backprojection from 500 directions.

It appears that the filtered backprojection reconstruction f is a very good approximation to the original object. The ordinary backprojection shows the circular nature of the object, but it appears to be a very poor reconstruction. This is confirmed by viewing these objects as graphs. Actually what we do is present graphs of the differences between the objects: the smaller the difference, the better the approximation. In the next figure, the left-hand graph is the difference between the original object and its reconstruction by filtered backprojection. The right-hand graph is the difference between the original object and its reconstruction by ordinary backprojection.

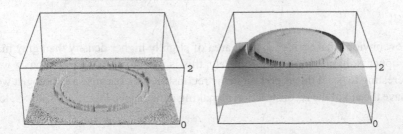

We can see that filtered backprojection does a much better job.

The original object actually represents a mathematical model of a simplified human brain. Such a model is called a phantom. In this phantom the densities are those of the actual tissue in the brain. The densest tissue is that of the skull, represented by the white outer ring. The gray matter has somewhat less than half the density of the skull. In actual CT we are interested in locating pathologies. It is easy to pick up the difference between gray matter and the skull. But the difference in density between normal tissue and diseased tissue can be very small. For example, there is only a 4.5% difference in density between a metastatic breast carcinoma and gray matter in the brain, as compared with a 50% difference between gray matter and the skull. In fact, the subtle difference in density between the gray matter and a carcinoma is almost impossible to detect by the unaided eye in the images we have been presenting. For that reason it would be interesting to perform an image enhancement process on the original object to enhance the contrast between normal and cancerous tissue. The next figure shows the original phantom after contrast enhancement.

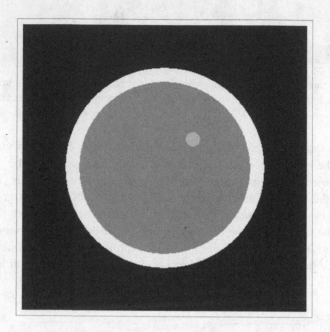

It is now obvious that there is a small area of slightly higher density than gray matter in the upper right quadrant. Unfortunately, this is indicative of a carcinoma. It would be interesting to see if this is present in the reconstruction, because a radiologist would only have the aid of the reconstruction in making a diagnosis. Figure 1.28 shows, left to

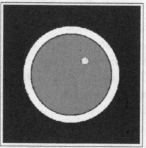

Figure 1.28. From left to right, original object, reconstruction by filtered backprojection, and reconstruction by ordinary backprojection with enhanced contrast.

right, density plots of the original object, the filtered backprojection reconstruction f, and the reconstruction by ordinary backprojection, all with contrast enhancement. We see that the reconstruction by filtered backprojection detects the carcinoma, whereas the ordinary backprojection does not. At this point we also can understand the comment made earlier that it is impossible to produce the detailed structure of f merely by looking at its x-ray data.

Go back to figure 1.27, which has a picture of the object before contrast enhancement. If you look carefully at the original object and its tomogram you may be able to see a faint trace of the carcinoma. This carcinoma is easily detected by the CT process, although the human eye requires some aid in the form of contrast enhancement to actually see it. This contrast enhancement is purely mathematical: it has no relation to the contrast-enhancing dyes used in some forms of radiology.

The carcinoma physically exists in the original object. Its effect on the Radon transform is very subtle and it would be very hard to detect in the sinogram, versus the same object without the carcinoma. The reason we cannot see the carcinoma with the naked eye is that there is such a small density difference between normal tissue and cancerous tissue. It is remarkable that this distinction can be detected by the process of tomography.

The contrast enhancement is not cheating, it merely emphasizes subtle differences in tissue. It is analogous to examining a picture in a dark room versus enhancing the picture with bright light from a lamp.

The next step is to see how well CT does with actual human organs.

1.7.2 Reconstruction of a Human Brain from X-ray Projections

We now present the reconstruction of an image of a human brain from its x-ray projections by the same filtered backprojection process used by CT scanners. The known data are the set of x-ray projections shown by the sinogram in figure 1.29, as may be collected by the detectors of a CT scanner.

Figure 1.30 shows the succession of filtered backprojections used in reconstructing the brain section. The x-ray projections were sampled in 1,000 equally spaced directions. The first backprojection and every fortieth backprojection thereafter are shown along with the reconstruction, which is the last backprojection (lower left).

Figure 1.31 presents a larger version of the reconstructed brain section.

We can show that this is a good reconstruction. First we compare the original brain image with the reconstruction (fig. 1.32).

In figure 1.33, we show graphs of the original brain section image, the filtered backprojection reconstruction, and the difference between the original image and the reconstruction.

We therefore see that the filtered backprojection process works very well, at least on objects no more complicated than human tissue.

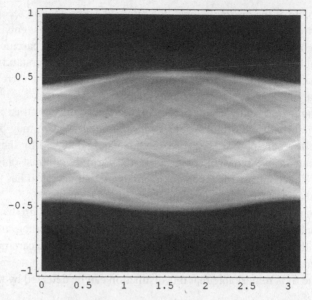

Figure 1.29. Sinogram showing the x-ray projections of a human brain.

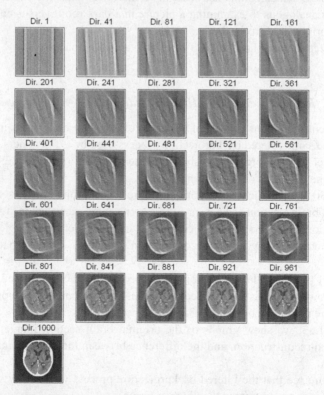

Figure 1.30. Snapshots of filtered backprojection reconstruction of a human brain (1,000 projections).

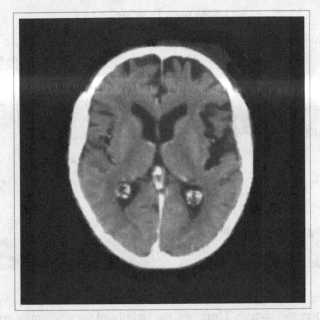

Figure 1.31. Filtered backprojection reconstruction of a human brain.

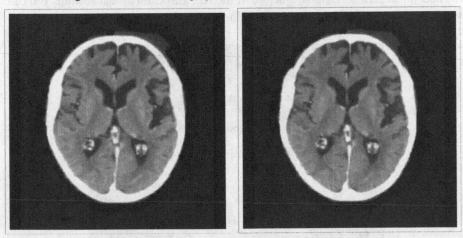

Figure 1.32. Comparison between original brain image and its filtered backprojection reconstruction. The original is on the left.

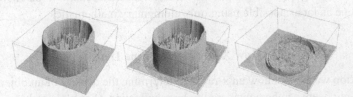

Figure 1.33. Another comparison. The original is on the left, the reconstruction is in the middle, and the difference between them is on the right.

1.8 Which Is Which?

We can now answer the question posed at the end of section 1.4 where we gave an example of x-ray data for a brain and an abdomen. Here are the sinograms representing these data.

 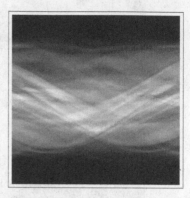

At this point we recognize the left-hand sinogram as that of the brain section discussed in the previous section. Therefore, the right-hand sinogram must correspond to the abdominal section. Here is a picture of the reconstruction of that section. You can see quite clearly the spinal column and kidneys and other anatomical features.

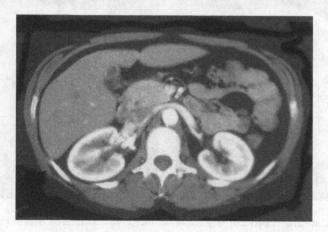

Many interesting questions about the Radon transform remain to be answered, but we have gone as far as possible using only elementary mathematics.

1.9 How X-rays Determine Density

In this section we explain how an x-ray can determine the density of an object along a line. An x-ray beam can be considered to be a collection of energetic photons moving along a straight line. As the x-ray beam traverses tissue some photons may be scattered or absorbed. Heavier tissue tends to scatter or absorb x-rays more than lighter tissue. Therefore, it is possible to get a good estimate of the total density along the line traversed by the x-ray beam by comparing the number of photons emitted by the x-ray source with the number of photons counted by the detector at the other end.

Actually the behavior of x-rays is more subtle. The attenuation of the x-ray beam due to absorption and scatter also depends on the energy of the beam and the atomic characteristics of the material being traversed. But the previous paragraph indicates that the total attenuation experienced by an x-ray beam traversing a line is informally related to the total density of the material along the line, that is, to the x-ray projection. At the risk of being slightly imprecise, we will treat the terms "attenuation" and "density" synonymously in this section.

This section is more technical than the rest of the chapter. However, I have tried to present the material so that a reader with just a small amount of mathematical knowledge can follow the argument. The only exception is a technical note that requires some knowledge of the calculus.

Figure 1.34 illustrates a classic experiment showing the phenomenon of **photon attenuation** in an x-ray beam. Identical plates consisting of a homogenous material are stacked together and an x-ray beam traverses the plates from left to right. As the beam exits each plate, the number of surviving photons decreases. Each plate can be examined to see the intensity of the x-ray beam as it exits the plate; in figure 1.34, note how fewer and fewer photons survive through each plate. Therefore, the number of photons that exit the last plate on the right is less than the number that entered on the left. This experiment was repeated by using many different materials for the plates. In each case it was observed that the number of photons that did not survive the trip over a small distance d is a fraction of the number beginning the trip. However, over a large distance, the relationship between the number of photons entering the material and the number exiting is more intricate. To investigate this relationship we introduce the idea

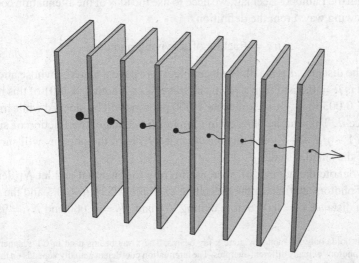

Figure 1.34. Photon attenuation in an x-ray beam.

of the ***attenuation coefficient*** of the material in question.[14] The attenuation coefficient is denoted by the symbol μ (μ is the symbol for the Greek letter mu). It is defined in the following way. Take the fraction of photons that did not survive the trip through a small distance d of the material. Then divide this fraction by d. The resulting quantity, μ, is the attenuation coefficient of the material. For example, if we observe that $10,000$ photons enter a 0.1-cm length of some material and that $9,500$ photons exit, then the fraction of photons that did not survive is

$$\frac{10,000 - 9,500}{10,000} = 0.05$$

If we now divide this fraction by the small distance $d = 0.1\ cm$ we arrive at the quantity μ:

$$\mu = \frac{0.05}{0.1}$$
$$= 0.5\ photons/cm$$

If we perform the same experiment over half the distance we would observe that $9,750$ photons survive the distance of $0.05\ cm$. Therefore the attenuation coefficient computes to

$$\mu = \frac{10,000 - 9,750}{10,000} \div 0.05. = 0.5$$

and we see that we get the same attenuation coefficient. The experiment illustrated in figure 1.34 leads to the Lambert-Beer law. One aspect of the Lambert-Beer law is that the attenuation coefficient is constant for homogeneous materials. Of course each material has a different attenuation coefficient.

To obtain the Lambert-Beer law, we need to use the idea of the attenuation coefficient in the following way. From the definition,

$$\mu d = \text{fraction of surviving photons,}$$

provided the distance d is small. For example, if we have a material whose attenuation coefficient is $\mu = 0.3$ and if an x-ray beam traverses a length $d = 0.01$ of this material, then $\mu d = 0.003 = \frac{3}{1,000}$, so 3 of every $1,000$ photons will be blocked while traversing the distance d. This can be phrased in terms of percentage: the fraction of surviving photons is $1 - \mu d = \frac{997}{1000}$ and this means that 99.7% of the photons will survive the passage of the small distance d.

Let N_0 denote the number of photons entering the material and let N_1 denote the number of photons exiting. In the preceding example where $\mu = 0.3$ and the photons traversed a distance $d = 0.01$ before exiting, we had $N_0 = 1,000$ and $N_1 = 997$.

---

[14] This result holds only for monoenergetic x-ray beams. The x-ray beams used in CT scanners actually consist of photons of many different energies. The attenuation coefficient actually depends on the material and the energy of the photons. However, there are ways of resolving this problem, and it is safe for the current discussion to assume that the attenuation coefficients depend only on the atomic number of the material.

We are more interested in what happens to N_1 when an x-ray beam traverses a long distance through inhomogeneous material. To investigate this, let us take a simple case in which an x-ray beam traverses a line containing 10 cm of tissue with density μ. The situation is depicted in the following figure.

$$\mu$$

In this investigation we use the law of exponents which states that

$$x^a x^b = x^{a+b}$$

and the fact that if

$$x^a = c$$

then

$$a = \log_x c$$

(i.e., the logarithm to the base x of a positive number c is precisely the power a such that $x^a = c$).

To continue this investigation, we need a simple mathematical relationship. Define the real number $g = 0.3679$ (those readers who have studied the calculus can check that g is approximately e^{-1}). The significance of this number is that it allows one to express the difference $1 - x$ as a power of g. That is, $1 - x \approx g^x$, if x is relatively small. This ability to express $1 - x$ as a certain power of g is advantageous later. When we use this relationship we can substitute any small number for x. For example if μ_1 is a fraction and d is a short distance, then the relationship $1 - x \approx g^x$ implies that $1 - \mu_1 d \approx g^{\mu_1 d}$. We will use this type of calculation several times.

This approximation can be verified in a completely rigorous way using the calculus, but let us illustrate the idea using a few values of x

x	g^x	$1 - x$
0.01	0.990 05	0.99
0.05	0.951 23	0.95
0.09	0.913 94	0.91

From this table we see evidence that the relationship $g^x \approx 1 - x$ is valid, at least if x is relatively small.

Remember that we are trying to see how many photons survive the trip through the 10-cm piece of material. N_0 photons enter at the left and N_1, the number of surviving photons, exit at the right.

To see what happens to photons traversing long distances, we need to break up a long distance into small pieces. We do this by breaking the entire 10-cm line into small pieces of size d. The actual length of d is not important, any small length will do. From the Lambert–Beer law we know that the photons that successfully pass through the first small segment of length d will contain only the fraction $(1 - \mu d)$ of the photons that were initially in the beam. The real number $(1 - \mu d)$ is always between 0 and 1 and hence is a fraction. It helps in the following discussion to think of the abstract fraction $(1 - \mu d)$ as a definite fraction such as $\frac{3}{100}$ or 0.456. Hence, multiplying a quantity N_0 by $(1 - \mu d)$ will always result in a smaller quantity N_1.

In particular, if we start with N_0 photons entering the segment, only $N_0 (1 - \mu d)$ will exit. We use the approximation, above with $x = \mu d$, to get

$$N_0 (1 - \mu d) \approx N_0 g^{\mu d}$$

as the number of photons, out of the initial number N_0, that have survived the passage over the segment of length d. Over the second segment of length d only the fraction $(1 - \mu d)$ of the $N_0 g^{\mu d}$ photons that enter will survive. The approximation then yields

$$(1 - \mu d)(N_0 g^{\mu d}) \approx N_0 g^{\mu d} g^{\mu d} = N_0 g^{2\mu d}.$$

In the same way, over the third small segment of length d we can check that approximately $N_0 g^{3\mu d}$ photons will survive. In general, we will have $N_0 g^{x\mu d}$ photons surviving past x adjacent small segments of length d. This pattern will persist until the beam traverses the entire 10-cm length of material. We note that after the last segment we have $x = \frac{10}{d}$, because, in a 10-cm line segment, there are exactly $\frac{10}{d}$ pieces of length d. At this point the surviving number of photons, which by definition is N_1 will be about

$$N_1 = N_0 g^{\left(\frac{10}{d}\right)\mu d} = N_0 g^{10\mu}$$

(Note how, in the end, there is no dependence of the answer on the particular length d.) Hence, we get $g^{10\mu} = \frac{N_1}{N_0}$. Take the logarithm to base g of this expression to get

$$10\mu = \log_g \left(\frac{N_1}{N_0} \right).$$

But the 10-cm line, being homogeneous, has a total density of 10μ/m so for homogeneous lines we get

$$\text{Total density} = \log_g \left(\frac{N_1}{N_0} \right).$$

If we have a line containing a nonhomogeneous amount of material, then we break the line up into pieces that are approximately homogeneous. We illustrate this with the simple case of a line containing 10 cm of tissue with density μ_1 followed by 5 cm of tissue with density μ_2 as depicted in the following figure.

μ_1 $\qquad\qquad\qquad\qquad\qquad\qquad\qquad\qquad$ μ_2

If N_0 photons enter at the left, then, as we saw above, $N_0 g^{10\mu_1}$ will survive the passage through the first 10-cm piece with density μ_1. We then use this number, $N_0 g^{10\mu_1}$, as the input number of photons to the 5-cm piece of density μ_2. This results in $N_1 = (N_0 g^{10\mu_1})g^{5\mu_2}$ as the number of photons which survive the entire passage. We can use the law of exponents to simplify this expression, thus arriving at

$N_1 = N_0 g^{10\mu_1 + 5\mu_2}$ and, hence, $g^{10\mu_1 + 5\mu_2} = \frac{N_1}{N_0}$. As before we take logarithms to get $10\mu_1 + 5\mu_2 = \log_g(\frac{N_1}{N_0})$. But the total density along the line is $10\mu_1 + 5\mu_2$, so again we arrive at

$$\text{total density} = \log_g \left(\frac{N_1}{N_0} \right), \qquad (1.1)$$

This result can be extrapolated to any number of adjacent homogeneous pieces. This is accurate to a good approximation for any piece of tissue that we may encounter.

Because CT scanners can produce these photon counts, we have solved the problem of how an x-ray beam can give the total density of material lying along a line. We use the CT scanner to obtain the number N_o (of photons emitted from the x-ray source) and the number N_1 (of photons that have survived to be counted at the detector) along a given line. We substitute these numbers into equation (1.1) which then provides us with the total density, or x-ray projection, along that line.

Those readers who have not studied the calculus can skip the following technical note.

Technical Note 1. *For those readers who know some calculus, we can derive the exact relationship between the line integral of the attenuation coefficients and the photon counts along the path of a monoenergetic x-ray beam. The **intensity of the beam** is the photon count at any point of the beam and we denote it by N. The attenuation of a monoenergetic x-ray beam is governed by the Lambert-Beer law, which is illustrated by the experiment shown in figure 1.34. The result of this experiment is that the attenuation of the x-ray beam is exponential relative to the distance traversed. More precisely, if an x-ray beam traverses a homogeneous material M along a straight line L, parametrized by distance s, then the intensity N of the beam at a distance s from the origin is given by*

$$N(s) = e^{-\mu s} \qquad (1.2)$$

for some positive constant μ. This relationship has been experimentally verified for many different materials and we accept it as an axiom. Equation (1.2) is the global version of the Lambert-Beer law. We can easily obtain a local version by differentiation:

$$\frac{dN}{ds} = -\mu N \qquad (1.3)$$

Going back to the experiment with the photographic plates from which we deduced the Lambert-Beer law, it is a reasonable assumption that (1) The experiment is independent of the direction of the line L: if we rotate the apparatus, then the same exponential law will arise, independent of the direction of rotation. (2) The local version of the Lambert-Beer law should hold for a nonhomogeneous medium as long as the intensity function is reasonably smooth.

Therefore, the constant μ in the Lambert-Beer law, at each point, is a property of the material and is independent of the position or orientation of the material, so it is reasonable to define a function $\mu = \mu(x, y)$ whose value at the the point located at (x, y) is the attenuation coefficient of the material located at that point. Furthermore, *the local form of the Lambert-Beer law is obeyed by μ. This means that if ds is the element of length along any line in the plane and if an x-ray beam is sent along this line, then the intensity of the beam satisfies the same differential equation as before:*

$$\frac{dN}{ds} = -\mu N \tag{1.4}$$

*The function μ is called the **linear attenuation coefficient** of the tissue (or other material) being scanned.*

Now let L be a line passing through the unknown object. We assume that the x-rays are being generated at a fixed point on the line exterior to the unknown object. We let N_0 be the number of photons that are produced by this x-ray source. Likewise an x-ray detector has been placed at another fixed point on the on the line and outside the unknown object. We let N_1 denote the number of photons that have survived the passage along L and are counted by this detector. By integrating equation (1.4) and using a simple change of variable from s to N we see that

$$\int_L \mu(s)\, ds = \int_{N_0}^{N_1} -\frac{1}{N} dN \tag{1.5}$$

$$= \ln\left(\frac{N_0}{N_1}\right)$$

Hence, the line integral of μ along L (i.e., the total density of μ along L) may be obtained directly from the photon counts according to equation (1.5). Therefore the collection of these integrals for all lines L in the plane form the Radon transform of μ. This is pursued in more detail in chapter 2.

Therefore, a CT scanner actually tries to reconstruct the linear attenuation coefficient function, not the density function. However, it is reasonable to suppose that a close connection exists between the attenuation coefficient and the density of the object. After all, the more matter present, the more chance there is to absorb or scatter a photon.

In practice there are some difficulties with this basic idea. The main difficulty is that the x-ray beams used in CT consist of photons at many different energy levels; they are not monochromatic as assumed above. Therefore, the attenuation coefficient is a function $\mu = \mu(x, E_0)$ of both position and the energy of the photon. A related problem is beam hardening, which means that less energetic photons tend to be absorbed or scattered earlier, thus causing the composition of the x-ray beam to consist of more energetic photons further down the line. Fortunately, there are corrective algorithms that take the number of initial photons and the number of exiting photons as input and can give as output a good estimate for $\ln\left(\frac{N_0}{N_1}\right)$ at a certain fixed energy level E_0. From this we can compute a good approximation to $\int_L \mu(x, E_0) dx$. Finally, radiologists are skilled in correlating the attenuation coefficients with the actual structure of the tissue, so CT scans turn out to be reliable diagnostic tools.

Herman's book [296] and the article by Cormack and Quinto [109] have more details on the interaction of radiation with matter.

1.10 Additional References and Results

1.10.1 History of Tomography in Medical Diagnosis

The most extensive history of tomography in medical diagnosis is contained in Webb's book [618]. The outline of this history that I have given here is taken mainly from [618] with supplementary information from Natterer [444], Natterer and Wubbeling [446], Herman [296], Cormack [104], and personal communications with colleagues.

The use of tomography in medicine is broadly classified into "classical tomography" and "computerized tomography." In classical tomography the desire is to image a two-dimensional section of the body. X-rays are transmitted through a neighborhood of this section in such a way as to produce as sharp a focus as possible on the desired section while blurring the surrounding tissue. This method was fairly popular in the early part of the twentieth century, but the results were of debatable resolution. One expert claimed that this method was "...an expensive way to obtain bad radiographs." (Webb [618], page 107). Because this method only shares its name with computerized tomography and does not utilize the Radon transform, we will not spend any more time in describing it. Readers interested in further details are referred to Webb [618].

According to Webb [618] there are "at least two (Frank and Takahashi) and possibly more pioneers whose work in the 1940s might be viewed as a direct precursor of CT." Frank was awarded a German patent for his ideas, but he used pure backprojection, which does not give reasonable reconstructions, as we saw in section 1.5. Kuhl and Edwards [366] also used backprojection with no better results.

Also according to Webb [618], the earliest implementation of computerized CT was developed by Korenblyum, Tetelbaum, and Tyutin [357] at the Kiev Polytechnic Institute in 1958. Their reconstruction was based on Tetelbaum's paper [599], which gave a method for inverting the Radon transform. In their paper [357] Korenblyum, Tetelbaum, and Tyutin outlined the mathematical procedure for fan beam scanning and described an experiment in which a 100×100 pixel image was reconstructed from x-ray projections. Their method involved recording the sinogram on film and then producing electrical signals from this film which were processed by an analogue computer. Webb [618] has diagrams of their apparatus. Korenblyum, Tetelbaum, and Tyutin described their work as experimental, and apparently, it was never clinically implemented. As of 1983, no further evidence of publications on CT were found in the Russian literature.

The Nobel prize for Physiology or Medicine was awarded jointly to A. M. Cormack and G. N. Hounsfield in 1979. Each of these laureates made separate and important contributions to the field of computerized tomography. The presentation speech for the 1979 Nobel prize for Physiology or Medicine was delivered by Professor Torgny Greitz of the Karolinska Medico-Chirurgical Institute. He stated that

Cormack realized that the problem of obtaining precise values for the tissue-density distribution within the body was a mathematical one. He found a solution and was able, in model experiments, to reconstruct an accurate cross-section of an irregularly shaped object. This was reported in two articles, in 1963 and 1964. Cormack's cross-section reconstructions were the first computerized tomograms ever made – although his "computer" was a simple desktop calculator.[15]

Greitz's comment about Cormack having made the first computerized tomograms may not be historically accurate (compare, the paragraph on Korenblyum, Tetelbaum, and Tyutin, above). However, the early Russian tomography never advanced beyond the simple experiment of Korenblyum and colleagues. On the other hand, Greitz went on to say:

> Publication of the first clinical results in the spring of 1972 flabbergasted the world. Up to that time, ordinary X-ray examinations of the head had shown the skull bones, but the brain had remained a gray, undifferentiated fog. Now, suddenly, the fog had cleared. Now, one could see clear images of cross-sections of the brain, with the brain's gray and white matter and its liquid-filled cavities. Pathological processes that previously could only be indicated by means of unpleasant – indeed, downright painful and not altogether risk-free examinations could now be rendered visible, simply and painlessly - and as clearly defined as in a section from an anatomical specimen.... Allan Cormack and Godfrey Hounsfield! Few laureates in physiology or medicine have, at the time of receiving their prizes, to the degree that you have, satisfied the provision in Alfred Nobel's will that stipulates that the prizewinner "shall have conferred the greatest benefit on mankind." Your ingenious new thinking has not only had a tremendous impact on everyday medicine; it has also provided entirely new avenues for medical research. It is my task and my pleasure to convey to you the heartiest congratulations of the Karolinska Institute and to ask you now to receive your insignia from His Majesty, the King.

Hounsfield, who worked for the EMI company in England, designed the first commercially available CT scanner in 1972. However, he traveled a long road to produce this machine. Hounsfield started his research in 1967. His initial experiment required nine days of radiation exposure to gather enough data to reconstruct one picture (I believe he used a cow's brain for this experiment). Then it required two and a half hours of computing time to produce the reconstructed image. By 1972 when EMI began marketing Hounsfield's CT scanner, the data collection time had been considerably reduced and it took only four and a half minutes of computing time to produce an image. Modern CT scanners work much faster still.

1.10.2 Further References

I can not make many suggestions for a reader who has no training in mathematics. One possibility is Cormack's article [102], which is very interesting and which describes his early research leading to his Nobel prize.

A few references are accessible to readers with a good background in undergraduate mathematics. Panton's 1981 article [468] still makes for very interesting reading. He discusses the history of tomography and graphically illustrates the backprojection

---

[15] The quotations by Greitz are from Nobel Lectures, Physiology or Medicine 1971–1980. Edited by Jan Lindsten. Singapore: World Scientific Publishing Co., 1992.

process. The remainder of his paper requires some advanced undergraduate mathematics. Other articles of interest are Shepp and Kruskal [556], Herman [296], Zalcman [625], Strichartz [585], Nievergelt [449,450], Deans [124], Kak and Slaney [328], Natterer and Wübbeling [446], and Epstein [150]. Also the next chapter of this book, which presents the mathematical derivation of the properties of the Radon transform, should be accessible to readers with a good background in undergraduate mathematics.

We should also mention the two interesting papers [400,401] by Louis, both with the title "Medical imaging: state of the art and future development." The first paper [400] is fairly elementary from the mathematical point of view but has very interesting informal descriptions of the process of data collection and reconstruction of images for medical diagnosis. This paper should be accessible to the reader with a good background in university-level mathematics. The second paper requires a more substantial mathematical background. It gives a very detailed overview of both the process and the mathematics of various forms of medical tomography. The emphasis in [401] is practical, with attention being paid to the development of algorithms for medical tomography. For the reader interested in the practical side of tomography, the articles [400,401] also provide an extensive list of references dealing with applications to medical imaging.

Hochbruck and Sautter have a fairly elementary article [313], in German, which has some interesting pictures of CT reconstructions. They also give a derivation of the Lambert-Beer law using a half-life argument. We assumed the Lambert-Beer law based on experimental evidence. The remainder of their article gives an algebraic reconstruction (ART) method for CT. ART methods are fairly simple but are not used much in CT. However, ART techniques are used in a different method of tomography called SPECT. In Chapter 5 we give a derivation of a generalized Radon transform which models SPECT tomography. However, we do not pursue ART methods, because they have little connection to mathematical analysis.

The next set of references require the reader to have had at least some graduate level mathematics courses.

First, I should mention the book by Gel'fand, Gindikin, and Graev [189], which is an almost breathtakingly beautiful introduction to the subject of mathematical tomography, written by some of the pioneers in the field.

The following books and articles delve deeply into various areas of the subject and require a substantial background in mathematics: Ehrenpreis [149], Gel'fand and Shilov [195], Gel'fand, Graev, and Vilenkin [194], Helgason [291], Natterer [444], Natterer and Wübbeling [446], Smith, Solmon, and Wagner [566], Kak and Slaney [328], and Ramm and Katsevich [513]. The articles [148, 147] by Ehrenpreis are a suggested accompaniment to reading his book [149].

The book by Lavrent'ev and Savel'ev [377] is a good reference for some of the functional analytic tools used in tomography.

The book by G. T. Herman [296] contains a wealth of information on the relation between line integrals and x-ray counts, photon statistics and data collection for CT, including correction schemes for beam hardening and polychromaticity. The book by S. R. Deans [124] also has a discussion of the relation between line integrals and x-ray counts. Both these references have many examples of practical uses of tomography besides CT. For a how-to guide to the practical side of tomography, consult Herman [296], Kak and Slaney [328], or Epstein [150]. The books by F. Natterer [444] and Ramm and Katsevich [513] have briefer derivations of the relation between line integrals and x-ray counts and briefer introductions to the applications but are valuable for the theoretical treatment of tomography.

There are a few survey articles accessible to a reader with a good background in graduate-level mathematics. Of particular note are the appendix [364], written by Kuchment and Quinto, of the Ehrenpreis book [149], the article [158] by Faridani titled "Introduction to the Mathematics of Computed Tomography," which appears in the book [606] edited by Uhlmann and the article by Smith, Solmon, and Wagner [566].

There are many references that give a very brief introduction to the Radon transform and its inversion. The translation of the Soviet 'Mathematical Encyclopedia' [612] (Vol. 9, Sto-Zyg, pp. 179–181) has an excellent short introduction to the Radon transform and tomography. Also, Khavin and Nikol'skij [351] have a one-paragraph treatment of the Radon transform; J. S. Walker [616] gives a succinct, somewhat specialized introduction to Radon transform in an appendix (pp. 380–400); Terras [597] defines the Radon transform early in the book (pp. 107–119) and has very succinct guided exercises in which the reader can develop Radon inversion; volume two of Terras [598] has a brief mention of applications to partial differential equation; and Dym and McKean [138] provide a short introduction to the Fourier inversion method for the Radon transform.

A few applications of Radon transforms are discussed in chapter 2, section 2.9.

1.10.3 Notes

Tomography and the theory of the Radon transform lie at the intersection of the mathematical fields of integral geometry and inverse problems. An area of integral geometry is concerned with geometric properties of objects that are determined by some set of integrals over portions of the objects. If these portions are straight lines, then we arrive at the Radon transform and hence tomography. The book by Gardner [185] contains a wealth of information on the integral geometric aspects of tomography. He also has a brief introduction to some aspects of analytic tomography. Also of interest in reference to integral geometry are the books of Nachbin [433], Santaló [543], and Ren [515].

To understand what is involved in the mathematical field of inverse problems, let us first describe the idea of a direct problem. A direct problem takes some given (input) data, performs some well defined operations on these data, and produces the results of this computation as output. We can symbolize a direct problem in the following way: let f represent the input data, let A represent the calculations to be performed on the input, and let g represent the output. The direct problem can be written

$$g = A(f)$$

A simple example of a direct problem is the computation of the values of a polynomial: if P is the polynomial $P(x) = x^2 + 1$, then we define $A(f) = P(f)$. For example, $A(2) = 2^2 + 1 = 5$. A more interesting direct problem is that of computing the Radon transform of a given function:

$$g(\theta, p) = Rf(\theta, p)$$

This means that we are given complete information about the object f and we must compute all the x ray projections of f.

An inverse problem can be written in the same form

$$g = A(f)$$

However, the roles of f and g are interchanged. Now it is g which is the input data and we are required to find f. In the case of the polynomial problem, a typical inverse problem would be to solve for x given that

$$17 = A(x) = x^2 + 1.$$

Solving this inverse problem would require finding all the solutions x to this equation. Likewise, the inverse Radon problem is to determine an unknown object by using its Radon transform as the input data.

2

The Radon Transform

2.1 Introduction

In chapter 1, we saw that the Radon transform is a fundamental tool for using x-rays in medical imaging. The value of the Radon transform of a function f on a line L was defined to be the total density of f along the line and could be represented as an integral of the form

$$\int_L f(x)\,dx \tag{2.1}$$

(Recall the discussion of x-ray projections and the Lambert-Beer law, chapter 1, section 1.9.) Furthermore, we described tomography as the study of reconstructing objects from their x-ray projections. Therefore, we can rephrase this by saying that *tomography* is the process of reconstructing an unknown function from its Radon transform.

In this chapter we rigorously define the Radon transform and develop its elementary properties. Consequently, we will be able to give a mathematical explanation for the filtered backprojection reconstruction method that was introduced in Chapter 1. We end the chapter with Radon's original inversion formula from his 1917 paper [508].

Because the field of tomography is very broad, and to accommodate those readers who are not mathematicians, I have tried to make this chapter accessible to readers who have had at least university-level courses in the calculus and linear algebra. These readers will be pleasantly surprised to learn that most of the basic theory of tomography follows from only two theorems that they will have studied in the calculus: the theorem on change of variables in a multiple integral and the theorem on iterating a multiple integral. However, all results are precisely stated, in n dimensions, and the proofs are mathematically rigorous.

There are practical and theoretical applications of the Radon transform on spaces of higher dimension than the plane, and with just a little effort we can extend the definition and study of the Radon transform to higher dimensions. Looking at the integral defining the Radon transform, equation (2.1), we see that an extension to higher dimensions can be accomplished by replacing the line L with some other domain. Therefore, we need to study integration on these domains. Let us begin by examining the mathematical description of lines in a plane.

A line in the plane is the locus of points described by a linear equation of the form $a_1x_1 + a_2x_2 = c$. The **Euclidean plane** $\mathbb{R}^2$ consists of all points x of the form (x_1, x_2), where x_1 and x_2 are real numbers. We now use the terms "point" and "vector" interchangeably because they are the same from the mathematical point of view. The inner product in $\mathbb{R}^2$ is denoted by $\langle a, x \rangle = a_1x_1 + a_2x_2$ for any two points $a, x \in \mathbb{R}^2$ (the expression $x \in A$ means x is an element of the set A or, equivalently, that x is in the set A). The inner product is sometimes called the dot product and is also denoted by $a \cdot x$. We can therefore rewrite the equation of the line in the form $\langle a, x \rangle = c$. Analytic geometry shows that this gives a line orthogonal to the vector $a = (a_1, a_2)$. When this equation is multiplied by a nonzero constant, we still get the same line. To normalize the situation we can demand that the vector a be a unit vector. It is traditional to denote unit vectors by the symbol θ and we denote the set of all unit vectors by S^1, which is called the **unit sphere** of $\mathbb{R}^2$. The unit sphere of $\mathbb{R}^2$ is just a fancy name for the circle of radius 1 centered at the origin. Shortly we will introduce unit spheres in any Euclidean space $\mathbb{R}^n$.

In the case that a is a unit vector θ, the equation of the line looks like

$$\langle \theta, x \rangle = c \tag{2.2}$$

and this determines the line orthogonal to the unit vector θ, which lies at a distance c from the origin. Also, if we are given any line in the plane that lies at a distance c from the origin, then there are precisely two unit vectors θ which satisfy the defining equation (2.2). If θ is one of these vectors then the other is $-\theta$ and the two corresponding equations are of the form:

$$\langle \theta, x \rangle = c$$
$$\langle -\theta, x \rangle = -c$$

From this we see that the distance c is a signed distance and can be any real number. In other words, specifying a unit vector θ and a real number c uniquely determines a line in the plane (although the vector and distance are not uniquely determined by the line).

From this discussion it is easy to generalize from the plane to *n-space*. Before doing this, note that we use the standard notation $\mathbb{R}$ for the set of real numbers and $\mathbb{C}$ for the set of complex numbers. Here is how we define Euclidean *n-space*:

Definition 2.1. $\mathbb{R}^n$ is the set of all ordered *n*-tuples of real numbers and is called **Euclidean n-space**. More precisely

$$\mathbb{R}^n = \{(x_1, x_2, \ldots, x_n) : \text{each } x_j \text{ is a real number}\}$$

Remark 2.2. Here we use the set descriptor notation: a set S which is characterized by a property P is written as $S = \{x : x \text{ has the property } P\}$. This is read as "the set of all x such that x has the property P".

The next definition is related and useful.

Definition 2.3. Given two sets A and B, the set of all ordered couples (a, b), where $a \in A$ and $b \in B$ is called the **Cartesian product** of A and B and is denoted by $A \times B$. Using ordered *n-tuples* we generalize this to

$$A_1 \times A_2 \times \cdots \times A_n = \{(x_1, x_2, \ldots, x_n) : x_j \in A_j, j = 1, \ldots, n\}$$

In the special case that each $A_j = \mathbb{R}$ we get Euclidean *n-space* and it makes much more sense to write this as $\mathbb{R}^n$ rather than $\overbrace{\mathbb{R} \times \mathbb{R} \times \cdots \times \mathbb{R}}^{n-factors}$

Example 2.4. $\mathbb{R}^1$ consists of all one-tuples, points of the form (x), where x is a real number. Therefore, except for the difference in notation, "x," versus "(x)," the real number system $\mathbb{R}$ and Euclidean 1-space $\mathbb{R}^1$ are identical.
Euclidean 2-space $\mathbb{R}^2$ is the set of ordered 2-tuples (more commonly known as ordered pairs or couples) of the form (x, y) where x, y are real numbers and Euclidean 3-space $\mathbb{R}^3$ is the set of ordered 3-tuples (triples) of the form (x, y, z) where x, y, z are real numbers. Therefore $\mathbb{R}^2$ represents the Euclidean plane and $\mathbb{R}^3$ represents the usual 3-dimensional Euclidean space. In higher dimensions, a typical point of $\mathbb{R}^n$ is denoted by the ordered n-tuple $(x_1, x_2, \ldots, x_n)$ where $x_1, x_2, \ldots, x_n$ are real numbers.

Remark 2.5. The symbol $\in$ denotes set membership, so writing $w \in S$ is the same as saying that w is an element of the set S. Therefore the expression $x \in \mathbb{R}^n$ means that x is a point of the Euclidean space $\mathbb{R}^n$ and hence there exist real numbers $x_1, x_2, \ldots, x_n$ such that $x = (x_1, x_2, \ldots, x_n)$. The context will always determine whether a particular variable is a vector in $\mathbb{R}^n$ or a real number. In this way we avoid the need to write vectors in boldface type or with arrows over them. Furthermore, this convention has the advantage of allowing one to see the similarities in the vector space structure of the real numbers and the Euclidean spaces.
In short, $x \in \mathbb{R}^n$ signals that x is a point or vector in n-dimensional Euclidean space.

Each vector in $\mathbb{R}^n$ can be assigned a length. This is done via the idea of inner product:

Definition 2.6. The **inner product** of two vectors $x, y \in \mathbb{R}^n$ is denoted by $x \cdot y$ or $\langle x, y \rangle$ and is defined by

$$\langle x, y \rangle = x \cdot y = \sum_{j=1}^{n} x_j y_j$$

We denote the **norm** or **length** of a point $x \in \mathbb{R}^n$ by $|x|$ and define it by

$$|x| = \sqrt{\langle x, x \rangle}$$
$$= \sqrt{\sum_{j=1}^{n} x_j^2}$$

The norm is a generalization of the idea of absolute value and in fact the norm of x and the absolute value of x coincide when $n = 1$.

In the case of $\mathbb{R}^n$, the set of unit vectors is denoted by S^{n-1}, which is called the **unit sphere** of $\mathbb{R}^n$. Concerning the sphere S^{n-1}, in $\mathbb{R}^2$ this is a circle, which has dimension 1, whereas $\mathbb{R}^2$ has dimension 2. In $\mathbb{R}^3$ the unit sphere is a usual sphere and has dimension 2, whereas $\mathbb{R}^3$ has dimension 3. In general the unit sphere of $\mathbb{R}^n$ has one less dimension than $\mathbb{R}^n$ itself: the dimension of S^{n-1} is $n-1$. This explains the superscript in the notation S^{n-1}. Regarding notation, θ will generally denote a point on the unit sphere of $\mathbb{R}^n$, not a real number (angle). However, on occasion we may use θ as a real number. As always, the context will clarify the usage.

2.1.1 Hyperplanes

A hyperplane is defined to be a set of points in $\mathbb{R}^n$ which is orthogonal to a specific unit vector θ and which lie a directed distance c from the origin. The precise definition is:

Definition 2.7. A **hyperplane** in $\mathbb{R}^n$ is the set of points $x \in \mathbb{R}^n$ satisfying an equation of the form

$$\langle \theta, x \rangle = c$$

where $\theta \in S^{n-1}$ is a unit vector and where c is a real number.

This definition makes sense. For example, in $\mathbb{R}^3$ if $\theta = (a_1, a_2, a_3)$, then the defining equation is of the form $\langle \theta, x \rangle = a_1 x_1 + a_2 x_2 + a_3 x_3 = c$. It is known from analytic geometry that such a set consists of the points in $\mathbb{R}^3$ which are orthogonal to θ and which lie a directed distance c from the origin. This set is therefore an ordinary plane and we see that definition 2.7 generalizes this concept to $\mathbb{R}^n$.

The equation $\langle \theta, x \rangle = 0$ is a rank 1 linear equation, so the set of points satisfying this equation forms an $n - 1$ dimensional vector space. Therefore, a hyperplane is just a parallel translate of such a vector space and therefore it also has dimension $n - 1$.

In $\mathbb{R}^2$ hyperplanes are lines, which are known to be one-dimensional. It is well known from calculus courses that equations of the form $\langle \theta, x \rangle = c$ in $\mathbb{R}^3$ determine planes that are two-dimensional. So in $\mathbb{R}^3$ hyperplanes are just ordinary planes. In $\mathbb{R}^n$ a hyperplane is $n - 1$ dimensional. As in $\mathbb{R}^2$ a hyperplane is uniquely determined by a unit vector θ and a signed distance c and we take the liberty of using the phrasing "the hyperplane $\langle \theta, x \rangle = c$" rather than the more precise but awkward expression "the hyperplane $\{x \in \mathbb{R}^n : \langle \theta, x \rangle = c\}$." We take similar liberties with other sets, using the defining equation as a synonym for the set itself.

Because unit vectors uniquely determine directions we find that the concepts of "unit vector," "direction," and "point of the unit sphere" are identical. In particular, we can write "$u \in S^{n-1}$" in lieu of "u is a unit vector." In the same way we can say "given a direction" in place of "given a unit vector."

The plan now is to define the Radon transform in higher dimensions by using the integral $\int_L f(x) \, dx$ with the line L replaced by a hyperplane. Before doing this we need to investigate hyperplanes and integration on hyperplanes.

Here are illustrations of typical hyperplanes in $\mathbb{R}^2$ and in $\mathbb{R}^3$. Both hyperplanes are defined by the equation $\langle x, \theta \rangle = p$. The distance from the origin to the hyperplane is p;

this is illustrated in the first figure which shows the hyperplane $\langle x, \theta \rangle = p$ in $\mathbb{R}^2$. This is of course the straight-line orthogonal to θ and a signed distance p from the origin. We have shown the unit sphere S^1 in $\mathbb{R}^2$, which is the unit circle in the figure. We have also shown the unit vector θ in two places: first, inside the sphere, and second, on the hyperplane. The second figure shows the analogous features in $\mathbb{R}^3$ but does not explicitly show the distance p. However, if you measured the distance from the origin to the plane in that diagram, the distance would be p.

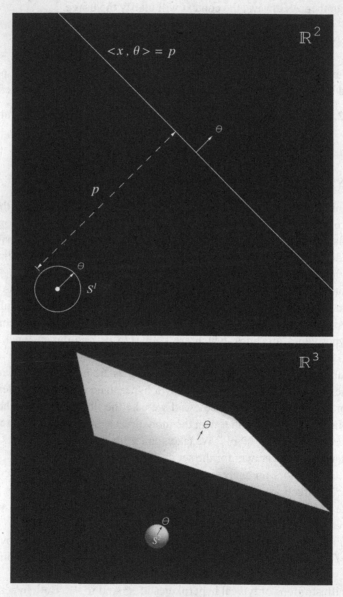

Along with hyperplanes, the following concept is useful.

Definition 2.8. If $\theta \in S^{n-1}$, then $\theta^\perp$ denotes the linear space orthogonal to θ, namely,

$$\theta^\perp = \{x \in \mathbb{R}^n : \langle x, \theta \rangle = 0\}$$

Remark 2.9. It is clear that $\theta^\perp$ is a linear space of dimension $n-1$ because it is defined by the rank 1 homogeneous linear equation $\langle x, \theta \rangle = 0$.

Remark 2.10. Occasionally we use the notation $\theta^\perp$ to represent a unit vector orthogonal to θ. The context will always resolve any ambiguity.

Remark 2.11. Recall that an **orthonormal basis** of a subspace V of $\mathbb{R}^n$ is a spanning set of unit vectors $\{\theta_1, \ldots, \theta_d\}$ in V with the property that these unit vectors are mutually orthogonal.

There is another way of looking at hyperplanes. If $a \in \mathbb{R}^n$ and U is a subset of $\mathbb{R}^n$ we use the notation $U + a = a + U$ to denote the set formed by adding a to arbitrary elements of U. The set $U - a$ is defined similarly.

Proposition 2.12 (hyperplane lemma). *A subset H of $\mathbb{R}^n$ is a hyperplane orthogonal to a unit vector θ lying a distance s from the origin if and only if $H = \theta^\perp + s\theta$.*

If H is the hyperplane $\langle \theta, x \rangle = s$ and if $\{\theta_1, \theta_2, \ldots, \theta_{n-1}, \theta\}$ is an orthonormal basis of $\mathbb{R}^n$ then

$$H = \{x_1\theta_1 + x_2\theta_2 + \cdots x_{n-1}\theta_{n-1} + s\theta : x_1, \ldots, x_{n-1} \in \mathbb{R}\} \tag{2.3}$$

Proof. By definition H is a hyperplane if and only if it is the set of points satisfying the equation $\langle \theta, x \rangle = s$ for some unit vector θ and some real number s. Define

$$V = H - s\theta \tag{2.4}$$

Because H is defined by the equation $\langle \theta, x \rangle = s$, we see that V is defined by the equation $\langle \theta, y \rangle = 0$. By Definition 2.8 this means that $V = \theta^\perp$ and equation (2.4) implies that $H = \theta^\perp + s\theta$. The proof of the converse, that any set of the form $\theta^\perp + s\theta$ is a hyperplane, is an easy variation of the preceding argument and is omitted.

If $H = \theta^\perp + s\theta$ then any element of H is of the form $u + s\theta$ where $u \in \theta^\perp$. Hence, if $\{\theta_1, \theta_2, \ldots, \theta_{n-1}, \theta\}$ is an orthonormal basis of $\mathbb{R}^n$ then it must be that $\{\theta_1, \theta_2, \ldots, \theta_{n-1}\}$ is an orthonormal basis for $\theta^\perp$, so u is of the form $x_1\theta_1 + x_2\theta_2 + \cdots + x_{n-1}\theta_{n-1}$ and the last statement follows immediately. ∎

Remark 2.13. The symbol ∎ is used to indicate the end of a proof.

2.2 Integration on Hyperplanes and the Definition of the Radon Transform

Almost all the basic theory of the Radon transform is a consequence of only two theorems from the calculus: the theorem on change of variables in an integral and the theorem on interchange of order of integration.

Before stating these theorems we need to develop some useful notation and to discuss the theory of integration.

We use the symbol

$$\int\limits_{\mathbb{R}^n} f(x)\,dx$$

to denote the (multiple) integral of the function $f(x) = f(x_1, x_2, \ldots, x_n)$ defined on $\mathbb{R}^n$. This avoids the awkward multiple integral notation:

$$\iint f(x, y)\,dx\,dy$$

$$\iiint f(u, v, w)\,du\,dv\,dw, \text{ and}$$

$$\int \cdots \int f(x_1, x_2, \ldots, x_n)\,dx_1\,dx_2 \cdots dx_n$$

and also allows one to see the many similarities in the theory of integration over different Euclidean spaces.

The type of integration studied in calculus courses is called **Riemann integration**. The theory of **Lebesgue integration** is an extension of Riemann integration in which any Riemann integrable function is also Lebesgue integrable and the Lebesgue and Riemann integrals have the same value. Therefore those readers who have studied the calculus already know many examples of Lebesgue integrable functions. Namely all the Riemann integrable functions studied in the calculus are also Lebesgue integrable, and there are many interesting Lebesgue integrable functions that are not Riemann integrable. Regardless, you can still read this chapter interpreting "integrable" to mean "Riemann integrable." However, care has been taken to ensure that all results are also valid for Lebesgue integrable functions. An example of a Lebesgue integrable function is given in example 2.36, and even those readers with only a background in the calculus will be able to appreciate this example.

All the basic properties of Riemann integrals extend to Lebesgue integrals. One of the most important of these properties is linearity. Lebesgue integration is a linear operator in the sense that if f and g are Lebesgue integrable functions and c is a constant, then

$$\int\limits_{\mathbb{R}^n} f(x) + g(x)\,dx = \int\limits_{\mathbb{R}^n} f(x)\,dx + \int\limits_{\mathbb{R}^n} g(x)\,dx$$

$$\int\limits_{\mathbb{R}^n} c \cdot f(x)\,dx = c \cdot \int\limits_{\mathbb{R}^n} f(x)\,dx$$

Four features of Lebesgue integration are particularly important for us. First, it is built into the definition of the Lebesgue theory that a function f is Lebesgue integrable if and only if its absolute value, $|f|$, is Lebesgue integrable. This feature becomes crucial in a few later arguments. For example, when we come to define the Fourier transform we need to integrate functions of the form $e^{-ix\omega}f(x)$ for real valued functions $f(x)$ and real constants ω. If f is known to be Lebesgue integrable, then so is $e^{-ix\omega}f(x)$ because these functions have the same absolute value.

The second important feature of Lebesgue integration is that it allows us as a legal value of both functions and integrals. Functions defined on $\mathbb{R}^n$ fall into three mutually exclusive classes: those for which it is impossible to compute a Lebesgue integral, those for which the Lebesgue integral can be computed but has an infinite value, and those possessing finite Lebesgue integrals. A function in the latter class is called **Lebesgue integrable**, and the set of all Lebesgue integrable functions on $\mathbb{R}^n$ forms a vector space denoted by $L^1(\mathbb{R}^n)$. When we say that f is integrable or Lebesgue integrable we mean that not only can we compute $\int_{\mathbb{R}^n} f(x)\,dx$, but also that $\int_{\mathbb{R}^n} |f(x)|\,dx < \infty$. The same meaning is implied when we say $f \in L^1(\mathbb{R}^n)$. Sometimes we simplify by saying "integrable" rather than "Lebesgue integrable."

The third important feature of Lebesgue integration is that it is fundamentally based on the notion of the measure of a set. We will not go into the general definition of the measure of a set. But we do note that if A is a set, then it may or may not have a measure. For those sets that possess measures we let the **measure** be denoted by the symbol $|A|$. Intuitively speaking, measure means length in $\mathbb{R}$, area in $\mathbb{R}^2$, and volume in $\mathbb{R}^3$. In $\mathbb{R}^n$ measure is the appropriate generalization of length, area, and volume to higher dimensions. However, we occasionally use the term "volume" as a synonym for measure, even in $\mathbb{R}^n$ with $n > 3$. Clearly, the measure of a set A depends on which Euclidean space contains A. For example the interval $[0, 1]$ has measure 1 as a subset of $\mathbb{R}$ and measure 0 as a subset of $\mathbb{R}^2$. In general, the context will determine which measure is being used. If it is necessary for clarity or emphasis we will say "the n-dimensional measure of A," otherwise we will just say "the measure of A." Concerning the symbol $|A|$, there is no possible confusion with the norm or absolute value symbol because these do not apply to sets. The somewhat awkward phrasing "measure of a measurable set" is made necessary because there are examples of sets that are not measurable. Nonmeasurable sets, albeit plentiful in theory, are hard to come by practically and they will not play a role in this book. All the sets familiar to us in the calculus, such as intervals, graphs of continuous functions, etc., are measurable and their measures in the Lebesgue sense are exactly what we would expect from our experience with the calculus. For example, the measure of a rectangle in $\mathbb{R}^2$ is exactly the product of the lengths of the base and the altitude. A two dimensional rectangle has measure zero in $\mathbb{R}^3$, but the measure of the region contained inside a sphere of radius R in $\mathbb{R}^3$ has a measure exactly equal to its traditional volume: $\frac{4}{3}\pi R^3$.

Remark 2.14. If A is a measurable set, then we use the notation

$$\int_A f(x)\,dx$$

to denote the integral of f restricted to A. Sometimes we write

$$\int\limits_{x \in A} f(x)\, dx$$

to emphasize the role of the variable of integration.

Sets of measure zero play a particularly useful role in the Lebesgue theory. Intuitively, a set of measure zero has a negligible amount of measure. A precise definition is possible, but all readers will either know the correct definition or else will be well served by this description.[1] One important thing about sets of measure zero is that if two Lebesgue integrable functions agree everywhere except on a set of measure zero, then they have the same integral. Therefore, sets of measure zero can be ignored as far as Lebesgue integration is concerned. Phrasing of the form "... agree except on a set of measure zero" occurs so often that we formalize the idea; we say that something happens **almost everywhere** if it happens except on a set of measure zero. We also say that a property is true for **almost all** x if it is true for all x except those in some set of measure zero.

Example 2.15. In $\mathbb{R}^2$, let f be the function which is identically 1 and let g be the function which is 1 except on the x_1 axis where it is zero. Then we can say that $f = g$ almost everywhere because the x_1 axis has measure zero in $\mathbb{R}^2$.

The fourth feature of the Lebesgue theory is that it allows an easy extension of the theory of integration to domains more general than $\mathbb{R}^n$. For example, there is a theory of Lebesgue integration on hyperplanes and on spheres. In the hyperplane and the sphere, Lebesgue integration is analogous to surface integration in Euclidean 3-space. Also, in the sphere, it is traditional to speak of the volume of a sphere when we really mean the measure. Of course the sphere, S^{n-1}, has n-dimensional measure zero as a subset of $\mathbb{R}^n$. But when we consider S^{n-1} as a space by itself and compute its volume via its $(n-1)$ dimensional Lebesgue measure, then we get a positive result. Before stating the formula for computing the volume of a sphere, we introduce the concept of a ball in $\mathbb{R}^n$. The open **unit ball** of $\mathbb{R}^n$ is denoted by the symbol B^n, or just B if the dimension is understood. It is defined by the inequality $|x| < 1$. The ball B^2 is a disk and the ball B^3 is the interior of the sphere S^2. In general, the unit sphere S^{n-1} is the boundary of the unit ball B^n. Occasionally people speak of the volume of the sphere when they actually mean the volume of the ball. This is clarified by noting that $|S^{n-1}|$ refers to the $(n-1)$-dimensional measure of the unit sphere and that this is different than the n-dimensional volume of the ball. We frequently will write B in place of B^n, if the dimension is known. However, we will always write S^{n-1} for the unit sphere.

---

[1] Okay, so here is the precise definition. First, a rectangle in $\mathbb{R}^n$ is a generalization of a rectangle in $\mathbb{R}^2$; it is a set of the form $R = [a_1, b_1] \times \cdots \times [a_n, b_n]$. We define $\mu(R) = (b_1 - a_1)(b_2 - a_2) \cdots (b_n - a_n)$ and call this the (n-dimensional) measure of the rectangle. Then we say that a set B has **Lebesgue measure** 0. If for every $\varepsilon > 0$ we can find a sequence of rectangles $R_1, R_2, \ldots, R_k, \ldots$ such that B is contained in the union of these rectangles and $\sum_{k=1}^{\infty} \mu(R_k) < \varepsilon$.

The following theorem details the volumes of spheres and balls in n-dimensional space. The proofs of these results may be found in Flanders [176] or Stroock [588].

Theorem 2.16.

$$\left| S^{n-1} \right| = \frac{2\pi^{n/2}}{\Gamma(n/2)}$$

$$\left| B^n \right| = \frac{1}{n} \left| S^{n-1} \right|$$

$$= \frac{\pi^{n/2}}{\Gamma\left(\frac{n+2}{2}\right)}$$

where Γ denotes the usual gamma function

$$\Gamma(x) = \int\limits_0^\infty e^{-t} t^{x-1} dt$$

defined for $x > 0$.[2]

The fact that $\frac{1}{n}\left| S^{n-1}\right| = \frac{\pi^{n/2}}{\Gamma\left(\frac{n+2}{2}\right)}$ is an easy consequence of the elementary properties of the gamma function.

We can easily compute the volumes of some spheres:

$$\left| S^1 \right| = 2\pi \quad \left| S^3 \right| = 2\pi^2$$
$$\left| S^2 \right| = 4\pi \quad \left| S^4 \right| = \tfrac{8}{3}\pi^2$$

The first two of these formulas should be very familiar: they give the circumference of a unit circle in $\mathbb{R}^2$ and the area of a unit sphere in $\mathbb{R}^3$. The last equation is the 4-dimensional volume of the 4-sphere in $\mathbb{R}^5$.

If $S^{n-1}(c, R)$ denotes the sphere with center c and radius R in $\mathbb{R}^n$, and if $B^n(c, R)$ denotes the open ball with center c and radius R in $\mathbb{R}^n$, then

$$\left| S^{n-1}(c, R) \right| = \left| S^{n-1}\right| R^{n-1}$$
$$\left| B^n(c, R) \right| = \left| B^n\right| R^n$$

For example, in $\mathbb{R}^3$ we get the familiar formulas $4\pi R^2$ for the area of the sphere and $\frac{4}{3}\pi R^3$ for the volume of a ball.

At this point we have given enough of an overview of the Lebesgue theory so that all readers can understand the statements and proofs that are developed in this chapter. From this point on we assume that the term *integrable* means Lebesgue integrable. This assumption should work even for readers who have never met a function that is Lebesgue integrable but not Riemann integrable. The only difference is that when those

---

[2] A fundamental identity for the gamma function is $\Gamma(x + 1) = x\Gamma(x)$ from which we easily derive $\Gamma(n) = (n - 1)!$ when n is a natural number. It can be shown that $\Gamma(\frac{1}{2}) = \sqrt{\pi}$. From these facts we can find $\Gamma(\frac{n}{2})$ for any positive integer n.

readers picture our results, they will picture the Riemann integrable functions that they studied in the calculus. This includes all continuous functions of bounded support. It also includes the so-called "pixel" functions, which are functions defined on subsquares of a fixed square and which are constant on each subsquare. These functions are useful in image processing.

So, what is the advantage of using the Lebesgue theory? First, most proofs become easier, because all the hard work has been done in developing the Lebesgue theory. Also, because many discontinuous functions arise in tomography, it is advantageous to use the Lebesgue theory, which handles these issues very smoothly. For example, it is quite possible to have a very nice Riemann integrable function f which is continuous almost everywhere on $\mathbb{R}^n$. In tomography we would like to integrate this function over all hyperplanes. But it is also quite possible that on some of these hyperplanes f will have discontinuities of positive measure and thus not be Riemann integrable. Also it is hard using just the Riemann theory to determine how many of these hyperplanes there are. On the other hand, it is a very simple matter to deal with this by using the Lebesgue theory.

Many good references are available for those readers interested in learning or reviewing the Lebesgue theory of integration, but we will mention only five: Hewitt and Stromberg [304], Strichartz [587], Evans and Gariepy [153], Kuttler [375], and Stroock [588]. The Stroock reference is a particularly focused introduction to the Lebesgue theory. You will find the theorem on change of variable and Fubini's theorem in these references.

Having finished this short tour of the Lebesgue theory of integration we can now precisely state the theorem on change of variables and Fubini's theorem on iterated integration.

Theorem 2.17 (Change of Variables in Integration).[3] *If g is a one-to-one, continuously differentiable function from $\mathbb{R}^n$ to $\mathbb{R}^n$, if f is an integrable function on $\mathbb{R}^n$, and if U is an open set in $\mathbb{R}^n$, then*

$$\int_{g(U)} f(x)\,dx = \int_U f(g(y))\,Jg(y)\,dy$$

The function Jg is the Jacobian determinant defined as follows.

Definition 2.18. Let g be a function from $\mathbb{R}^n$ to $\mathbb{R}^n$ which possesses first partial derivatives of all orders. Define the matrix

$$g'(x) = \begin{bmatrix} \frac{\partial g_1}{\partial x_1} & \frac{\partial g_1}{\partial x_2} & \cdots & \frac{\partial g_1}{\partial x_n} \\ \frac{\partial g_2}{\partial x_1} & \frac{\partial g_2}{\partial x_2} & \cdots & \frac{\partial g_2}{\partial x_n} \\ \vdots & \vdots & & \vdots \\ \frac{\partial g_n}{\partial x_1} & \frac{\partial g_n}{\partial x_2} & \cdots & \frac{\partial g_n}{\partial x_n} \end{bmatrix}$$

---

[3] This can also be called the theorem on substitution. It gives the effect of the substitution $x = g(y)$ on the integral $\int_{g(U)} f(x)\,dx$.

and define the *Jacobian* of g to be the function

$$Jg(x) = \left| \det \left(g'(x) \right) \right|$$

The theorem on interchange of order of integration is called Fubini's theorem and the following convention is useful there and in several other instances later on. It x is a variable element of $\mathbb{R}^n$ then we define

$$x' = (x_1, \ldots, x_k) \text{ and}$$
$$x'' = (x_{k+1}, \ldots, x_n)$$

Therefore, we can write $x = (x', x'')$ and, in the same way, a function f of n variables can be written $f(x', x'')$. When we encounter this notation we must keep in mind that $x' \in \mathbb{R}^k$ actually represents a *k-tuple* of numbers and $x'' \in \mathbb{R}^{n-k}$ represents an $n - k$-tuple, so (x', x'') represents an *n-tuple* of numbers, a typical element of $\mathbb{R}^n$. The context will generally clarify what the correct value of k is. However, when we use the symbol x' in connection with a variable x without mention of k, then we will always assume that $k = n - 1$, that is, that $x' = (x_1, \ldots, x_{n-1})$. In the same way, the differential $dx' = dx_1 \cdots dx_{n-1}$ denotes the volume element on $\mathbb{R}^{n-1}$.

Theorem 2.19 (Fubini's theorem on changing the order of integration). *If f is an integrable function on $\mathbb{R}^n$, then the function assigning $x'' \in \mathbb{R}^{n-k}$ to $\int_{x' \in \mathbb{R}^k} f(x', x'') \, dx'$ is integrable. Hence $\int_{x' \in \mathbb{R}^k} f(x', x'') \, dx'$ exists for almost all x'' in $\mathbb{R}^{n-k}$. Furthermore*

$$\int_{\mathbb{R}^n} f(x) \, dx = \int_{x'' \in \mathbb{R}^{n-k}} \left(\int_{x' \in \mathbb{R}^k} f(x', x'') \, dx' \right) dx''$$

$$= \int_{x' \in \mathbb{R}^k} \left(\int_{x'' \in \mathbb{R}^{n-k}} f(x', x'') \, dx'' \right) dx'$$

Conversely, if the function assigning x'' to $\int_{x' \in \mathbb{R}^k} |f(x', x'')| \, dx'$ is integrable on $\mathbb{R}^{n-k}$, then f is integrable on $\mathbb{R}^n$ and the previous equation holds.

Except for the notational convention on multiple integrals and the terminology about a function existing "almost everywhere," the theorem on change of variable in an integral and Fubini's theorem are close enough to the corresponding theorems in the calculus to make the validity of the theorems plausible. The interested reader can find the proofs in the references listed above theorem 2.17.

Remark 2.20. Let f be integrable on $\mathbb{R}^n$. In the following figure we see $\mathbb{R}^n$ sliced into hyperplanes $\langle \theta, x \rangle = p$. These hyperplanes are parametrized by p and we can compute the integral of f over each such hyperplane to obtain a function $I(p)$ of the single variable p. The geometric meaning of Fubini's theorem is that $I(p)$ is integrable over the real line and its integral is precisely the multiple integral $\int_{\mathbb{R}^n} f(x) \, dx$.

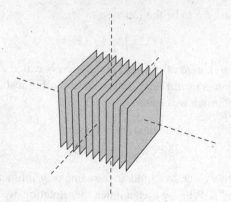

Fubini's theorem: dissection of $\mathbb{R}^n$ into hyperplanes.

Remark 2.21. In the Lebesgue theory it is quite possible to have an integrable function that is finite everywhere, yet which is not integrable on some hyperplanes. A consequence of Fubini's theorem is that there cannot be too many such hyperplanes where the integral is infinite, compare proposition 2.30 and example 2.36.

An important special case of the change of variables theorem occurs when g is a linear transformation with determinant ± 1 and where the open set U is all of $\mathbb{R}^n$. Then by linear algebra we obtain that $g(U) = \mathbb{R}^n$ also. If we let A denote the linear transformation and apply the theorem with $g = A$ we then get

Corollary 2.22. *A linear transformation with determinant ± 1 leaves integrals invariant:*

$$\int_{\mathbb{R}^n} f(x)\, dx = \int_{\mathbb{R}^n} f(A(y))\, dy.$$

The **translation** of a function f by a vector a is the function whose value at x is $f(x - a)$ and is denoted by f_a. A translation is of the form $f \circ T$ where $T(x) = x - a$. It is obvious that the Jacobian of T is identically equal to 1 and we immediately have the following important result:

Corollary 2.23 (Invariance of integration under translations). *Integration is invariant under translations: if a is a fixed point in $\mathbb{R}^n$, then*

$$\int_{\mathbb{R}^n} f(x)\, dx = \int_{\mathbb{R}^n} f_a(x)\, dx = \int_{\mathbb{R}^n} f(x - a)\, dx.$$

A matrix is said to be orthogonal if $AA^t = I$. This happens if and only if the columns of A form an orthonormal basis of $\mathbb{R}^n$. A linear transformation is said to be an orthogonal transformation if it preserves lengths, that is, if $\|T(x)\| = \|x\|$ for all $x \in \mathbb{R}^n$. It is easily shown that a linear transformation is orthogonal if and only if its associated matrix is orthogonal. Orthogonal matrices give rise to linear transformations with determinant ± 1. This easy observation from the defining equation is the basis of the next corollary.

Corollary 2.24. *If $\{\theta_1, \theta_2, \ldots, \theta_{n-1}, \theta_n\}$ is an orthonormal basis for $\mathbb{R}^n$, then*

$$\int_{\mathbb{R}^n} f(x)\,dx = \int_{\mathbb{R}^n} f(y_1\theta_1 + y_2\theta_2 \cdots + y_{n-1}\theta_{n-1} + y_n\theta_n)\,dy \qquad (2.6)$$

Proof. If $B = \{\theta_1, \theta_2, \ldots, \theta_{n-1}, \theta_n\}$ is an orthonormal basis for $\mathbb{R}^n$, then its matrix A is orthogonal, so by the previous corollary

$$\int_{\mathbb{R}^n} f(x)\,dx = \int_{\mathbb{R}^n} f(A(y))\,dy$$

If we now let $y = (y_1, y_2, \ldots, y_{n-1}, y_n)$ be an arbitrary element of $\mathbb{R}^n$, then $A(y) = y_1\theta_1 + y_2\theta_2 \cdots + y_{n-1}\theta_{n-1} + y_n\theta_n$ and hence we see that equation (2.6) is correct. ∎

Remark 2.25. This corollary is the first instance of a useful general principle. Sometimes it is helpful to re-express the argument $x = (x_1, x_2, \ldots, x_n)$ in a more natural coordinate system. If $\{\theta_1, \theta_2, \ldots, \theta_{n-1}, \theta_n\}$ is an orthonormal basis of $\mathbb{R}^n$, then the expression

$$x = y_1\theta_1 + y_2\theta_2 \cdots + y_{n-1}\theta_{n-1} + y_n\theta_n$$

is just as valid as

$$x = (x_1, x_2, \ldots, x_n)$$

In this case we can consider the values $y_1, y_2, \cdots y_n$ to be new coordinates for x. In a sense the symbols "$\theta_j +$" play the same role as the commas "$,$" do in the standard coordinate system.

An important consequence of these theorems is the technique of using polar coordinates in integrals. Most readers will be familiar with the change of variable to polar coordinates in the plane: if f is integrable on the plane, then

$$\int_{\mathbb{R}^2} f(x)\,dx = \int_0^{2\pi} \int_0^\infty f(r(\cos\alpha, \sin\alpha))r\,dr\,d\alpha \qquad (2.7)$$

and conversely, if the integral on the right exists, then f is integrable on the plane. The vector $(\cos\alpha, \sin\alpha)$ is just a typical unit vector, so the outer integration is actually integration over the circle S^1. So using $\theta = (\cos\alpha, \sin\alpha)$ we can rewrite the previous equation as

$$\int_{\mathbb{R}^2} f(x)\,dx = \int_{S^1} \int_0^\infty f(r\theta)r\,dr\,d\theta$$

We will not go into the details, but theorem 2.17, on change of variables, and Theorem 2.19, Fubini's theorem, validate the following generalization of the polar coordinate formula, equation (2.7), to $\mathbb{R}^n$.

$$\int_{\mathbb{R}^n} f(x)\,dx = \int_{S^{n-1}} \int_0^\infty f(r\theta) r^{n-1}\,dr\,d\theta \tag{2.8}$$

Note that in going to higher dimensions the sphere becomes S^{n-1} whereas the power of r changes to r^{n-1}. A special case occurs when f is a function only of $|x|$ in which case the function is constant with respect to θ. The integration over the sphere therefore gives the volume of the sphere, which is denoted by $|S^{n-1}|$. The formula then becomes

$$\int_{\mathbb{R}^n} f(|x|)\,dx = \left| S^{n-1} \right| \int_0^\infty f(r) r^{n-1}\,dr \tag{2.9}$$

$$= \frac{2\pi^{n/2}}{\Gamma(n/2)} \int_0^\infty f(r) r^{n-1}\,dr$$

The books by Strichartz [587] and Stroock [588] contain derivations of the polar coordinate substitution in n-dimensional Euclidean space.

According to the plan for generalizing the Radon transform, we need to be able to integrate functions f defined on $\mathbb{R}^n$ over hyperplanes $\langle \theta, x \rangle = s$. We therefore introduce the following definition.

Definition 2.26 (*Integration over hyperplanes*). Given a hyperplane H defined by $\langle \theta, x \rangle = s$ and a function f defined on $\mathbb{R}^n$, let $\{\theta_1, \theta_2, \ldots, \theta_{n-1}\}$ be an orthonormal basis of $\theta^\perp$. Then, letting $x' = (x_1, \ldots, x_{n-1}) \in \mathbb{R}^{n-1}$, we define the *hyperplane integral of* f over H by

$$\int_{\langle \theta, x \rangle = s} f(x)\,dx = \int_{x' \in \mathbb{R}^{n-1}} f(x_1\theta_1 + x_2\theta_2 + \cdots + x_{n-1}\theta_{n-1} + s\theta)\,dx'$$

Remark 2.27. If $\{\theta_1, \theta_2, \ldots, \theta_{n-1}, \theta\}$ is an orthonormal basis of $\mathbb{R}^n$, then it is immediate that $\{\theta_1, \theta_2, \ldots, \theta_{n-1}\}$ is an orthonormal basis of $\theta^\perp$. The hyperplane lemma then shows that the argument in the integrand is a typical element of the hyperplane $\langle \theta, x \rangle = s$. This justifies the definition. We will frequently make use of an orthonormal basis of $\mathbb{R}^n$ whose last vector is θ. It will go without saying, from now on, that the first $n-1$ vectors in such a basis are an orthonormal basis of $\theta^\perp$.

We are now in a position to generalize the preliminary definition of the Radon transform given earlier.

Definition 2.28 (*The Radon transform*). The **Radon transform** is an operator R defined on $L^1(\mathbb{R}^n)$ whereby for any integrable function f on $\mathbb{R}^n$, the function Rf is defined for $\theta \in S^{n-1}$ and $s \in \mathbb{R}$ by

$$Rf(\theta, s) = \int_{\langle \theta, x \rangle = s} f(x) \, dx$$

whenever the integral exists. The left-hand side is read "the value of the Radon transform of the function f on the hyperplane $\langle \theta, x \rangle = s$." An alternate expression is

$$Rf(\theta, s) = \int_{y \in \theta^{\perp}} f(y + s\theta) \, dy$$

where this integral is just a convenient expression for the hyperplane integral

$$\int_{x' \in \mathbb{R}^{n-1}} f(y_1 \theta_1 + y_2 \theta_2 \ldots + y_{n-1} \theta_{n-1} + s\theta) \, dx'$$

($\{\theta_1, \theta_2, \ldots, \theta_{n-1}\}$ being an orthonormal basis for $\theta^{\perp}$).

For a fixed $\theta \in S^{n-1}$ we also define the **Radon projection (orthogonal to** θ**)**, $R_\theta f$, by

$$R_\theta f(s) = Rf(\theta, s)$$

Remark 2.29. The domain of the Radon transform of a specific function is $S^{n-1} \times \mathbb{R}$, because any element of $S^{n-1} \times \mathbb{R}$ is of the form (θ, s) with $\theta \in S^{n-1}$ and $s \in \mathbb{R}$. We define the symbol $Z^n = S^{n-1} \times \mathbb{R}$ and call it the **cylinder** (of dimension n). A moment's visualization of the product $S^1 \times \mathbb{R}$ justifies the name. The reason for the symbol Z is that "cylinder" is spelled "zylinder" in German. For those readers who know some topology we note that Helgason [291] lets the domain be P_n, the set of all hyperplanes in $\mathbb{R}^n$. The space P_n is homeomorphic to real projective space $\mathbb{R} \, \mathbb{P}^n$ and there is a natural double covering of P_n by Z^n.

The following questions now arise.

1. Under what conditions will the integral in Definition 2.26 exist? This is the same as asking "when does $Rf(\theta, s)$ exist?"
2. If we choose a different basis, will we arrive at the same answer for the hyperplane integral?
3. The hyperplane defined by $\langle \theta, x \rangle = s$ is the same as the one defined by $\langle -\theta, x \rangle = -s$. Will the integrals be the same? This is the same as asking if the Radon transform is an even function on the cylinder.

These questions are answered in the next few results.

Proposition 2.30 (The hyperplane integration theorem). *Let* $\theta \in S^{n-1}$. *If* $f \in L^1(\mathbb{R}^n)$, *then* $\int_{\langle \theta, x \rangle = s} f(x) \, dx$ *exists for almost all* $s \in \mathbb{R}$ *and is independent of the* *orthonormal basis used to define the hyperplane. Furthermore*

$$\int_{s=-\infty}^{\infty} \int_{\langle \theta,x \rangle=s} f(x)\,dx\,ds = \int_{\mathbb{R}^n} f(w)\,dw \tag{2.10}$$

Proof. Let $\{\theta_1, \theta_2, \ldots, \theta_{n-1}, \theta\}$ be an orthonormal basis of $\mathbb{R}^n$. The linear transformation that changes the coordinates of $\mathbb{R}^n$ from one orthonormal basis to another is orthogonal. Applying this idea to the standard basis and the given basis, Corollary 2.24, shows that

$$\int_{\mathbb{R}^n} f(x)\,dx = \int_{\mathbb{R}^n} f(y_1\theta_1 + y_2\theta_2 \cdots + y_{n-1}\theta_{n-1} + y_n\theta)\,dy$$

Now apply Fubini's theorem to the integral on the right, in the y-coordinate system. This gives two results. First, that for almost every s,

$$\int_{y' \in \mathbb{R}^{n-1}} f(y_1\theta_1 + y_2\theta_2 \cdots + y_{n-1}\theta_{n-1} + s\theta)\,dy'$$

is integrable, which proves the statement that $\int_{\langle \theta,x \rangle=s} f(x)\,dx$ exists for almost all $s \in \mathbb{R}$, and, second, that equation (2.10) is valid.

To show the independence of the orthonormal basis used in definition 2.26 let $A = \{\alpha_1, \alpha_2, \ldots, \alpha_{n-1}\}$ and $B = \{\beta_1, \beta_2, \ldots, \beta_{n-1}\}$ be orthonormal bases of $\theta^\perp$ and let C be the $(n-1) \times (n-1)$ matrix which is the matrix of the change of basis from A to B. Then it is easy to show that C is an orthogonal matrix and that if $v = x_1\alpha_1 + x_2\alpha_2 + \cdots x_{n-1}\alpha_{n-1} + s\theta$ is an element of the hyperplane $\langle \theta, x \rangle = s$, then we also have $v = y_1\beta_1 + y_2\beta_2 + \cdots + y_{n-1}\beta_{n-1} + s\theta$ where

$$y' = Cx' \tag{2.11}$$

(as usual $y' = (y_1, \ldots, y_{n-1})$, $x' = (x_1, \ldots, x_{n-1})$). If we apply the change of variable theorem on $\mathbb{R}^{n-1}$ to the integral

$$\int_{\langle \theta,x \rangle=s} f(x)\,dx = \int_{x' \in \mathbb{R}^{n-1}} f(x_1\alpha_1 + x_2\alpha_2 + \cdots + x_{n-1}\alpha_{n-1} + s\theta)\,dx'$$

with the change of variables (2.11) then we also get

$$\int_{\langle \theta,x \rangle=s} f(x)\,dx = |\det(C)| \int_{y' \in \mathbb{R}^{n-1}} f(y_1\beta_1 + y_2\beta_2 \cdots + y_{n-1}\beta_{n-1} + s\theta)\,dy'$$

But C is orthogonal, so $|\det(C)| = 1$ and this proves the hyperplane integral is independent of the orthonormal basis used in the definition. ∎

Corollary 2.31. $\displaystyle \int_{\langle \theta,x \rangle=s} f(x)\,dx = \int_{\langle -\theta,x \rangle=-s} f(x)\,dx$

The proof is really easy; just apply the definitions.

Corollary 2.32 (The Radon transform is even). *The Radon transform has the following symmetry property.*

$$Rf(-\theta, -s) = Rf(\theta, s)$$

Proof. This is a direct consequence of the definition of the Radon transform and the previous corollary. ∎

The next corollary tells us that the values of the Radon projection of an integrable function exist as finite real numbers for almost all, if not all, s. It is an immediate consequence of the hyperplane integration theorem and its proof is omitted.

Corollary 2.33. *If f is an integrable function on $\mathbb{R}^n$ and if $\theta \in S^{n-1}$ is a fixed direction, then $R_\theta f(s)$ is defined for almost all s in $\mathbb{R}$. Furthermore*

$$\int_{s \in \mathbb{R}} R_\theta f(s)\, ds = \int_{x \in \mathbb{R}^n} f(x)\, dx$$

This also shows that the projection of an integrable function is again integrable.

The next corollary is also an immediate consequence of the hyperplane integration theorem.

Corollary 2.34. *The integral defining $Rf(\theta, s)$ is independent of the particular orthonormal basis used in the definition of the Radon transform.*

Remark 2.35. The next two figures give an intuitive idea of Proposition 2.30 and the interplay between Fubini's theorem and the change of variables theorem. The first figure illustrates Fubini's theorem, showing the domain of f sliced into hyperplanes parallel to the usual coordinate axes. The second picture shows the same set of hyperplanes rotated to be orthogonal to a unit vector θ. The essence of the proposition is that f will be integrable on almost every hyperplane and that the value of $\int_{x \in \mathbb{R}^n} f(x)\, dx$ is independent of this rotation.

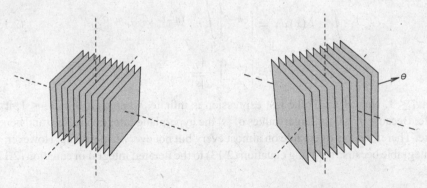

So if f is integrable on $\mathbb{R}^n$, then its restriction to almost every hyperplane is integrable. However, as the next example shows, this does not imply that f is integrable on *every* hyperplane.

Example 2.36. Example of an integrable function f on $\mathbb{R}^n$ which is finite everywhere, yet for which there exists a hyperplane on which f is not integrable.

Construction: let

$$f(x) = \left(x_1^2 + \cdots + x_{n-1}^2\right)^{-\frac{n-1-\sqrt{|x_n|}}{2}} \quad \text{if } \left(x_1^2 + \cdots + x_{n-1}^2\right) \le 1 \text{ and } 0 < |x_n| \le 1$$

with f defined to be zero elsewhere.

Let $\theta = (0, 0, \ldots, 0, 1)$. Note that by the definition of θ the hyperplane $\langle \theta, x \rangle = s$ simplifies to $x_n = s$ and we can apply the hyperplane integration theorem to obtain

$$\int_{\mathbb{R}^n} f(x)\, dx = \int_{-\infty}^{\infty} \int_{\langle \theta, x \rangle = s} f(x)\, dx\, ds \tag{2.12}$$

$$= \int_{-1}^{1} \int_{|x'| \le 1} \left(x_1^2 + \cdots + x_{n-1}^2\right)^{-\frac{n-1-\sqrt{|s|}}{2}} dx'\, ds$$

The restriction of the x' variable to the unit ball in $\mathbb{R}^{n-1}$ is justified by the definition of the function f.

Focus attention on the inner integral. This integration is being performed over $\mathbb{R}^{n-1}$. Use the polar coordinate substitution (2.9), but of course applied with $n-1$ in place of n. We then have

$$x' = r'\theta'$$

where

$$r' = \sqrt{x_1^2 + \cdots + x_{n-1}^2}$$

and where $\theta' \in S^{n-2}$, which is the unit sphere of $\mathbb{R}^{n-1}$. Therefore, when we use the formula for polar coordinates we must use S^{n-2} and the power r^{n-2}. This yields

$$\int_{\langle \theta, x \rangle = s} f(x)\, dx = \left| S^{n-2} \right| \int_{0}^{1} r'^{-(n-1-\sqrt{|s|})} r'^{n-2} dr' \tag{2.13}$$

$$= \left| S^{n-2} \right| \frac{1}{\sqrt{|s|}}$$

for $|s| \le 1$. Now if $s = 0$ the last expression is infinite, whereas if $0 < |s| \le 1$, it is finite. Because f is 0 for larger values of $|s|$ the hyperplane integral is also 0 and hence finite. Therefore f is integrable on almost every but not every hyperplane. However, f is integrable because applying equation (2.13) to the iterated integral in equation (2.12) gives

$$\int_{\mathbb{R}^n} f(x)\, dx = \int_{-1}^{1} \left| S^{n-2} \right| \frac{1}{\sqrt{|s|}} ds = 4 \left| S^{n-2} \right| < \infty.$$

Remark 2.37. Hence, there are integrable functions f for which there is a direction θ and a point p such that $Rf(\theta, p)$ does not exist. Nonetheless, it is possible to show that $Rf(\theta, n)$ does exist almost everywhere when f is an integrable function. Proposition 2.30 was a result in that direction and the full result is proved in Chapter 3, Theorem 3.35. In fact it is not necessary for f to be integrable for $Rf(\theta, p)$ to exist as a finite Lebesgue integral almost everywhere. Theorem 3.35 shows that one only needs to have the integrability of $(1 + |x|)^{-1} f(x)$ for $Rf(\theta, p)$ to exist almost everywhere, and Theorem 3.36 shows that this is a necessary and sufficient condition when f is nonnegative.

At this point the following question may have occurred to you: Why do we integrate over hyperplanes rather than lines in the definition of the Radon transform?

Indeed lines generalize to any linear space, so we could have decided to integrate over lines instead of hyperplanes. In chapter 3 we define what is called the *k-plane transform*. This takes a function on $\mathbb{R}^n$ and integrates over translates of k-dimensional subspaces. If $k = n - 1$, then this is precisely the Radon transform as we defined it. If $k = 1$ this integrates over lines and is called the *x-ray transform*. In dimension 2, there is no practical difference between the Radon transform and the x-ray transform, whereas in higher dimensions there are significant differences.

We have chosen the term Radon transform for the operator that integrates a function over hyperplanes for reasons both of tradition and expediency.

The field of mathematics called integral geometry is closely related to mathematical tomography. One aspect of integral geometry is the study of properties of manifolds that can be determined from integration over submanifolds. Clearly, the Radon transform fits neatly into this aspect of integral geometry. The reader may refer to the books by Gardner [185], Nachbin [433], and Santaló [543] for an introduction to the ideas of integral geometry. Santaló's book presents a more classical form of integral geometry. Ren [515] also treats this more classical form of integral geometry. A typical question in this field is Buffon's needle problem: If one randomly drops a needle of length N on a ruled sheet of paper, where the lines have distance D from each other and $N \leq D$, then what is the probability of the needle touching a line? The answer is $\frac{2N}{\pi D}$.

2.3 Properties of the Radon Transform

All functions in this section are assumed to be integrable on $\mathbb{R}^n$.

The first two results are identical with corollaries 2.33 and 2.32 of the hyperplane integration theorem (Proposition 2.30) but are repeated here for convenience.

Proposition 2.38 (Existence of Radon projections). *If f is an integrable function on $\mathbb{R}^n$ and if $\theta \in S^{n-1}$ is a fixed direction, then $R_\theta f(s)$ is defined for almost all s in $\mathbb{R}$. Furthermore*

$$\int_{s \in \mathbb{R}} R_\theta f(s)\, ds = \int_{x \in \mathbb{R}^n} f(x)\, dx.$$

Proposition 2.39 (Even symmetry of the Radon transform). *The Radon transform has the following symmetry property:*

$$Rf(-\theta, -s) = Rf(\theta, s)$$

This is to say that Rf is an even function on $S^{n-1} \times \mathbb{R}$.

Because integration is a linear operator, it follows that the Radon transform is also a linear operator. Therefore, we immediately have Proposition 2.40.

Proposition 2.40 (Linearity of the Radon transform). *For any functions $f, g \in L^1(\mathbb{R}^n)$ and any constants a, b we have*

$$R(af + bg)(\theta, s) = aRf(\theta, s) + bRg(\theta, s)$$

Two important transformations of Euclidean space are translations and orthogonal linear transformations. The next two results show the effect of these on the Radon transform. First, the reader should recall the definition of composition of functions: $f \circ g(x) = f(g(x))$. We recall that the translation of a function f by a vector a is the function whose value at x is $f(x - a)$ and is denoted by f_a. Because a translation shifts the graph of a function in the direction a the next result is sometimes called the *shift theorem* for the Radon transform.

Proposition 2.41 (Translation-Shift Theorem).

$$Rf_a(\theta, s) = Rf(\theta, s - \langle \theta, a \rangle)$$

In words, the Radon projection of the translation of f by a is the translation of the Radon projection of f by $\langle \theta, a \rangle$.

Proof. If $\{\theta_1, \theta_2, \ldots, \theta_{n-1}, \theta\}$ is an orthonormal basis of $\mathbb{R}^n$, then

$$Rf_a(\theta, s) = \int_{x' \in \mathbb{R}^{n-1}} f_a(x_1\theta_1 + x_2\theta_2 + \cdots + x_{n-1}\theta_{n-1} + s\theta) \, dx' \qquad (2.14)$$

$$= \int_{x' \in \mathbb{R}^{n-1}} f(x_1\theta_1 + x_2\theta_2 + \cdots + x_{n-1}\theta_{n-1} + s\theta - a) \, dx'$$

By the orthonormality of the basis we can expand the vector a in the following way: define $b_k = \langle a, \theta_k \rangle$ for $k = 1, \ldots, n - 1$. Then,

$$a = \sum_{k=1}^{n-1} b_k\theta_k + \langle a, \theta \rangle \theta$$

so we can rewrite equation (2.14) as

$$Rf_a(\theta, s)$$
$$= \int_{x' \in \mathbb{R}^{n-1}} f((x_1 - b_1)\theta_1 + \cdots + (x_{n-1} - b_{n-1})\theta_{n-1} + (s - \langle a, \theta \rangle)\theta) \, dx'.$$

Because of the translation invariance of integration on Euclidean spaces, in this case on $\mathbb{R}^{n-1}$, we can treat $(x_j - b_j)$ as variables. Hence the last expression is the hyperplane integral of f on the hyperplane $\langle x, \theta \rangle = s - \langle a, \theta \rangle$. This is the same as $Rf(\theta, s - \langle a, \theta \rangle)$ and the proof is finished. ∎

Proposition 2.42 (Rotation theorem). *Let T be an orthogonal transformation of $\mathbb{R}^n$. Then*

$$R(f \circ T)(\theta, s) = Rf(T(\theta), s)$$

Proof. If $\{\theta_1, \theta_2, \ldots, \theta_{n-1}, \theta\}$ is an orthonormal basis of $\mathbb{R}^n$, then

$$R(f \circ T)(\theta, s) = \int_{x' \in \mathbb{R}^{n-1}} (f \circ T)(x_1\theta_1 + x_2\theta_2 + \cdots + x_{n-1}\theta_{n-1} + s\theta) \, dx'$$

$$= \int_{x' \in \mathbb{R}^{n-1}} f(x_1 T(\theta_1) + x_2 T(\theta_2) + \cdots + x_{n-1} T(\theta_{n-1}) + s T(\theta)) \, dx'.$$

We used the linearity of T to expand the linear combination in this equation. But the orthogonality of the transformation T implies that $\{T(\theta_1), T(\theta_2), \ldots, T(\theta_{n-1}), T(\theta)\}$ is an orthonormal basis of $\mathbb{R}^n$, so by the definition of hyperplane integration, the last expression is precisely $\int_{\langle T(\theta), x \rangle = s} f(x) \, dx = Rf(T(\theta), s)$. ∎

A *radial function* is one of the form $f(|x|)$, where f is a function of a single real variable. Radial functions are precisely those that are invariant under all rotations. Therefore, the next result is an immediate consequence of the rotation theorem.

Corollary 2.43 (Radial function theorem). *If f is a radial function, then its Radon transform is direction independent. More precisely: if $f(x) = g(|x|)$ and f is integrable on $\mathbb{R}^n$, then for any directions $\theta, \eta \in S^{n-1}$ we have*

$$Rf(\theta, p) = Rf(\eta, p)$$

for almost all p.

Note that $f = g(|x|)$ is integrable on $\mathbb{R}^n$ if and only if $g(r)r^{n-1}$ is integrable on the real line, as the polar coordinate theorem shows.

Remark 2.44. The following precise integral formula for the Radon transform of a radial function $f(x) = g(|x|)$ is a special case of theorem 3.31, Chapter 3.

$$Rf(\theta, p) = \frac{2\pi^{(n-1)/2}}{\Gamma((n-1)/2)} \int_p^\infty g(r) \left(r^2 - p^2\right)^{\frac{n-3}{2}} r \, dr.$$

The rotation theorem can be generalized to arbitrary nonsingular linear transformations but requires some more machinery. First, recall the fundamental theorem of calculus: if φ is continuous on $\mathbb{R}$ and a is any real number, then

$$\frac{d}{dx} \int_a^x \varphi(t)\,dt = \varphi(x).$$

Writing the left-hand side in terms of limits leads to

$$\lim_{h \to 0} \frac{\int_a^{x+h} \varphi(t)\,dt - \int_a^x \varphi(t)\,dt}{h} = \lim_{h \to 0} \frac{\int_x^{x+h} \varphi(t)\,dt}{h}$$

so we end up with

$$\varphi(x) = \lim_{h \to 0} \frac{1}{h} \int_x^{x+h} \varphi(t)\,dt$$

for any continuous function h. A fundamental result in the Lebesgue theory is that this equation is true almost everywhere for integrable functions. Precisely we have lemma 2.45.

Lemma 2.45 (Lebesgue's differentiation theorem). *If φ is integrable on $\mathbb{R}$, then*

$$\varphi(x) = \lim_{h \to 0} \frac{1}{h} \int_x^{x+h} \varphi(t)\,dt$$

almost everywhere.

The proof of this lemma may be found in various sources, for example, Kuttler [375], Corollary 14.9, or Stein [581], corollary I.1.

Next, we define a transform auxiliary to the Radon transform.

Definition 2.46. If f is integrable on $\mathbb{R}^n$ define

$$Mf(\theta, s) = \int_{\langle \theta, x \rangle \le s} f(x)\,dx$$

Lemma 2.47. *Let $\theta \in S^{n-1}$, let $s \in \mathbb{R}$ and let f be integrable on $\mathbb{R}^n$. Then*

$$\frac{\partial}{\partial s} Mf(\theta, s) = Rf(\theta, s)$$

Proof. Because θ is a unit vector it is easy to see that an orthogonal change of variables will reduce us to the case where θ is the standard unit vector $(0, \ldots, 0, 1)$. In this case $\langle \theta, x \rangle = x_n$ and the domain of integration for $Mf(\theta, s)$ is $\{x_n \le s\}$. The integral defining $Mf(\theta, s)$ can be calculated by Fubini's theorem as follows. First, define

$$\varphi(x_n) = \int_{x' \in \mathbb{R}^{n-1}} f(x', x_n)\,dx'$$

Then φ represents the inner integral in the application of Fubini's theorem:

$$
Mf(\theta, s) = \int_{x_n \leq s} f(x) \, dx
$$

$$
= \int_{-\infty}^{s} \int_{x' \in \mathbb{R}^{n-1}} f(x', x_n) \, dx' dx_n
$$

$$
= \int_{-\infty}^{s} \varphi(x_n) \, dx_n
$$

Furthermore, Fubini's theorem guarantees that φ is integrable on the real line. Therefore,

$$
\lim_{h \to 0} \frac{Mf(\theta, s+h) - Mf(\theta, s)}{h} = \lim_{h \to 0} \frac{\int_{-\infty}^{s+h} \varphi(x_n) \, dx_n - \int_{-\infty}^{s} \varphi(x_n) \, dx_n}{h}
$$

$$
= \lim_{h \to 0} \frac{1}{h} \int_{s}^{s+h} \varphi(x_n) \, dx_n
$$

$$
= \varphi(s)
$$

almost everywhere, the last step being justified by the integrability of φ and Lebesgue's differentiation theorem. This establishes the existence of the partial derivative and the equation

$$
\frac{\partial}{\partial s} Mf(\theta, s) = \varphi(s)
$$

The proof is completed by the easy observation that $\varphi(s) = \int_{\langle \theta, x \rangle = s} f(x) \, dx = Rf(\theta, s)$. ∎

Now we can prove the result about linear transformations.

Proposition 2.48 (Linear transformation theorem). *Let T be a nonsingular linear transformation of $\mathbb{R}^n$ with transpose denoted by T^*. Then*

$$
R(f \circ T)(\theta, s) = \frac{1}{|\det T|} \frac{1}{|(T^*)^{-1}(\theta)|} Rf\left(\frac{(T^*)^{-1}(\theta)}{|(T^*)^{-1}(\theta)|}, \frac{s}{|(T^*)^{-1}(\theta)|} \right)
$$

Remark 2.49. If T is an orthogonal transformation and θ is a unit vector, then $(T^*)^{-1}(\theta)$ is also a unit vector. However, this is not true for a general linear transformation, and this accounts for the more complicated form here versus the form in the rotation theorem.

Proof. Recall that the defining property of the transpose is that for all $x, y \in \mathbb{R}^n$ we have

$$\langle Tx, y \rangle = \langle x, T^*y \rangle.$$

Let D_s be the set of all points x in $\mathbb{R}^n$ which satisfy the inequality $\langle \theta, x \rangle \leq s$. Also let E_s be the set of all points y in $\mathbb{R}^n$ which satisfy the inequality $\langle (T^*)^{-1}(\theta), y \rangle \leq s$. Then we can prove that the image of D_s under T satisfies the equation

$$T(D_s) = E_s. \tag{2.15}$$

In fact equation (2.15) is equivalent to $D_s = T^{-1}(E_s)$, because T is nonsingular. But $x \in D_s$ if and only if $\langle \theta, x \rangle \leq s$, which of course happens if and only if $\langle \theta, T^{-1}Tx \rangle = \langle (T^{-1})^*\theta, Tx \rangle \leq s$. By the definition of E_s, this last inequality holds if and only if $Tx \in E_s$. This is equivalent to $x \in T^{-1}(E_s)$ and this establishes equation (2.15).

Now apply the change of variables theorem with the substitution defined by T. Note that the Jacobian of T is just $|\det T|$, so we get

$$\int_{T(D_s)} f(y)\,dy = \int_{D_s} f(T(x))\,JT(x)\,dx \tag{2.16}$$

$$= |\det T| \int_{D_s} (f \circ T)(x)\,dx.$$

Note that $(T^*)^{-1}(\theta)$ is not necessarily a unit vector. This is not a problem, however, because the set $T(D_s)$ is the same as the set E_s and it is easy to see that E_s is identical with the set of points $y \in \mathbb{R}^n$ such that

$$\left\langle \frac{(T^*)^{-1}(\theta)}{|(T^*)^{-1}(\theta)|}, y \right\rangle \leq \frac{s}{|(T^*)^{-1}(\theta)|}.$$

Here we see that $\frac{(T^*)^{-1}(\theta)}{|(T^*)^{-1}(\theta)|}$ is a unit vector. Therefore, equation (2.16) leads to

$$\int_{\left\langle \frac{(T^*)^{-1}(\theta)}{|(T^*)^{-1}(\theta)|}, y \right\rangle \leq \frac{s}{|(T^*)^{-1}(\theta)|}} f(y)\,dy = |\det T| \int_{\langle \theta, x \rangle \leq s} (f \circ T)(x)\,dx.$$

This can be rephrased in terms of the operator M as

$$Mf\left(\frac{(T^*)^{-1}(\theta)}{|(T^*)^{-1}(\theta)|}, \frac{s}{|(T^*)^{-1}(\theta)|} \right) = |\det T|\, M(f \circ T)(\theta, s).$$

If we now differentiate both sides of this equation with respect to s and use lemma 2.47 then we get

$$\frac{1}{|(T^*)^{-1}(\theta)|} Rf\left(\frac{(T^*)^{-1}(\theta)}{|(T^*)^{-1}(\theta)|}, \frac{s}{|(T^*)^{-1}(\theta)|} \right) = |\det T|\, R(f \circ T)(\theta, s)$$

which establishes the conclusion. ∎

A dilation of a function $f(x)$ is a function of the form $f(\gamma x)$, where γ is a nonzero real number. Therefore we define an operator δ_γ by $\delta_\gamma f(x) = f(\gamma x)$ and we call δ_γ the **dilation operator** with factor γ.

Proposition 2.50 (Dilation theorem). *Let $\theta \in S^{n-1}$, let $s \in \mathbb{R}$ and let f be integrable on $\mathbb{R}^n$. Then if $\gamma \neq 0$ we have*

$$R(\delta_\gamma f)(\theta, s) = \gamma^{1-n} Rf(\theta, \gamma s)$$

Proof. (sketch). If $\gamma > 0$, this is an easy consequence of the previous theorem with $T(x) = \gamma x$. In this case the matrix of T is γI and hence $\det T = \gamma^n$, $T = T^*$, and $T^{-1} = \frac{1}{\gamma} T$. If $\gamma < 0$ we can use the fact that $\gamma = sgn(\gamma)|\gamma|$ to achieve the same result. ∎

In Chapter 1 we gave the name "convolution" to the filtering process required for reconstructing a function from its projections. In this chapter we can precisely define the concept.

Definition 2.51 (Convolution). If f and g are integrable functions on $\mathbb{R}^n$, then we define the **convolution** $f * g$ of f and g by

$$(f * g)(x) = \int_{\mathbb{R}^n} f(x - y) g(y) \, dy$$

If F and G are integrable functions on the cylinder Z^n then we define their convolution as

$$(F * G)(\theta, s) = \int_{t \in \mathbb{R}} F(\theta, s - t) g(t) \, dt$$

Note that the integration in this case is only with respect to the second variable.

The question arises as to which functions can be convolved successfully. By using Fubini's theorem and the change of variables theorem we can show that the convolution of integrable functions is again integrable. To show this let f and g be integrable functions on $\mathbb{R}^n$ and define $h = f * g$ according to definition 2.51. We now use Fubini's theorem on $\mathbb{R}^n \times \mathbb{R}^n$ as follows: in the current context x' and x'' will both denote elements of $\mathbb{R}^n$ so (x', x'') will be a typical element of $\mathbb{R}^n \times \mathbb{R}^n$. We can then write definition 2.51 in the form

$$h(x') = \int_{x'' \in \mathbb{R}^n} f(x' - x'') g(x'') \, dx''$$

Let

$$K = \int_{\mathbb{R}^n} |f(u)| \, du$$

Because f is integrable, K is a finite real number. Then, note that in the following integration x'' is constant,

$$\int_{\mathbb{R}^n} \left| f(x' - x'') g(x'') \right| dx' = \left| g(x'') \right| \int_{\mathbb{R}^n} \left| f(x' - x'') \right| dx'$$

$$= \left| g(x'') \right| \int_{\mathbb{R}^n} \left| f(x') \right| dx' = K \left| g(x'') \right|.$$

Here, we used the invariance of integration under translations, which was proved in corollary 2.23, to drop the term x'' in the integral. We thus have

$$\int_{\mathbb{R}^n} \left(\int_{\mathbb{R}^n} \left| f(x' - x'') g(x'') \right| dx' \right) dx'' = \int_{\mathbb{R}^n} K \left| g(x'') \right| dx''$$

$$= K \int_{\mathbb{R}^n} \left| g(x'') \right| dx'' < \infty$$

by the hypothesis that g is integrable. Thus the function $x'' \mapsto \int_{\mathbb{R}^n} |f(x' - x'') g(x'')| dx'$ is integrable for all $x' \in \mathbb{R}^n$. But by the absolute integrability of the Lebesgue theory this means that also the function $x'' \mapsto \int_{\mathbb{R}^n} f(x' - x'') g(x'') dx'$ is integrable, so by the second part of Fubini's theorem, the function $(x', x'') \mapsto f(x' - x'') g(x'')$ is integrable on $\mathbb{R}^{2n}$. Now, by the first part of Fubini's theorem, the function that takes x' to $\int_{\mathbb{R}^n} f(x' - x'') g(x'') dx''$ is integrable and this is precisely the function $h = f * g$. We thus have proved that the convolution is integrable.

Now that we know that convolutions of integrable functions are integrable, we can investigate the effect of convolution on the Radon transform.

Proposition 2.52 (Convolution theorem for the Radon transform). *If f and g are integrable on $\mathbb{R}^n$, then so is $f * g$ and*

$$R(f * g)(\theta, s) = Rf(\theta, s) * Rg(\theta, s)$$

Proof. Before beginning the proof, recall our convention that anytime there is a variable of the form $u = (u_1, \ldots, u_n) \in \mathbb{R}^n$ we let $u' = (u_1, \ldots, u_{n-1}) \in \mathbb{R}^{n-1}$.

We start by taking an orthonormal basis $\{\theta_1, \theta_2, \ldots, \theta_{n-1}, \theta\}$ of $\mathbb{R}^n$ containing θ. Then we use the definition of the Radon transform and the definition of convolution to get

$$R(f * g)(\theta, s)$$

$$= \int_{x' \in \mathbb{R}^{n-1}} (f * g)(x_1\theta_1 + x_2\theta_2 + \cdots + x_{n-1}\theta_{n-1} + s\theta) dx'$$

$$= \int_{\mathbb{R}^{n-1}} \int_{\mathbb{R}^n} f(x_1\theta_1 + x_2\theta_2 + \cdots + x_{n-1}\theta_{n-1} + s\theta - w) g(w) dw dx'$$

Make the orthogonal change of variables

$$w = y_1\theta_1 + y_2\theta_2 + \cdots + y_{n-1}\theta_{n-1} + t\theta$$

in which the differential dw becomes $dy'dt$. This is possible by corollary 2.24 because the integrand in the inner integral is Lebesgue integrable for almost all w. From this we get

$$R(f*g)(\theta, s) \tag{2.17}$$

$$= \int\limits_{\mathbb{R}^{n-1}} \int\limits_{\mathbb{R}} \int\limits_{\mathbb{R}^{n-1}} f((x_1 - y_1)\theta_1 + \cdots + (x_{n-1} - y_{n-1})\theta_{n-1} + (s - t)\theta)$$

$$\times g(y_1\theta_1 + y_2\theta_2 + \cdots + y_{n-1}\theta_{n-1} + t\theta)\, dy'dt\, dx'$$

The integration in the innermost integral is translation invariant, so we can replace $x_j - y_j$ by x_j. Also, because the integrand in equation (2.17) is integrable, Fubini's theorem implies that we can interchange the order of integration. Taking this into account yields

$$R(f*g)(\theta, s) = \int\limits_{\mathbb{R}} \int\limits_{\mathbb{R}^{n-1}} \int\limits_{\mathbb{R}^{n-1}} f(x_1\theta_1 + \cdots + x_{n-1}\theta_{n-1} + (s - t)\theta)\, dx'$$

$$\times g(y_1\theta_1 + y_2\theta_2 + \cdots + y_{n-1}\theta_{n-1} + t\theta)\, dy'dt$$

The innermost integral is

$$\int\limits_{\mathbb{R}^{n-1}} f(x_1\theta_1 + \cdots + x_{n-1}\theta_{n-1} + (s - t)\theta)\, dx'$$

and we recognize this as $Rf(\theta, s - t)$, which is independent of y and hence factors through the integration with respect to the differential dy'. Hence, the previous equation becomes

$$\int\limits_{\mathbb{R}} \left[\int\limits_{\mathbb{R}^{n-1}} g(y_1\theta_1 + y_2\theta_2 + \cdots + y_{n-1}\theta_{n-1} + t\theta)\, dy' \right] Rf(\theta, s - t)\, dt$$

$$= \int\limits_{\mathbb{R}} Rg(\theta, t)\, Rf(\theta, s - t)\, dt$$

$$= Rf(\theta, s) * Rg(\theta, s)$$

Here we used the fact that the integral in brackets is precisely $Rg(\theta, t)$. The proof is complete. ∎

An important connection exists between the Radon transform and the Fourier transform.

Definition 2.53 (Fourier transform). If f is an integrable function on $\mathbb{R}^n$ then we define

$$(\mathcal{F}f)(y) = (2\pi)^{-\frac{n}{2}} \int\limits_{\mathbb{R}^n} f(x)\, e^{-i\langle x,y\rangle}\, dx$$

The function $\mathcal{F}f$ is called the **Fourier transform** of f and is also denoted by $\widehat{f}$. If *expr* denotes a long expression, then we may use the notation $(expr)^{\wedge}$ to denote the Fourier transform of *expr*.

Note that the Fourier transform can be defined on any Euclidean space. If it is necessary to emphasize the dimension, then we will write $\mathcal{F}_n$, but usually the context determines the correct dimension.

Remark 2.54. Some authors use $\widehat{f}$ to denote the Radon transform of f and the notation $\widetilde{f}$ to denote the Fourier transform. However, it is more common outside the field of tomography to use $\widehat{f}$ for the Fourier transform and we follow this convention.

Remark 2.55. The absolute value of the integrand in the definition of the Fourier transform is $|f(x)|$, since $|e^{-i\langle x, y\rangle}| = 1$. Because we know that f is integrable, this shows that the Fourier transform of an integrable function exists for all y in $\mathbb{R}^n$.

For our purposes we need just one property of the Fourier transform: the **uniqueness property of the Fourier transform**: If $\mathcal{F}f = 0$, then $f = 0$ almost everywhere.

The Fourier transform has many other interesting and useful properties. Readers who have not encountered the Fourier transform can find a proof of the uniqueness property in the first few pages of Stein and Weiss [583]. This reference has many more applications of the Fourier transform as do Dym and McKean [138] and Hewitt and Stromberg [304], among numerous excellent references.

The relation between the Radon transform and the Fourier transform is sometimes, called the Fourier slice theorem and sometimes, the projection theorem. We call it the **slice-projection theorem**.

Theorem 2.56 (*Slice projection*). *If f is integrable on $\mathbb{R}^n$ and if θ is a unit vector, then*

$$\mathcal{F}_1 R_\theta f(s) = (2\pi)^{\frac{n-1}{2}} \mathcal{F}_n f(s\theta) \tag{2.18}$$

Proof. By the hypothesis that f is integrable on $\mathbb{R}^n$ we see that the n-dimensional Fourier transform on the right of equation (2.18) is defined and that by the existence of projections theorem, proposition 2.38, the one-dimensional Fourier transform on the left of the equation is defined.

If we now look at the definition of the Fourier transform, the integration over $\mathbb{R}^n$ can be carried out by integrating over any set of parallel hyperplanes. In particular, this could be done over the set of hyperplanes orthogonal to θ. This is a consequence of Fubini's theorem and the result is

$$\mathcal{F}_n f(s\theta) = (2\pi)^{-\frac{n}{2}} \int_{\mathbb{R}} \left[\int_{\langle \theta, x\rangle = p} f(x) e^{-i\langle x, s\theta\rangle} dx \right] dp$$

But if x is in the hyperplane $\langle \theta, x \rangle = p$, then $\langle x, s\theta \rangle = s \langle x, \theta \rangle = sp$. Hence, the exponential term is constant in the inner integral, so

$$\mathcal{F}_n f(s\theta) = (2\pi)^{-\frac{n}{2}} \int_{\mathbb{R}} \left[\int_{\langle \theta, x \rangle = p} f(x) \, dx \right] e^{-isp} dp$$

We recognize the inner integral to be the Radon transform, hence,

$$\mathcal{F}_n f(s\theta) = (2\pi)^{-\frac{n}{2}} \int_{\mathbb{R}} R_\theta f(p) \, e^{-isp} dp$$

$$= (2\pi)^{-\frac{n-1}{2}} \mathcal{F}_1 R_\theta f(s)$$

The constant $(2\pi)^{-\frac{n-1}{2}}$ arises because the one-dimensional Fourier transform needs to borrow the factor $(2\pi)^{-\frac{1}{2}}$ from $(2\pi)^{-\frac{n}{2}}$. ∎

The name of the slice projection theorem comes from the fact, just established, that restricting the Fourier transform of f to the slice of $\mathbb{R}^n$ along the direction θ yields the Fourier transform of the Radon projection orthogonal to θ.

Using the properties of the Fourier transform stated above, we can give the following important uniqueness result.

Theorem 2.57 (Uniqueness theorem for the Radon transform). *An integrable function is uniquely determined by its Radon transform: if f and g are integrable functions on $\mathbb{R}^n$ with*

$$Rf(\theta, s) = Rg(\theta, s) \tag{2.19}$$

for all unit vectors θ and all $s \in \mathbb{R}$, then

$$f(x) = g(x)$$

for almost all $x \in \mathbb{R}^n$.

Proof. Let $h(x) = f(x) - g(x)$. By the linearity of the Radon transform and the hypothesis (2.19) we see that

$$R_\theta h = 0$$

for all unit vectors θ, from which it is immediately known that the Fourier transform of $R_\theta h$ is also zero for all θ. By the slice-projection theorem we see that for all $\theta \in S^{n-1}$ and all $s \in \mathbb{R}$ we have

$$\mathcal{F}_n h(s\theta) = 0$$

But any $x \in \mathbb{R}^n$ is of the form $s\theta$ (e.g., if $x \neq 0$, then we can take $s = |x|$ and $\theta = \frac{x}{|x|}$), so we end up with $\mathcal{F}_n h(x) = 0$ for all $x \in \mathbb{R}^n$. From this and the uniqueness property of the Fourier transform it follows that $h(x) = 0$ for almost all $x \in \mathbb{R}^n$. The definition of h then implies the conclusion of the theorem. ∎

Note that the uniqueness theorem does not give an inversion formula for the Radon transform. An inversion formula would accept the Radon transform as input data and would yield the original function as output. Those readers who are familiar with the inverse Fourier transform will naturally wonder whether this can be applied to give an explicit inversion formula for the Radon transform. This can be done but, unfortunately, not directly. One problem is that not every integrable function has an integrable Fourier transform. Even such a simple integrable function as the characteristic function of the unit square has a nonintegrable Fourier transform.

We will reproduce Radon's original inversion formula from his 1917 paper in section 2.6. Some readers may want to consult the excellent introductory paper of Strichartz [585] in which a formal inversion[4] of the Radon transform via the Fourier transform is presented. In a like manner Nievergelt [449, 450] gives an elementary inversion formula. Nievergelt's formula is closely related to the filtered backprojection method that is discussed in section 2.7.

Remark 2.58. There are many other inversion formulas and uniqueness results for Radon transforms. These are presented in chapter 3.

2.4 A Homogeneous Extension of the Radon Transform

It is useful to extend the domain of Rf from the cylinder $Z = S^{n-1} \times \mathbb{R}$ to $(\mathbb{R}^n \setminus \{0\}) \times \mathbb{R}$. This yields a different parameterization of hyperplane integrals. If the extension is done in such a way as to make the Radon transform homogeneous of degree -1, then this extension corresponds to the Radon transform developed by Gel'fand, Graev, and Vilenkin [194], compare chapter 3 of Gel'fand and Shilov [195].

Definition 2.59. A function $F(\xi, s)$ defined on $(\mathbb{R}^n \setminus \{0\}) \times \mathbb{R}$ is said to be **homogeneous** of degree k if $F(a\xi, as) = a^k F(\xi, s)$ for $a > 0$.

Given a function $h(\theta, s)$ defined on $S^{n-1} \times \mathbb{R}$, we can extend h to $(\mathbb{R}^n \setminus \{0\}) \times \mathbb{R}$ by defining $H(\xi, s) = |\xi|^k h(\frac{\xi}{|\xi|}, \frac{s}{|\xi|})$ for $\xi \in \mathbb{R}^n \setminus \{0\}, s \in \mathbb{R}$. It is easy to check that H is an extension of h which is homogeneous of degree k.

We can extend the Radon transform in this way:

Definition 2.60. If $\xi \in \mathbb{R}^n \setminus \{0\}$, then we define $Gf(\xi, s) = \frac{1}{|\xi|} Rf(\frac{\xi}{|\xi|}, \frac{s}{|\xi|})$.

The proof of the next result follows directly from the definitions and the evenness of the Radon transform.

Proposition 2.61. *The extended Radon transform G has the following symmetry property: Gf is an even function.*

$$Gf(-\xi, -s) = Gf(\xi, s)$$

---

[4] A "formal" inversion derives the inversion formula without regard to the validity of integrations and other limit operations.

Also G is homogeneous of degree −1: if a is a nonzero real number then

$$Gf(a\xi, as) = |a|^{-1} Gf(\xi, s)$$

Finally, G agrees with R on the unit sphere: if $\theta \in S^{n-1}$ then

$$Gf(\theta, s) = Rf(\theta, s)$$

The proof of the next result is omitted because it involves techniques beyond the prerequisites for this chapter.

Theorem 2.62. *If $(\xi, s) \in (\mathbb{R}^n \setminus \{0\}) \times \mathbb{R}$, and $\xi_n \neq 0$, then*

$$Gf(\xi, s) = \frac{1}{|\xi_n|} \int_{\mathbb{R}^{n-1}} f\left(x_1, \ldots, x_{n-1}, \frac{1}{\xi_n}\left(s - (\xi_1 x_1 + \cdots \xi_{n-1} x_{n-1})\right)\right) dx'.$$

It can be shown that the integral defining $Gf(\xi, s)$ gives the hyperplane integral of f over the hyperplane defined by the equation $\langle x, \xi \rangle = s$. In the particular case that ξ is a unit vector θ, we obtain the following.

Corollary 2.63. *If $\theta \in S^{n-1}$, then $Gf(\theta, s) = Rf(\theta, s)$.*

Therefore, Gf is a homogeneous extension of degree −1 of the Radon transform from the cylinder to $(\mathbb{R}^n \setminus \{0\}) \times \mathbb{R}$. The operator G is the version of the Radon transform studied by Gel'fand and coworkers [194, 195].

Remark 2.64. One interpretation of this theorem is that we have a new parametrization of the Radon transform. Let ξ be a unit vector. Formerly we computed the hyperplane integral over $\langle \xi, x \rangle = s$ by parametrizing the hyperplane by

$$x_1 \alpha_1 + x_2 \alpha_2 + \cdots + x_{n-1} \alpha_{n-1} + s\xi$$

where the α_j values are chosen so that $\{\alpha_1, \alpha_2, \ldots, \alpha_{n-1}, \xi\}$ is an orthonormal basis of $\mathbb{R}^n$. The Gel'fand, Graev, and Vilenkin formulation yields the alternate parametrization

$$\left(x_1, \ldots, x_{n-1}, \frac{1}{\xi_n}\left(s - (\xi_1 x_1 + \cdots + \xi_{n-1} x_{n-1})\right)\right).$$

2.5 Examples of the Radon Transform

2.5.1 The Radon Transform on $\mathbb{R}^2$

Let f be an integrable function on $\mathbb{R}^2$. For any hyperplane (i.e., line) of the form $\langle x, \theta \rangle = s$ in $\mathbb{R}^2$, we can find a real number α between 0 and 2π such that $\theta = (\cos\alpha, \sin\alpha)$. Define $\theta^\perp = (-\sin\alpha, \cos\alpha)$. Note that $\{\theta, \theta^\perp\}$ form an orthonormal basis of $\mathbb{R}^2$ and that $\{\theta^\perp\}$ forms a basis of the linear space (line through the origin in this case) which is parallel to the hyperplane $\langle x, \theta \rangle = s$. Then any point on this hyperplane can be expressed as

$$t\theta^\perp + s\theta = (s\cos\alpha - t\sin\alpha, s\sin\alpha + t\cos\alpha)$$

where t is a real number. According to the definition of integration over hyperplanes, (2.26) with $\theta_1 = \theta^\perp$, we get

$$
\begin{aligned}
Rf(\theta, s) &= \int\limits_{\langle x, \theta \rangle = s} f(x)\, dx \\
&= \int\limits_{\mathbb{R}} f\left(t\theta^\perp + s\theta\right) dt \\
&= \int\limits_{-\infty}^{\infty} f(s \cos\alpha - t \sin\alpha, s \sin\alpha + t \cos\alpha)\, dt.
\end{aligned}
$$

This last expression is useful in practical calculations of the Radon transform on $\mathbb{R}^2$.

2.5.2 The Radon Transform on $\mathbb{R}^3$

Let f be an integrable function on $\mathbb{R}^3$. Any unit vector $\theta \in \mathbb{R}^3$ can be expressed in spherical coordinates as $\theta = (\sin\phi \cos\beta, \sin\phi \sin\beta, \cos\phi)$ for some real numbers ϕ and β. We define

$$
\begin{aligned}
\theta_1 &= (\cos\phi \cos\beta, \cos\phi \sin\beta, -\sin\phi) \\
\theta_2 &= (-\sin\beta, \cos\beta, 0)
\end{aligned}
$$

Then it is easy to check that $\{\theta_1, \theta_2, \theta\}$ is an orthonormal basis of $\mathbb{R}^3$ and that $\{\theta_1, \theta_2\}$ is an orthonormal basis of the linear space $\theta^\perp$ orthogonal to θ. Therefore, if f is an integrable function on $\mathbb{R}^3$, then f restricted to the hyperplane $\langle x, \theta \rangle = s$ has the form

$$
f(r\theta_1 + t\theta_2 + s\theta),
$$

where r and t are real numbers. We can then expand this when we integrate to get the Radon transform:

$$
Rf(\theta, s) = \int\limits_{\langle x, \theta \rangle = s} f(x)\, dx = \int\limits_{-\infty}^{\infty} \int\limits_{-\infty}^{\infty} f(r\theta_1 + t\theta_2 + s\theta)\, dr\, dt
$$

$$
\begin{aligned}
= \int\limits_{-\infty}^{\infty} \int\limits_{-\infty}^{\infty} f(&r \cos\phi \cos\beta - t \sin\beta + s \sin\phi \cos\beta, r \cos\phi \sin\beta \\
&+ t \cos\beta + s \sin\phi \sin\beta, -r \sin\phi + s \cos\phi)\, dr\, dt
\end{aligned}
$$

2.5.3 The Radon Transform of Some Common Functions

In this section we derive formulas for the Radon transform of Gaussian functions, balls and squares. When we talk about the Radon transform of a set we really mean the Radon transform of the function which is identically 1 on the set with other values equal to zero. Such a function is called the ***characteristic function*** of the set.

Proposition 2.65 (Gaussian functions). *Let $a > 0$ and let $\gamma_a(x) = \exp(-a|x|^2)$ for $x \in \mathbb{R}^n$. Then*

$$R\,\gamma_n(\theta, s) = \left(\frac{\pi}{a}\right)^{\frac{n-1}{2}} \exp\left(-as^2\right)$$

Proof. It is clear from the definition that γ_a is a radial function. By the radial function theorem, corollary 2.43, it is enough to show that $R\,\gamma_a(e_n, s) = \left(\frac{\pi}{a}\right)^{\frac{n-1}{2}} \exp(-as^2)$. Here e_n is the *n*th standard unit vector and the hyperplane is $x_n = 0$, so

$$R\,\gamma_a(\theta, s) = \int_{x' \in \mathbb{R}^{n-1}} \exp\left(-a|x'|^2 - as^2\right) dx'$$

$$= \exp\left(-as^2\right) \int_{-\infty}^{\infty} \exp\left(-ax_1^2\right) dx_1 \cdots \int_{-\infty}^{\infty} \exp\left(-ax_{n-1}^2\right) dx_{n-1}$$

We used the laws of exponents and also Fubini's theorem (to iterate the integration). These integrals over $(-\infty, \infty)$ are all the same and there are $n - 1$ factors. The result now follows from the classical result that $\int_{-\infty}^{\infty} \exp(-ax_1^2)\,dx_1 = \sqrt{\frac{\pi}{a}}$. ∎

Proposition 2.66. *Let β_r be the characteristic function of the ball of radius r centered at the origin of $\mathbb{R}^n$. Then*

$$R\beta_r(\theta, p) = \begin{cases} \dfrac{\left(\pi(r^2 - p^2)\right)^{\frac{n-1}{2}}}{\Gamma\left(\frac{n+1}{2}\right)} & \text{if } |p| \leq r \\ 0 & \text{if } |p| > r \end{cases}$$

In particular, in dimension $n = 2$ we have

$$R\beta_r(\theta, p) = \begin{cases} 2\sqrt{r^2 - p^2} & \text{if } |p| \leq r \\ 0 & \text{if } |p| > r \end{cases}$$

and in dimension $n = 3$

$$R\beta_r(\theta, p) = \begin{cases} \pi\left(r^2 - p^2\right) & \text{if } |p| \leq r \\ 0 & \text{if } |p| > r \end{cases}$$

Proof. The ball of radius r centered at the origin can be defined as the set of points in $\mathbb{R}^n$ satisfying the equation

$$|x| < r \tag{2.20}$$

It is easy to see from the symmetry of the ball that its characteristic function is radial, so we can use the radial function theorem, corollary 2.43. This means that we can find the Radon transform for all directions by looking only at hyperplanes orthogonal to a standard unit vector. Such a hyperplane has the form $x_n = p$, so equation (2.20) becomes

$$\sqrt{x_1^2 + \cdots x_{n-1}^2 + p^2} \le r \text{ which is equivalent to}$$

$$\sqrt{x_1^2 + \cdots x_{n-1}^2} \le \sqrt{r^2 - p^2}$$

This is valid, of course, only for $|p| \le r$. For other values of p an empty intersection exists between the ball and the hyperplane, so the Radon transform will be zero for $|p| > r$. The last equation is that of a ball of radius $\sqrt{r^2 - p^2}$ in $\mathbb{R}^{n-1}$. The hyperplane integral of the characteristic function then becomes the volume of this $(n-1)$ ball which is

$$\left(\sqrt{r^2 - p^2}\right)^{n-1} \frac{\pi^{\frac{n-1}{2}}}{\Gamma\left(\frac{n+1}{2}\right)}$$

(compare theorem 2.16 for the volume of a ball). Hence, we see that $R\beta_r(\theta, p) = \frac{(\pi(r^2 - p^2))^{\frac{n-1}{2}}}{\Gamma(\frac{n+1}{2})}$ if $|p| \le r$ with the value 0 for other p.   ∎

In the next result we use a slight abuse of language in which the unit vector θ and its angle share the same symbol.

Proposition 2.67. *Let σ_a be the characteristic function of the square of side $2a$ centered at the origin ($a > 0$). Then,*

if $\theta \in \left(0, \frac{\pi}{4}\right]$,

$$R\sigma_a(\theta, p)$$
$$= \begin{cases} 2a\sec\theta & \text{if} & |p| \le a(\cos\theta - \sin\theta) \\ \frac{a\cos\theta + a\sin\theta - |p|}{\sin\theta\cos\theta} & \text{if} & a(\cos\theta - \sin\theta) < |p| \le a(\cos\theta + \sin\theta) \\ 0 & \text{elsewhere} \end{cases}$$

if $\theta \in \left(\frac{\pi}{4}, \frac{\pi}{2}\right)$,

$$R\sigma_a(\theta, p)$$
$$= \begin{cases} 2a\csc\theta & \text{if} & |p| \le a(\sin\theta - \cos\theta) \\ \frac{a\cos\theta + a\sin\theta - |p|}{\sin\theta\cos\theta} & \text{if} & a(\sin\theta - \cos\theta) < |p| \le a(\cos\theta + \sin\theta) \\ 0 & \text{elsewhere} \end{cases}$$

if $\theta \in \left(\frac{\pi}{2}, \frac{3\pi}{4}\right]$,

$$R\sigma_a(\theta, p)$$
$$= \begin{cases} 2a\csc\theta & \text{if} & |p| \le a(\sin\theta + \cos\theta) \\ \frac{|p| - a\sin\theta + a\cos\theta}{\sin\theta\cos\theta} & \text{if} & a(\sin\theta + \cos\theta) < |p| \le a(\sin\theta - \cos\theta) \\ 0 & \text{elsewhere} \end{cases}$$

if $\theta \in \left(\frac{3\pi}{4}, \pi\right)$,

$R\sigma_a(\theta, p)$

$$= \begin{cases} 2a \,|\sec\theta| & if & |p| \le a(\sin\theta + \cos\theta) \\ |p| \dfrac{a\sin\theta + a\cos\theta}{\sin\theta\cos\theta} & y' & a(\sin\theta + \cos\theta) \le |p| \le a(\sin\theta - \cos\theta) \\ 0 & elsewhere \end{cases}$$

Remark 2.68. The translation theorem for the Radon transform can be used with this result to obtain the Radon transform for squares not centered at the origin. The linear transformation theorem can also be applied to the current result to obtain the Radon transform of rectangles and parallelograms.

Remark 2.69. If $\theta = 0, \frac{\pi}{2}$, or π, then it is easy to check that $R\sigma_a(\theta, p) = 2a$ if $|p| \le a$ and is zero elsewhere.

Sketch of the proof. First let θ be in the range from 0 to $\frac{\pi}{4}$. By definition of characteristic function, σ_a has the value 1 on the square $[-a, a] \times [-a, a]$ and the value 0 elsewhere. Therefore, every line integral will be the length of the intersection of the line with the square and these integrals will determine the Radon transform. The following diagram illustrates a typical case.

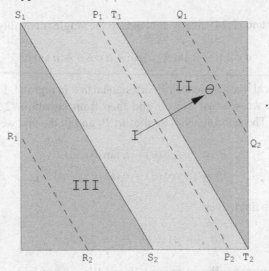

In the diagram we see three typical regions of the square: region I, region II, and region III. We also see the unit vector θ issuing from the origin and five lines orthogonal to θ. Two of these lines, $S_1 S_2$ and $T_1 T_2$ are boundaries of region I. All the lines satisfy the equation

$$x\cos\theta + y\sin\theta = p \tag{2.21}$$

where p is the signed distance of the line from the origin.

First, every line orthogonal to θ in region I has a value of p satisfying

$$|p| < a\cos\theta - a\sin\theta. \tag{2.22}$$

This is because the extremes for p in this region are attained at the lines $T_1 T_2$ and $S_1 S_2$. But it is evident from the diagram that $T_2 = (a, -a)$ so equation (2.21) implies that

$$p = a \cos \theta - a \sin \theta$$

for points on the line $T_1 T_2$. A similar calculation with $S_1 = (-a, a)$ shows that

$$p = -a \cos \theta + a \sin \theta$$

for points on the line $S_1 S_2$. This is exactly the negative of the value for the line $T_1 T_2$, so we then have the desired bounds in (2.22).

In a similar fashion we can determine that in region II all the lines orthogonal to θ satisfy

$$a \cos \theta - a \sin \theta \le p < a \cos \theta + a \sin \theta \qquad (2.23)$$

and that in region III

$$-a \cos \theta - a \sin \theta < p \le -a \cos \theta + a \sin \theta. \qquad (2.24)$$

Inequalities (2.23) and (2.24) clearly coalesce to the single inequality

$$a \cos \theta - a \sin \theta \le |p| < a \cos \theta + a \sin \theta \qquad (2.25)$$

Now take a typical line segment $P_1 P_2$ orthogonal to θ in region I. Let $P_1 = (x_1, y_1)$. From the diagram we see that $y_1 = a$ and then from equation (2.21) we get $x_1 = p \sec(\theta) - \tan(\theta)$. The same analysis applies to P_2 and in the end we arrive at

$$P_1 = (p \sec(\theta) - a \tan(\theta), a)$$
$$P_2 = (p \sec(\theta) + a \tan(\theta), -a).$$

An easy calculation then gives

$$R\sigma_a(\theta, p) = |P_1 P_2| = 2a \sec \theta.$$

The same reasoning applies in regions II and III; we can find expressions for the endpoints of segments $Q_1 Q_2$ and $R_1 R_2$, compute the lengths, and thus obtain the value of the Radon transform.

Piecing all this together gives the desired result for θ in the range 0 to $\frac{\pi}{4}$.

Next let θ be in the range from $\frac{\pi}{4}$ to $\frac{\pi}{2}$. As the reader can easily show by manipulating the preceding diagram, a rotation through negative 90° followed by a reflection about the x axis will transform the calculation of the Radon transform into the range 0 to $\frac{\pi}{4}$. Using the linear transformation theorem, we observe that the effect of this transformation is to interchange sin with cos (and, of course, as a consequence, sec with csc). Making these replacements in the formula for $R\sigma_a$ in the range 0 to $\frac{\pi}{4}$ yields the formula stated for the range $\frac{\pi}{4}$ to $\frac{\pi}{2}$.

To obtain the Radon transform in the range $\frac{\pi}{2}$ to π we again take a linear transformation, this time a rotation through negative 90°. The reader can check that the stated formulas follow from the case of the range 0 to $\frac{\pi}{2}$.

By the symmetry theorem, Proposition 2.39, there is no need to do separate calculations for θ in the range π to 2π because, if δ is any direction with an angle between π and 2π, $-\delta$ has an angle between 0 and π. ∎

2.6 Inversion, Reconstruction, and Approximate Identities

In this and the next few sections we investigate the possibility of recovering a function from its Radon transform. This was treated informally in chapter 1 where the Radon transform was obtained via x-rays through the function. We saw that an approximate inversion formula was possible by using a filtered version of backprojection. In section 2.7 we will precisely define the backprojection operator and derive a theorem justifying the filtered backprojection process. Any such process that yields an exact or approximate version of the unknown function based on the knowledge of its Radon transform is called "reconstruction of a function from its Radon transform" or, in brief, *reconstruction*.

Inversion refers to the possibility of finding a linear operator S with the property that $S(Rf) = f$. If we can do this inversion, then we get an exact reconstruction of f because knowing the formula for f and the values of Rf clearly lead to a formula for f. The general problem of inversion will be treated in chapter 3, although in Section 2.8 of this chapter we present the inversion method developed by Radon in his 1917 paper [508]. In this chapter we also present a rigorous development of the filtered backprojection method for the approximate inversion of the Radon transform.

But first we need a result on approximate identities. This result was used without proof or reference by Radon, so we present a detailed discussion and derivation.

2.6.1 Approximate Identities

The lemma on approximate identities shows how a bounded, continuous, real valued function can be approximated by certain convolutions. It is a useful tool in both inversion and reconstruction.

The linear space $L^1(\mathbb{R})$ of Lebesgue integrable real valued functions can be considered as a commutative algebra if multiplication is defined as convolution. However, this algebra has no multiplicative identity; there is no element ϕ with the property that $f * \phi = f$ for all f in the algebra. We can get around this inconvenience by introducing a family of functions φ_ε, defined for $\varepsilon > 0$, such that $f * \phi_\varepsilon \approx f$ for all f in the algebra. The meaning of the approximation will be made precise in the next lemma. Such a family of functions is called an *approximate identity*. However, do such approximate identities exist? The next lemma shows that they do. Subsequent results show their utility, which may not be apparent just now.

Lemma 2.70 (Approximate identity theorem). *Let φ be a nonnegative, integrable function defined on $(-\infty, \infty)$ such that*

$$\int_{-\infty}^{\infty} \varphi(x)\,dx = 1 \qquad (2.26)$$

For each $\varepsilon > 0$ define

$$\varphi_\varepsilon(x) = \frac{1}{\varepsilon}\varphi\left(\frac{x}{\varepsilon}\right) \qquad (2.27)$$

Then for any continuous, bounded function f, we have

$$\lim_{\varepsilon \to 0} f * \varphi_\varepsilon(x) = f(x)$$

for all real numbers x.

Proof. Because f is assumed to be bounded, we can choose an an upper bound M for $|f|$. Because the integral in equation (2.26) is finite, for any real number $\delta > 0$ we can find a real number $a > 0$ such that

$$\int_{|x|>a} \varphi(x)\,dx < \frac{\delta}{4M}. \qquad (2.28)$$

This is a consequence of the fact that the tails of an absolutely convergent integral must tend to zero. Then using the substitution $x = \frac{y}{\varepsilon}$ we get

$$\int_{|y|>\varepsilon a} \varphi_\varepsilon(y)\,dy < \frac{\delta}{4M}. \qquad (2.29)$$

The same substitution shows that

$$\int_{-\infty}^{\infty} \varphi_\varepsilon(y)\,dy = 1, \qquad (2.30)$$

so, letting x represent a fixed real number, we have

$$f(x) = \int_{-\infty}^{\infty} f(x)\varphi_\varepsilon(y)\,dy \qquad (2.31)$$

(Note that $f(x)$ is constant with respect to the integration; equation (2.31) then follows immediately from equation (2.30)). Use this equation together with the definition of the convolution to get

$$f * \varphi_r(x) - f(x) = \int_{-\infty}^{\infty} f(x-y)\varphi_\varepsilon(y)\,dy - \int_{-\infty}^{\infty} f(x)\varphi_\varepsilon(y)\,dy$$

$$= \int_{-\infty}^{\infty} (f(x-y) - f(x))\varphi_\varepsilon(y)\,dy$$

Break up this last integral into two pieces obtaining

$$f * \varphi_\varepsilon(x) - f(x) = \int_{|y|>\varepsilon a} (f(x-y) - f(x))\varphi_\varepsilon(y)\,dy \qquad (2.32)$$

$$+ \int_{-\varepsilon a}^{\varepsilon a} (f(x-y) - f(x))\varphi_\varepsilon(y)\,dy$$

For the first piece we can use the boundedness of f and equation (2.29) to get

$$\left| \int_{|y|>\varepsilon a} (f(x-y) - f(x))\varphi_\varepsilon(y)\,dy \right| \le 2M \int_{|y|>\varepsilon a} \varphi_\varepsilon(y)\,dy \qquad (2.33)$$

$$< \frac{\delta}{2}$$

For the second piece we can use the second mean value theorem for integrals[5] to get the existence of a real number $c = c(\varepsilon, a)$ in the interval $[-\varepsilon a, \varepsilon a]$ such that

$$\int_{-\varepsilon a}^{\varepsilon a} (f(x-y) - f(x))\varphi_\varepsilon(y)\,dy = (f(x-c) - f(x)) \int_{-\varepsilon a}^{\varepsilon a} \varphi_\varepsilon(y)\,dy \qquad (2.34)$$

From this and equation (2.30) it follows immediately that

$$\left| \int_{-\varepsilon a}^{\varepsilon a} (f(x-y) - f(x))\varphi_\varepsilon(y)\,dy \right| \le |f(x-c) - f(x)| \qquad (2.35)$$

because the nonnegativity of φ and equation (2.30) shows that

$$\left| \int_{-\varepsilon a}^{\varepsilon a} \varphi_\varepsilon(y)\,dy \right| \le 1$$

---

[5] The statement and proof of the second mean value theorem for integrals can be found in many elementary calculus books (e.g., Salas, Hille, and Etgen [540]).

Going back to equation (2.32) and using equations (2.33) and (2.35), we get

$$\left| f * \varphi_\varepsilon (x) - f (x) \right| \le \frac{\delta}{2} + |f (x - c) - f (x)| \tag{2.36}$$

Because f is continuous at x, there exists some $\eta > 0$, η depending on δ, such that $|c| < \eta$ implies $|f (x - c) - f (x)| < \frac{\delta}{2}$. If we now choose any $\varepsilon < \frac{\eta}{a}$ we then have, because $c = c(\varepsilon, a)$ is in the interval $[-\varepsilon a, \varepsilon a]$, that $|c| < \eta$ so indeed we get $|f (x - c) - f (x)| < \frac{\delta}{2}$. Using this back in equation (2.36) shows that for any $\delta > 0$ there exists an $\eta > 0$ such that whenever $\varepsilon < \frac{\eta}{a}$ we have $|f * \varphi_\varepsilon (x) - f (x)| < \delta$. This proves that $\lim_{\varepsilon \to 0} f * \varphi_\varepsilon (x) = f (x)$. ∎

Remark 2.71. The approximate identity theorem is valid on $\mathbb{R}^n$ with

$$\varphi_\varepsilon (x) = \frac{1}{\varepsilon^n} \varphi \left(\frac{x}{\varepsilon} \right). \tag{2.37}$$

Remark 2.72. The second mean value theorem for integrals is usually stated to require the hypothesis that both functions in the integrand are continuous. However, the proof easily extends to the case that φ is a nonnegative integrable function.

Remark 2.73. For those readers who are familiar with the L^p spaces we mention that a much more general result is possible. We can change the hypotheses so that f is now assumed to be an L^p function on $\mathbb{R}^n$. Then, if the convergence of $f * \varphi_\varepsilon$ is taken with respect to the L^p norm, we get the same result as the lemma. See Stein and Weiss [583], theorem 1.18, for details.

2.7 Backprojection, Filtered Backprojection, and Reconstruction

We now investigate approximate methods for reconstructing a function from its Radon transform. These are also called "approximate inversion methods."

In chapter 1 we gave an intuitive introduction to the method of *filtered backprojection* for reconstructing a function from its Radon transform. This is also called the *convolution reconstruction method*. We now formalize the definition of backprojection, which is an important constituent of the filtered backprojection method. Before defining backprojection, let us review the procedure in chapter 1.

In chapter 1 we defined the backprojection $R_\theta^\# g$ of a function $g(\theta, p)$ defined on the cylinder Z^n. The function $R_\theta^\# g$ was obtained by smearing the values of $g(\theta, p)$ backward in the direction orthogonal to θ. In this smearing process all points x on the hyperplane $\langle x, \theta \rangle = p$ take on the value $g(\theta, p)$. In other words $R_\theta^\# g(x) = g(\theta, \langle x, \theta \rangle)$. This is a precise formula of the function $R_\theta^\# g$ that was described informally in chapter 1. In chapter 1 we then averaged these backprojections over all possible directions.

In this chapter we are more interested in the averaged backprojections than in backprojections in specific directions. Therefore, we do not bother to define the functions $R_\theta^\# g$; we go directly to the definition of the averaged backprojections. The resulting function is denoted by $R^\# g$ and is a function defined on $\mathbb{R}^n$. We know from elementary calculus that the average of a function over an interval is proportional to the integral of the function. In our situation we do not care about the proportionality constant and we merely define the backprojection as the integral of the function $g(\theta, \langle x, \theta \rangle)$ over the unit sphere as follows:

Definition 2.74 (Backprojection). If $g : S^{n-1} \times \mathbb{R} \to \mathbb{C}$ we define the *backprojection* of g to be the function $R^{\#}g : \mathbb{R}^n \to \mathbb{C}$ defined by

$$R^{\#}g(x) = \int\limits_{S^{n-1}} g(\theta, \langle x, \theta \rangle) \, d\theta$$

whenever the integral is defined. The backprojection operator is also called the *formal adjoint* of the Radon transform.

We need to use the concept of *measurable function* in a few places from now on. We will not give details on the definition of measurable functions; these can be found in the references for the Lebesgue theory: Hewitt and Stromberg [303], Strichartz [587], Evans and Gariepy [153], Kuttler [375], and Stroock [588]. However, we will say that measurable functions are in the same relation to functions as measurable sets are to sets (compare the discussion on measurable sets, in section 2 of this chapter.) Just as there are many nonmeasurable sets, there are many nonmeasurable functions. On the other hand, just as nonmeasurable sets are hard to find in practice, the same is true of nonmeasurable functions. All the common functions encountered in tomography: integrable, continuous, pixel functions, etc., are measurable.

We will also need to use an extension of Fubini's theorem from $\mathbb{R}^k \times \mathbb{R}^p$ to $S^{n-1} \times \mathbb{R}^n$. This can be justified and a proof of a very general version of Fubini's theorem is given in the references cited for the Lebesgue theory. We will call this the *extended Fubini theorem*. The original Fubini theorem applied to integrable functions, so all the integrals came out to be finite. This is also true of the extended Fubini theorem, but the extended theorem also allows interchanges in the order of integration for any nonnegative measurable functions. The only thing is that in this case some or all of the integrals may be infinite.

The backprojection operator is called the formal adjoint of the Radon transform because it is the dual of the Radon transform in a sense analogous to the way the adjoint of a linear transformation on Euclidean space is dual to the original transformation.[6] Although this duality does not play a role in this chapter, in subsequent chapters it plays a fruitful and even indispensable role. For now we just explain why backprojection is called the formal adjoint of the Radon transform.

In linear algebra the adjoint of a linear transformation $S : \mathbb{R}^m \to \mathbb{R}^n$ is defined to be the unique linear transformation $S^* : \mathbb{R}^n \to \mathbb{R}^m$ which satisfies the equation

$$\langle Sx, y \rangle = \langle x, S^*y \rangle$$

---

[6] Compare, Theorem 3.104 in chapter 3.

for all $x \in \mathbb{R}^m$ and $y \in \mathbb{R}^n$. This equation expresses the duality between the operators S and S^*. We wish to explore the possibility of a similar duality between the Radon transform and the backprojection operator.

Because the inner product in $\mathbb{R}^n$ is defined as a certain sum and because integrals are limits of sums, it is natural to define an inner product on $L^1(\mathbb{R}^n)$ by the formula

$$\langle f, g \rangle = \int_{\mathbb{R}^n} f(x) g(x) \, dx$$

We can define a similar inner product on the cylinder: if $u = u(\theta, p)$ and $w = w(\theta, p)$ then we define

$$\langle u, w \rangle = \int_{\mathbb{R}} \int_{S^{n-1}} u(\theta, p) w(\theta, p) \, d\theta dp$$

In analogy with the case S and S^* we would like to show that

$$\langle Rf, w \rangle = \left\langle f, R^{\#} w \right\rangle$$

Unfortunately, there are certain combinations of functions f and w for which the integrals defining the inner product do not have finite values. This is the reason that $R^{\#}$, the backprojection operator, is called the formal adjoint and not the adjoint of the Radon transform. However, we do have theorem 2.75.

Theorem 2.75 (Formal Adjoint Theorem). $\langle Rf, w \rangle = \langle f, R^{\#} w \rangle$.

Proof. Because we do not need this result in this chapter we proceed rather formally. We do the proof for nonnegative measurable functions, not caring much if the inner products sometimes come out to be infinite. In chapter 3 we revisit this result and derive a more general and precise version: theorem 3.29.

Because we are dealing with nonnegative measurable functions, the extended Fubini theorem provides the justification for the various changes in the order of integration.

$$\left\langle f, R^{\#} w \right\rangle = \int_{\mathbb{R}^n} f(x) R^{\#} w(x) \, dx \qquad (2.38)$$

$$= \int_{\mathbb{R}^n} f(x) \int_{S^{n-1}} w(\theta, \langle x, \theta \rangle) \, d\theta dx$$

$$= \int_{S^{n-1}} \int_{\mathbb{R}^n} f(x) w(\theta, \langle x, \theta \rangle) \, dx d\theta$$

In the inner integral, θ is fixed so we can use corollary 2.33 to the hyperplane integration theorem to obtain

$$\int_{\mathbb{R}^n} f(x)\,w\,(\theta,\langle x,\theta\rangle)\,dx = \int_{p\in\mathbb{R}}\int_{\langle\theta,x\rangle=p} f(x)\,w\,(\theta,\langle x,\theta\rangle)\,dx\,dp$$

$$= \int_{p\in\mathbb{R}}\int_{\langle\theta,x\rangle=p} f(x)\,w\,(\theta,p)\,dx\,dp$$

$$= \int_{p\in\mathbb{R}} w\,(\theta,p)\int_{\langle\theta,x\rangle=p} f(x)\,dx\,dp$$

$$= \int_{p\in\mathbb{R}} w\,(\theta,p)\,Rf\,(\theta,p)\,dp$$

Substituting this result back into equation (2.38) we get

$$\left\langle f, R^{\#}w\right\rangle = \int_{S^{n-1}}\int_{p\in\mathbb{R}} w\,(\theta,p)\,Rf\,(\theta,p)\,dp\,d\theta$$

$$= \langle Rf, w\rangle$$

∎

In chapter 1 we investigated the direct use of backprojection for the inversion of the Radon transform. Recall that even for a simple object consisting of the characteristic function of three disks, the direct use of backprojection gave very poor results, as indicated in the accompanying figure:

Original object   Back Projection   Difference between
                   (500 directions)   object & back projection

The reason for this poor result is indicated by the following result.

Theorem 2.76 (The Backprojection Theorem). *If f is a nonnegative Lebesgue mea-surable function on $\mathbb{R}^n$ then we have*

$$R^{\#}Rf = T * f$$

where the function T is defined by

$$T(x) = \left|S^{n-2}\right|\frac{1}{|x|}$$

Recall that $|S^{n-2}|$ denotes the $n-2$-dimensional measure of the unit $n-2$ sphere.

The proof of this theorem depends on some concepts beyond the prerequisites for this chapter, so we refer the interested reader to chapter 3, theorem 3.67, for a more general result or to Natterer [444]. The convolution $T * f$ differs only by a multiplicative constant from an important operator called the Riesz potential which is studied in detail in chapter 3.

Remark 2.77. Lebesgue integration is very powerful, although there are some integrals that would be useful, but which lie outside the scope of Lebesgue integration. For example $\int_{-1}^{1} \frac{1}{x} dx$ is not Lebesgue integrable. However we can assign a meaning to this integral via the equation

$$\int_{-1}^{1} \frac{1}{x} dx = \lim_{\varepsilon \to 0} \int_{[-1,-\varepsilon] \cup [\varepsilon,1]} \frac{1}{x} dx \qquad (2.39)$$

In this simple example, the limit evaluates to zero. However, many interesting examples of this type of integration exist. Integrals of the form (2.39) are called Cauchy principal value integrals. They are sometimes denoted by the symbol $(p.v.) \int_A f(x) \, dx$, but if it is obvious that the singularity is being treated by the Cauchy method, then we usually drop the prefix $(p.v.)$. Cauchy principal value integrals are a subset of the more general class of singular integrals.

The integral defining the convolution $T * f$ is, in general, a singular integral. However, we will show in chapter 3 that if the function $(1 + |x|)^{-1} f(x)$ is integrable, then the integral defining the convolution $T * f (x)$ exists as a finite Lebesgue integral for almost all x (compare, theorem 3.64, chapter 3). In Chapter 3 we also show that the condition that $(1 + |x|)^{-1} f(x)$ is integrable is essentially the most general condition under which the integral defining the convolution $T * f (x)$ is not a singular integral (see chapter 3, theorem 3.65). Furthermore, $(1 + |x|)^{-1} f (x)$ is integrable if, and essentially only if, the Radon transform of f exists almost everywhere (see chapter 3, theorem 3.35).

Remark 2.78. This remark requires some more advanced knowledge of Fourier transforms.

The function T is not integrable, although it is a tempered function. Readers who are conversant with tempered distributions will know that any tempered function has a Fourier transform that is a tempered distribution. In the case of T it is possible to show that its Fourier transform is also a tempered function: $\widehat{T}(\xi) = 2^{\frac{n}{2}} \pi^{\frac{n}{2}-1} \frac{1}{|\xi|^{n-1}}$.

The convolution of integrable functions is again integrable. Therefore, if g and f are integrable functions, then $g * f$ has a Fourier transform and the convolution theorem for the Fourier transform gives the following formula:

$$(g * f)^{\wedge} (\xi) = (2\pi)^{\frac{n}{2}} \widehat{g}(\xi) \widehat{f}(\xi).$$

However, if we replace g by the nonintegrable function T, then the formula $(T * f)^{\wedge}(\xi) = (2\pi)^{\frac{n}{2}} \widehat{T}(\xi) \widehat{f}(\xi) = 2^n \pi^{n-1} |\xi|^{1-n} \widehat{f}(\xi)$, is not true for all integrable functions f. The study of which functions do give rise to this formula and the consequences for inverting the Radon transform are pursued in chapter 3, section 3.8. However, we do indicate later in this chapter how one may be led to an exact inversion formula for the Radon transform via the backprojection theorem.

The next result, theorem 2.79, is the mathematical foundation of the filtered back-projection method which was proposed in chapter 1 to remedy the deficiency of the backprojection method. Theorem 2.79 rigorously supports the experimental evidence in chapter 1 that applying backprojection to a suitably filtered Radon transform will yield a good reconstruction of the original function.

Theorem 2.79 (Filtered Backprojection). *Let f be integrable on $\mathbb{R}^n$ and let w be a bounded measurable function on the cylinder $Z^n = S^{n-1} \times \mathbb{R}$. Then we have*

$$R^{\#}(w) * f = R^{\#}(w * Rf) \tag{2.40}$$

Proof. Use the definition of convolution on $\mathbb{R}^n$, the definition of backprojection, and the extended Fubini theorem to obtain

$$\left(R^{\#}(w) * f\right)(x) = \int_{\mathbb{R}^n} \left(R^{\#}(w)(x - y)\right) f(y)\,dy \tag{2.41}$$

$$= \int_{\mathbb{R}^n} \left(\int_{S^{n-1}} w\left(\theta, \langle(x - y),\, \theta\rangle\right) d\theta\right) f(y)\,dy$$

$$= \int_{S^{n-1}} \left(\int_{\mathbb{R}^n} w\left(\theta, \langle x, \theta\rangle - \langle y, \theta\rangle\right) f(y)\,dy\right) d\theta$$

The use of the Fubini theorem is justified because w is a bounded measurable function on S^{n-1} and f is an integrable function on $\mathbb{R}^n$ so the function

$$w\left(\theta, \left(\langle x, \theta\rangle - \langle y, \theta\rangle\right)\right) f(y) \tag{2.42}$$

is integrable on the cylinder $Z^n = S^{n-1} \times \mathbb{R}$.

Now fix $x \in \mathbb{R}^n$ and $\theta \in S^{n-1}$ and define the function h on $\mathbb{R}^n$ by

$$h(y) = w\left(\theta, \left(\langle x, \theta\rangle - \langle y, \theta\rangle\right)\right) f(y)$$

The function h is the restriction of the function defined in (2.42) to the set where both x and θ are fixed constants. As in the case of (2.42), h is an integrable function on the cylinder $\mathbb{R}^n$. Therefore by corollary 2.33 we have

$$\int_{\mathbb{R}} R_{\theta} h(s)\,ds = \int_{\mathbb{R}^n} h(y)\,dy \tag{2.43}$$

But $R_\theta h(s)$ is the integral of h over the hyperplane $\langle y, \theta \rangle = s$. On this hyperplane we have $w(\theta, (\langle x, \theta \rangle - \langle y, \theta \rangle)) = w(\theta, (\langle x, \theta \rangle - s))$ which is independent of y so

$$R_\theta h(s) = \int_{\langle y,\theta \rangle = s} h(y)\,dy = \int_{\langle y,\theta \rangle = s} w(\theta, (\langle x, \theta \rangle - \langle y, \theta \rangle))\, f(y)\,dy$$

$$= w(\theta, (\langle x, \theta \rangle - s)) \int_{\langle y,\theta \rangle = s} f(y)\,dy$$

$$= w(\theta, (\langle x, \theta \rangle - s))\, Rf(\theta, s)$$

Thus, from equation (2.43) we get

$$\int_{y \in \mathbb{R}^n} w(\theta, (\langle x, \theta \rangle - \langle y, \theta \rangle))\, f(y)\,dy \tag{2.44}$$

$$= \int_\mathbb{R} w(\theta, (\langle x, \theta \rangle - s))\, Rf(\theta, s)\,ds$$

which is valid for any (θ, x) in the cylinder Z^n. Substituting (2.44) into (2.41) and taking into account the definition of backprojection and the definition of convolution we get

$$\left(R^\#(w) * f\right)(x) = \int_{S^{n-1}} \left(\int_\mathbb{R} w(\theta, (\langle x, \theta \rangle - s))\, Rf(\theta, s)\,ds\right) d\theta$$

$$= \int_{S^{n-1}} (w * Rf)(\theta, \langle x, \theta \rangle)\,d\theta$$

$$= R^\#(w * Rf)(x)$$

which proves the theorem. ∎

Equation (2.40) is the basis of the filtered backprojection method for approximately reconstructing a function from its Radon transform.

Define $A = R^\#(w)$. Then A is called a **filter** or **point spread function** and w is called a **reconstruction kernel**. This is the reason the name "filtered backprojection" is given to this procedure. If we can establish a family of such filters A which form an approximate identity, then the approximate identity lemma shows that, in the limit, $A * f$ approaches the function f. Therefore, we can choose a specific element A of the approximate identity such that $A * f$ approximates f to any desired degree of accuracy.

Equation (2.40) becomes

$$A * f = R^\#(w * Rf)$$

and we can now think of the left-hand side, $A * f$, as an approximation to the unknown function f. However, both Rf (obtained from the x-ray data) and w are known.

Therefore, the approximation to the unknown function f may be computed from the known data via this equation.

Let α be a general procedure which operates on a function f and produces an approximation $\alpha(f)$ to f. Let us consider this idea when α represents some approximate, linear, translation invariant reconstruction algorithm. Paraphrasing a remark of K. T. Smith ([562], page 12) we note that if the approximation procedure results in the same reconstruction in Cincinnati and San Francisco, then it must be translation invariant. Since it is reasonable to assume that a CT scanner works the same in both Cincinnati and San Francisco, we see that α is a linear and translation invariant operator. It is known that linear and translation invariant operators must be of the form $\alpha(f) = A * f$. The precise statement of the theorem connecting linear translation invariant operators to convolutions may be found in Stein and Weiss [583], theorem 3.16. Then theorem 2.79 shows that as long as $A = R^{\#}(w)$ for some w defined on the cylinder, then the only type of linear, translation invariant reconstruction process that we could possibly obtain from the Radon transform is the filtered backprojection method.

In practice the kernel w can be chosen to be a bounded continuous function such that when its backprojection A is convolved with f, then $A * f$ does give a good approximation to f.

Two common kernels (actually families of kernels) are the Ramachandran-Lakshminarayanan kernel, w_{LR} [510], given by

$$w_{LR}(\theta, s) = \frac{b^2}{4\pi^2} \left(\frac{\cos(bs) - 1}{b^2 s^2} + \frac{\sin(bs)}{bs} \right)$$

(with the value at 0 being $\frac{1}{8} \frac{b^2}{\pi^2}$ to preserve continuity) and the Shepp-Logan kernel, w_{SL} [557],

$$w_{SL}(\theta, s) = \frac{b^2}{2\pi^3} \frac{\frac{\pi}{2} - bs \sin(bs)}{\left(\frac{\pi}{2}\right)^2 - b^2 s^2}$$

(with the value at $\pm \frac{\pi}{2b}$ being $\frac{b^2}{2\pi^4}$). Natterer's book [444], chapter 5, justifies the fact that a proper choice of filter together with an appropriate choice of the parameter b does allow the filtered backprojection method to give a good approximate inversion.

Nievergelt [449], (compare, Nievergelt [450]) gives a very elementary derivation of the filtered backprojection process on $\mathbb{R}^2$ for the special case where the kernel w_b is defined on $S^1 \times \mathbb{R}$ by

$$w_b(\theta, s) = \begin{cases} \frac{1}{\pi b^2} & \text{if } |s| \le b \\ \frac{1}{\pi b^2} \left(1 - \frac{|s|}{\sqrt{s^2 - b^2}} \right) & \text{if } |s| > b \end{cases}$$

He is able to prove that

$$R^{\#} w_b(x) = \begin{cases} \frac{1}{\pi b^2} & \text{if } |x| \le b \\ 0 & \text{if } |x| > b \end{cases}$$

Now recall remark 2.73 about the extension of approximate identities to $\mathbb{R}^n$. It is easy to check that if we let $\varphi_b = R^\# w_b$ and if we let $\varphi = \varphi_1$, then

$$\varphi_b(x) = \frac{1}{b^2}\varphi\left(\frac{x}{b}\right)$$

and hence the family φ_b is an approximate identity on $\mathbb{R}^2$. The filtered backprojection theorem then yields

$$
\begin{aligned}
f &= \lim_{b\to 0} f * \varphi_b \\
&= \lim_{b\to 0} R^\#(w_b) * f \\
&= \lim_{b\to 0} R^\#(w_b * Rf)
\end{aligned}
$$

which provides both an inversion formula and an approximate inversion formula for the Radon transform. Nievergelt's kernel is not used in practical computerized tomography, probably because of engineering concerns in which the Ramachandran-Lakshminarayanan and Shepp-Logan kernels give better performance under noisy conditions. These kernels are designed to eliminate or taper off high frequencies in the unknown function. The most obvious such approach is to accept all frequencies in a given bandwidth and to reject higher frequencies. This gives rise to the ideal low-bandpass filter which is the kernel used by Nievergelt. Despite its forbidding name, the ideal low bandpass filter is a very simple concept mathematically. It merely multiplies the Fourier transform by the characteristic function of a disk. The radius of this disk is the bandwidth B. Therefore, the ideal low-bandpass filter cuts off any frequencies greater than B, but allows lower frequencies to pass through unchanged. Although the ideal low-bandpass filter is the simplest filter from the mathematical point of view, it does not give the best performance relative to noise reduction. The Ramachandran-Lakshminarayanan and Shepp-Logan kernels are designed to give better reconstructions than the ideal low-bandpass filter.

On the other hand, the Ramachandran-Lakshminarayanan, Shepp-Logan and the ideal low-bandpass filter kernels are not approximate identities. If w represents any of these kernels and if we define $\varphi = R^\# w$, then it can be shown that

$$\varphi\left(\frac{x}{b}\right) = \sqrt{b}\,\frac{1}{b^n}\varphi_1\left(\frac{|x|}{b}\right)$$

so we are off by a factor of $\sqrt{b}$ from having an approximate identity. Therefore, the kernels based on restricting bandwidth are appropriate for approximate reconstructions but not exact inversion.

We can also use the backprojection theorem in conjunction with remark 2.78 to obtain an exact inversion formula for the Radon transform on $\mathbb{R}^n$.

We mentioned in Remark 2.78 that the Fourier transform convolution theorem does not generally apply to the convolution $T * f$. However, there is a large class of functions for which this convolution theorem is true. For those functions we will have

$$(T * f)^\wedge (\xi) = 2^n \pi^{n-1} \frac{1}{|\xi|^{n-1}} \widehat{f}(\xi). \tag{2.45}$$

The last logarithm to the proposed inversion theorem is the operation Λ defined by letting Λg be the inverse Fourier transform of $|\xi| g'(\xi)$. Composing Λ ($n - 1$ times) gives Λ^{n-1} with the property

$$\left(\Lambda^{n-1} g\right)^\wedge = |\xi|^{n-1} \widehat{g}(\xi).$$

Theorem 2.80. *If f is a function satisfying equation (2.45), then we have the following inversion formula for the Radon transform:*

$$\frac{1}{2^n \pi^{n-1}} \Lambda^{n-1} R^{\#} R f = f$$

The class of functions for which this is valid includes rapidly decreasing differentiable functions and compactly supported integrable functions. These functions include all the functions of practical interest in tomography.

Proof. We start by taking the Fourier transform of $\Lambda^{n-1}(R^{\#} R f)$. From the various definitions we then have

$$\left(\Lambda^{n-1}\left(R^{\#} R f\right)\right)^\wedge = |\xi|^{n-1} \left(R^{\#} R f\right)^\wedge (\xi)$$
$$= |\xi|^{n-1} (T * f)^\wedge (\xi)$$

The last step comes from the backprojection theorem: $R^{\#} R f = T * f$. Continuing the calculation by using equation (2.45) we get

$$\left(\Lambda^{n-1}\left(R^{\#} R f\right)\right)^\wedge = |\xi|^{n-1} \left(2^n \pi^{n-1} \frac{1}{|\xi|^{n-1}} \widehat{f}(\xi)\right)$$
$$= 2^n \pi^{n-1} \widehat{f}(\xi)$$

Taking the inverse Fourier transform yields the desired inversion formula. ∎

Of course, we need to describe the class of functions for which this formula is valid and to justify the convolution theorem in equation (2.45). This is done, in greater generality, in chapter 3, sections 3.6, 3.7, and 3.8. In particular, see corollary 3.80.

Practical application of the filtered backprojection method is something of an art. An appropriate choice of the filter along with the parameter b and the (necessarily finite) number of directions is crucial to getting a useful reconstruction. A discussion of this topic is beyond the scope of this book. However, the books by Herman [296], Kak and Slaney [328], Natterer [444], and Natterer and Wübbeling [446] are useful in this regard as are the papers by Bracewell and Riddle [65], Chang and Herman [85], Lewitt, Bates, and Peters [389], Ramachandran and Lakshminarayanan [510], Shepp and Logan [557], Smith [562, 563], and Smith and Keinert [564].

2.8 Inversion of the Radon Transform in $\mathbb{R}^2$ – Radon's Proof

Having shown the approximate inversion of the Radon transform in the preceding section, we now present Radon's method for inverting the Radon transform in $\mathbb{R}^2$. This is one of several major results presented by Johann Radon in his ground-breaking paper of 1917 [508]. The reader who is interested in reading the original may find it convenient to look in the following sources which have reprints of his paper: *75 Years of Radon Transform* [211] or Helgason [277].[7] These reprints are in the original German. Furthermore, Deans [124] and Parks [509] have English translations of the paper. The translation by Parks also has corrections to misprints in the original paper.

More general inversion results are given in Chapter 3.

Before proceeding to the proof let us remark that Radon observed that an inversion formula could be developed very quickly via the use of Abel integral equations. Radon was not convinced of the rigor of this approach, but Gel'fand, Gindikin, and Graev [189] have a modern, rigorous, and beautifully concise development by using this approach. I highly recommend the proof in [189] to the more advanced reader. The approach here is longer to respect the more modest prerequisites assumed for this chapter.

Radon defines for every point $P = (x, y)$ and for functions f defined on $\mathbb{R}^2$:

$$\overline{f}_P(r) = \frac{1}{2\pi} \int_0^{2\pi} f(x + r\cos\phi, y + r\sin\phi)\, d\phi, \qquad (2.46)$$

$$F(p, \phi) = \int_{-\infty}^{\infty} f(p\cos\phi - s\sin\phi, p\sin\phi + s\cos\phi)\, ds \qquad (2.47)$$

$$\overline{F}_P(q) = \frac{1}{2\pi} \int_0^{2\pi} F(x\cos\phi + y\sin\phi + q, \phi)\, d\phi \qquad (2.48)$$

In current terminology $F(p, \phi)$ is called the Radon transform, namely, $F(p, \phi) = Rf(\theta, p)$, where $\theta = (\cos\phi, \sin\phi)$.

The operators $\overline{f}_P$ and $\overline{F}_P$ are mean value operators on the original function and its Radon transform, respectively. Note that these are actually functions of (x, y) as well as the stated variable. The line $\langle \theta, x \rangle = p$ is a tangent line to the circle of radius p centered at the origin. As ϕ varies, $F(p, \phi)$ runs through the line integrals on all tangent lines to the circle of radius p centered at the origin. More generally one can show that the set of lines defined by the equation

$$x\cos\phi + y\sin\phi = x_0\cos\phi + y_0\sin\phi + q$$

generates the set of tangent lines to the circle of radius q centered at (x_0, y_0) as ϕ ranges from 0 to 2π. Therefore the mean value function $\overline{F}_P(q) = \overline{F}_P(q, x, y)$ is obtained by integrating the Radon transform over all lines tangent to the circle of radius q centered at (x, y). This is a dual operation to the Radon transform: in one case we integrate

---

[7] Only the first edition [277], not the second edition [291] of Helgason's book *The Radon Transform* has the reprint of Radon's 1917 paper.

tangent lines to produce a value at a point, whereas in the other case we integrate over points to produce the value at a tangent line. There is another aspect to this duality. In section 2.7 we defined the backprojection operator as the formal adjoint of the Radon transform. We note here that, in terms of the backprojection operator, $\overline{F}_P(q)$ is the backprojection of f translated by $-q$ in the first variable.

Radon made the following assumptions about the functions f that he dealt with:

(a_1) f is continuous on $\mathbb{R}^2$.

(b_1) $\displaystyle\int_{\mathbb{R}^2} \frac{|f(x,y)|}{\sqrt{x^2+y^2}}dxdy < \infty$

(c_1) $\displaystyle\lim_{r\to\infty} \overline{f}_P(r) = 0$

Remark 2.81. The conditions (a_1)–(c_1) are somewhat artificial, although they are certainly fulfilled by compactly supported[8] continuous functions. More generally, a continuous function which is $O(|x|^{-N})$ with $N > 2$ will satisfy these conditions.[9] This growth condition is the hypothesis of an inversion formula presented by Helgason for infinitely differentiable functions (Helgason [291], theorems 3.1 and 6.2). We have a generalization to measurable functions which are bounded near the origin and which satisfy this growth condition (chapter 3, corollary 3.77).

Functions which are $O(|x|^{-N})$ with $N > 2$ are integrable. There are functions that fail to be integrable yet which satisfy Radon's criteria: let $f(x) = \dfrac{1}{|x|^{\frac{3}{2}}}$ if $|x| \geq 1$ with $f(x) = 1$ elsewhere. Then f is easily shown to be a continuous function which is not integrable, yet which satisfies Radon's conditions (a_1)–(c_1) and hence has a Radon transform that can be inverted by the methods of this section.

In chapter 3 we derive a very general inversion formula for the Radon transform on a class of function that contains nonintegrable functions. The main property of functions in this class is that the Riesz potential of the order $n-1$ exists almost everywhere. It is interesting to note that Radon's condition (b_1) for a continuous function f is equivalent to f having a Riesz potential of the order $n-1$, which is finite almost everywhere. The interested reader can look at chapter 3, sections 3.7 and 3.8, especially theorems 3.64 and 3.65. The advanced reader will find an interesting update and generalization of Radon's theorem in Madych, Radon's inversion formulas, Trans, AMS, 356 (2004) 4475–4491.

Radon stated and presented proofs of the following theorems.

Theorem I. $F(p, \phi)$ *is defined almost everywhere in the sense that for any given circle in the plane, those points on the circle on whose tangent lines the integral defining F does not exist form a set of measure zero on the circle.*

Theorem II. *The integral defining $\overline{F}_P(q)$ is absolutely convergent for all P and q.*

---

[8] A function f defined on $\mathbb{R}^n$ is said to be **compactly supported** if the set of points where $f(x) \neq 0$ is contained in a bounded region of $\mathbb{R}^n$.

[9] The "big O" notation is defined as follows: we say that $f(x) = O(|x|^N)$ if there is a constant C such that $|f(x)| \leq C|x|^N$ for all $x \in \mathbb{R}^n$.

Theorem III. *The value of f is uniquely determined by $\overline{F}_P$ and can be found by the following computation:*

$$f(P) = -\frac{1}{\pi} \int_0^\infty \frac{d\overline{F}_P(q)}{q} \qquad (2.49)$$

This Stieltjes integral can be also be expressed in the form

$$f(P) = \frac{1}{\pi} \lim_{\varepsilon \to 0} \left(\frac{\overline{F}_P(\varepsilon)}{\varepsilon} - \int_\varepsilon^\infty \frac{\overline{F}_P(q)}{q^2} dq \right) \qquad (2.49')$$

We will now present proofs of theorems I to III. The presentation follows Radon's paper very closely, although we give more details.

Radon observes but does not prove that the double integral

$$\int_{x^2+y^2>q^2} \frac{f(x,y)}{\sqrt{x^2+y^2-q^2}} dx dy \qquad (2.50)$$

is absolutely convergent. The proof can be done by choosing any real number a larger than $|q|$ and breaking the integral into two pieces, the first I_1 with domain $q^2 < x^2 + y^2 < a^2$ and the second, I_2, with domain $x^2 + y^2 \geq a^2$.

Noting that $\sqrt{\frac{r^2}{r^2-q^2}}$ is a decreasing function for $r > q$, it is easy to see that there is a constant such that

$$\frac{1}{\sqrt{r^2-q^2}} \leq K\frac{1}{r}$$

for $r \geq a$ (just take $K = \sqrt{\frac{a^2}{a^2-q^2}}$). Then letting $r = \sqrt{x^2+y^2}$ and substituting in the integral I_2 we get

$$|I_2| \leq \int_{x^2+y^2\geq a^2} \frac{|f(x,y)|}{\sqrt{x^2+y^2-q^2}} dx dy$$

$$\leq K \int_{x^2+y^2\geq a^2} \frac{|f(x,y)|}{\sqrt{x^2+y^2}} dx dy < \infty$$

by condition (b_1).

In the integral I_1 change to polar coordinates:

$$|I_1| \leq \int_0^{2\pi} \int_q^a \frac{|f(r\cos\phi, r\sin\phi)|}{\sqrt{r^2-q^2}} r \, dr \, d\phi$$

After the change of variable $u^2 = r^2 - q^2$ we obtain

$$|I_1| \leq \int_0^{2\pi} \int_0^{\sqrt{q^2-q^2}} \left| f\left(\sqrt{u^2 + q^2} \cos \phi, \sqrt{u^2 + q^2} \sin \phi \right) \right| du d\phi$$

which is finite because of the continuity of f. This proves the desired absolute convergence of the integral (2.50).

We now make the change of coordinates, from x, y to s, ϕ variables ($s \geq 0$, $0 \leq \phi \leq 2\pi$):

$$x = q \cos \phi - s \sin \phi$$
$$y = q \sin \phi + s \cos \phi$$

in the integral (2.50). The Jacobian is

$$\left| \det \begin{bmatrix} -\sin \phi & -q \sin \phi - s \cos \phi \\ \cos \phi & q \cos \phi - s \sin \phi \end{bmatrix} \right| = s$$

(recall that $s \geq 0$) and we note that

$$\begin{bmatrix} x \\ y \end{bmatrix} = \begin{bmatrix} \cos \phi & -\sin \phi \\ \sin \phi & \cos \phi \end{bmatrix} \begin{bmatrix} q \\ s \end{bmatrix}$$

Because the matrix is orthogonal we see that $x^2 + y^2 = q^2 + s^2$ so the integral (2.50) transforms to

$$\int_0^{2\pi} \int_0^{\infty} f(q \cos \phi - s \sin \phi, q \sin \phi + s \cos \phi) \, ds d\phi \qquad (2.51)$$

Now replace ϕ by $\phi + \pi$ and s by $-s$. We can also replace q by $-q$ since the integral (2.50) depends only on q^2. Then equation (2.51) transforms to

$$\int_0^{2\pi} \int_{-\infty}^{0} f(q \cos \phi - s \sin \phi, q \sin \phi + s \cos \phi) \, ds d\phi$$

and adding these results and dividing by 2 gives

$$\int_{x^2+y^2>q^2} \frac{f(x, y)}{\sqrt{x^2 + y^2 - q^2}} dx dy \qquad (2.52)$$

$$= \frac{1}{2} \int_0^{2\pi} \int_{-\infty}^{\infty} f(q \cos \phi - s \sin \phi, q \sin \phi + s \cos \phi) \, ds d\phi$$

We note that the inner integral is now $F(q, \phi)$ so we get the following backprojection type of result:

$$\int\limits_{x^2+y^2>q^2} \frac{f(x, y)}{\sqrt{x^2 + y^2 - q^2}} dx dy = \frac{1}{2} \int\limits_{0}^{2\pi} F(q, \phi) d\phi \qquad (2.53)$$

$$= \pi \overline{F}_0(q)$$

(the last step following from the definition of the mean value function $\overline{F}_P$ when P is the origin).

From Fubini's theorem and the convergence of the integral (2.52) we see that the inner integral in (2.52), which is $F(q, \phi)$, exists for almost all ϕ which proves theorem I. Theorem II is proved similarly.

We now give Radon's proof of the identity (2.49′) in theorem III.

Introduce polar coordinates in equation (2.50) and use the definition of $\overline{f}_0(r)$ to get

$$\overline{F}_0(q) = \frac{1}{\pi} \int\limits_{q}^{\infty} \int\limits_{0}^{2\pi} \frac{f(r\cos\phi, r\sin\phi)}{\sqrt{r^2 - q^2}} r d\phi dr \qquad (2.54)$$

$$= \int\limits_{q}^{\infty} 2\overline{f}_0(r) \frac{r}{\sqrt{r^2 - q^2}} dr$$

We note that there is either a misprint or a mistake in Radon's paper at the point where this equation is stated (page 265, equation (2) of [508]).[10,11] However, Radon does use the correct form in the remainder of his proof of theorem III.

Radon notes that the conditions (a_1)–(c_1) on the function f are invariant under rigid motions of the plane, so it suffices to prove equation (2.49′) for $P = 0$. Using

[10] Radon makes a change to polar coordinates in the following equation appearing on page 264 of his paper [508]

$$\int\limits_{x^2+y^2>q^2} \frac{f(x, y)}{\sqrt{x^2 + y^2 - q^2}} dx dy$$

obtaining, at the bottom of the same page

$$\int\limits_{q}^{\infty} dr \int\limits_{0}^{2\pi} \frac{f(r\cos\phi, r\sin\phi)}{\sqrt{r^2 - q^2}} d\phi$$

This should be

$$\int\limits_{q}^{\infty} r \, dr \int\limits_{0}^{2\pi} \frac{f(r\cos\phi, r\sin\phi)}{\sqrt{r^2 - q^2}} d\phi$$

[11] In his translation [509] of Radon's paper, P. C. Parks has included several more corrections.

equation (2.54), the right-hand side of equation (2.49') becomes

$$\frac{1}{\pi} \lim_{\varepsilon \to 0} \left(\frac{\overline{F}_0(\varepsilon)}{\varepsilon} - \int_{\varepsilon}^{\infty} \frac{\overline{F}_0(q)}{q^2} dq \right) \tag{2.55}$$

$$= \frac{2}{\pi} \lim_{\varepsilon \to 0} \left(\frac{1}{\varepsilon} \int_{\varepsilon}^{\infty} \frac{r \overline{f}_0(r)}{\sqrt{r^2 - \varepsilon^2}} dr - \int_{\varepsilon}^{\infty} \int_{q}^{\infty} \frac{r \overline{f}_0(r)}{\sqrt{r^2 - q^2}} dr \frac{1}{q^2} dq \right)$$

At this point Radon suggests computing the iterated integral in reverse order and then, without any detail, states that the value of equation (2.55) becomes

$$\frac{2}{\pi} \lim_{\varepsilon \to 0} \int_{\varepsilon}^{\infty} \frac{\overline{f}_0(r)}{r \sqrt{r^2 - \varepsilon^2}} dr$$

which expression yields, he claims, $\overline{f}_0(0) = f(0,0)$, "wie unschwer zu zeigen ist" ([508], page 266). In fact, this expression is wrong and should be $\frac{2}{\pi} \varepsilon \lim_{\varepsilon \to 0} \int_{\varepsilon}^{\infty} \frac{\overline{f}_0(r)}{r\sqrt{r^2-\varepsilon^2}} dr$. It is probably better, therefore, to provide a few details.

First, it is easy to check, via a simple trigonometric substitution, that

$$\int_{\varepsilon}^{r} \frac{1}{q^2 \sqrt{r^2 - q^2}} dq = \frac{1}{r^2} \frac{\sqrt{r^2 - \varepsilon^2}}{\varepsilon}$$

We can use this relation to see that when the order is switched in the iterated integral in equation (2.55) we obtain:

$$\int_{\varepsilon}^{\infty} r \overline{f}_0(r) \left(\int_{\varepsilon}^{r} \frac{1}{q^2 \sqrt{r^2 - q^2}} dq \right) dr$$

$$= \int_{\varepsilon}^{\infty} r \overline{f}_0(r) \left(\frac{1}{r^2} \frac{\sqrt{r^2 - \varepsilon^2}}{\varepsilon} \right) dr$$

$$= \frac{1}{\varepsilon} \int_{\varepsilon}^{\infty} \overline{f}_0(r) \frac{\sqrt{r^2 - \varepsilon^2}}{r} dr$$

For $r > \varepsilon > 0$ we certainly have $r^2 - \varepsilon^2 \leq r^2$ and from this it follows easily that $\frac{\sqrt{r^2-\varepsilon^2}}{r} \leq \frac{r}{\sqrt{r^2-\varepsilon^2}}$. Hence we obtain the estimate

$$\int_{\varepsilon}^{\infty} |\overline{f}_0(r)| \frac{\sqrt{r^2 - \varepsilon^2}}{r} dr \leq \int_{\varepsilon}^{\infty} |\overline{f}_0(r)| \frac{r}{\sqrt{r^2 - \varepsilon^2}} dr = \left| \frac{1}{2} \overline{F}_0(\varepsilon) \right| < \infty$$

by equation (2.54) and theorem II. This shows, by Fubini's theorem, that the double integral associated to the iterated integral is absolutely convergent and that the change in the order of integration is justified. Using this result back in equation (2.55) yields

$$\frac{1}{\pi} \lim_{\varepsilon \to 0} \left(\frac{\overline{F}_0(\varepsilon)}{\varepsilon} - \int_{\varepsilon}^{\infty} \frac{\overline{F}_0(q)}{q^2} dq \right)$$

$$= \frac{2}{\pi} \lim_{\varepsilon \to 0} \left(\frac{1}{\varepsilon} \int_{\varepsilon}^{\infty} \frac{r \overline{f}_0(r)}{\sqrt{r^2 - \varepsilon^2}} dr - \frac{1}{\varepsilon} \int_{\varepsilon}^{\infty} \overline{f}_0(r) \frac{\sqrt{r^2 - \varepsilon^2}}{r} dr \right)$$

$$= \frac{2}{\pi} \lim_{\varepsilon \to 0} \left(\int_{\varepsilon}^{\infty} \overline{f}_0(r) \frac{\varepsilon}{r \sqrt{r^2 - \varepsilon^2}} dr \right)' \qquad (2.56)$$

If we now define $\varphi_\varepsilon(r) = \frac{2}{\pi} \varepsilon \frac{1}{r \sqrt{r^2 - \varepsilon^2}}$ on $[\varepsilon, \infty)$, with $\varphi_\varepsilon(r) = 0$ elsewhere, we see that we can interpret the right-hand side of equation (2.49') as the convolution

$$\lim_{\varepsilon \to 0} \overline{f}_0 * \varphi_\varepsilon(0)$$

and this suggests applying the approximate identity theorem, lemma 2.70. We can indeed do this because if we define $\varphi(r) = \varphi_0(r)$ then it is easy to verify that:

1. φ satisfies the hypotheses of lemma 2.70.
2. $\varphi_\varepsilon(r) = \frac{1}{\varepsilon} \varphi\left(\frac{r}{\varepsilon}\right)$
3. $\overline{f}_0$ satisfies the conditions of lemma 2.70 because from condition (a_1) that f is continuous it follows easily that the mean value function $\overline{f}_0$ is also continuous. This in conjunction with condition (c_1) that $\lim_{r \to \infty} \overline{f}_0(r) = 0$ yields that $\overline{f}_0$ is also bounded.

Therefore, we can apply lemma 2.70 to conclude that

$$\overline{f}_0(0) = \lim_{\varepsilon \to 0} \overline{f}_0 * \varphi_\varepsilon(0)$$

Finally we observe, from equation (2.46), that

$$\overline{f}_0(0) = \frac{1}{2\pi} \int_0^{2\pi} f(0,0) d\phi$$

$$= f(0,0)$$

Putting together all these facts we see that we have established equation (2.49'):

$$f(P) = \frac{1}{\pi} \lim_{\varepsilon \to 0} \left(\frac{\overline{F}_P(\varepsilon)}{\varepsilon} - \int_{\varepsilon}^{\infty} \frac{\overline{F}_P(q)}{q^2} dq \right) \qquad (2.57)$$

We can now use this result to verify equation (2.49) in theorem III. Integration by parts in $\int_0^\infty \frac{dF_P(q)}{q}$ yields

$$\lim_{q \to \infty} \frac{F_P(q)}{q} - \lim_{\varepsilon \to 0} \frac{F_P(\varepsilon)}{\varepsilon} + \int_0^\infty \frac{F_i(q)}{q^2} dq$$

It turns out that $\lim_{q \to \infty} \frac{F_P(q)}{q} = 0$ (the reader can refer to Radon's paper for the details about this point). Combining this result with equation (2.57) yields formula (2.49).

Remark 2.82. Radon did not use approximate identities explicitly. He basically stated that it is obvious that $\int_\varepsilon^\infty \frac{2}{\pi} \overline{f}_0(r) \frac{\varepsilon}{r \sqrt{r^2 - \varepsilon^2}} dr$ converges to $\overline{f}_0(0)$ as $\varepsilon \to 0$. But this is only obvious if you know about approximate identities and analyze this integral in the form of a convolution with an approximate identity, as we did previously.

2.9 Additional References and Results

The historical remarks, further references, and applications are of general interest. The section on further results is for readers with a very good background in advanced mathematics. Here, one may find many results about the Radon transform beyond the basic results presented in this chapter. References but not proofs are provided for these more advanced results.

2.9.1 Historical Remarks

According to Cormack [104], "the first person I know of who tackled Radon's problem was the great Dutch physicist H. A. Lorentz." Lorentz actually solved the problem of inverting the three-dimensional Radon transform before Radon's 1917 paper [508]. Cormack went on to say "We have no idea why Lorentz thought of the problem, or what his method of proof was, and the only reason we know of his work is that the above result was attributed to him by Bockwinkel who used the result in a long paper on the propagation of light in biaxial crystals." Since Bockwinkel's paper [56] appeared in 1906, it seems that the first mathematical result in tomography along with the first application far outdated Radon's paper of 1917.

Radon's pioneering 1917 paper [508] contained the seeds of many of the currently active areas of mathematical research in tomography. In the years before 1970, his results were either "rediscovered" or generalized by various researchers. Uhlenbeck [605] inverted the n-dimensional Radon transform for odd n in 1925. In 1927 Mader [411] succeeded in finding an inversion formula for the Radon transform in all dimensions (compare, John [322]). In 1936 Cramér and Wold [110] inverted the Radon transform in the context of mathematical statistics; x-ray projections are abstractly the same as marginal distributions. Also in 1936 Ambartsumian [17] "rediscovered" the Radon transform and gave an interesting application to stellar astronomy; see the discussion below. Also of interest during the period before 1970 are the papers of

John [323–326], Gel'fand, Graev, and Vilenkin [194], Helgason [269], and Ludwig [405]. After 1970, there was an explosion of papers in the field, both theoretical and practical.

Curiously, many practical implementations of tomography were done without any knowledge of Radon's work. Apparently, it was not until 1964 that Radon's contribution was acknowledged in the practical CT literature. An early practical implementation of tomography was developed in 1936 by the astronomer Ambartsumian [17]. He obtained the same inversion formula as Radon [508], using essentially the same method. Remarkably, he did this without knowledge of Radon's paper, which he learned of two years after his publication. The Nobel prize for Physiology or Medicine was awarded jointly to G. N. Hounsfield and A. M. Cormack in 1979. Each of these laureates made separate and important contributions to the field of computerized tomography. Apparently, however, neither Hounsfield nor Cormack were aware of Radon's contributions. According to Deans [124], Page 5, "... Marr (1982) notes that Pincus (1964) apparently was the first person to develop a reconstruction algorithm with knowledge of the available material in the mathematics literature including Radon's 1917 paper." We cannot quote from Marr's paper directly because the symposium volume [434] containing his paper was never published. Unfortunately, many other interesting and valuable papers were contained in this unpublished volume. In our bibliography these references are Marr [426] and Pincus [494].

There are more details on the history of computerized tomography in Chapter 1, section 1.10.1.

One of the earliest applications of tomography and the earliest development of the filtered backprojection method both occurred in astronomy. These are the application of Ambartsumian [17], already mentioned, and the development of filtered backprojection by Bracewell and Riddle [65].

In his 1936 paper [17] Ambartsumian considered the following problem: it is relatively easy to compute how fast a star is receding from or approaching to the earth. Astronomers refer to this as the radial velocity of the star. However, it is not obvious how to compute the motion of the star in other than the radial direction. The astronomer and physicist A. S. Eddington proposed the next best thing. He challenged astronomers to find a method of computing the number of stars with a specific velocity in three dimensions. Ambartsumian paraphrased Eddington's challenge in the following way: "is it possible to find the distribution function $\varphi(\xi, \eta, \delta)$ of the components of stellar space velocities in the solar neighborhood from radial velocities alone ... ? [18]"[12]

Ambartsumian solved this problem, in both two and three dimensions, by showing that what-we-now-call the Radon transform of the distribution function $\varphi(\xi, \eta, \delta)$ of the components of stellar space velocities is expressible in terms of the radial velocities of the stars. As remarked above, he then obtained the same inversion formula as Radon [508] by using essentially the same method.

Medical applications of tomography reconstruct an object in ordinary Euclidean space. However, Ambartsumian's idea shows that one can use tomography to reconstruct

---

[12] Actually $\varphi(\xi, \eta, \delta)$ is the probability density function of the sample space, consisting of the velocity vectors of observable stars in the solar neighborhood.

the velocity distribution in velocity space just from the knowledge of radial velocities. He published the result only for the distribution of stellar velocities, but one could obtain the distribution of velocities of any system of particles knowing only the radial velocities. The idea of using tomography in medical applications did not appear until later.

Bracewell and Riddle [65] gave the earliest implementation of the filtered back-projection method while using tomographic methods to reconstruct a microwave map of the sun. Their input data, corresponding to x-rays in medical applications, were derived from strip integrals of microwave activity on the sun. These strip integrals were obtained by using radio telescopes with long, narrow apertures. The projections derived from such a telescope vary by angle as the earth rotates or as the orientation of the telescope is changed. Thereby a sample of the values of the Radon transform is obtained.

2.9.2 Further References

The reader may want to consult the references already presented in chapter 1, section 1.10.2, some of which are repeated here.

We presented an application to computerized tomography in diagnostic radiology in chapter 1. There are many other practical applications of tomography: the books by Deans [124], Herman [296], and Natterer [444] contain numerous detailed descriptions of applications of tomography to science, medicine, and technology. Natterer's paper [441], which is in German, gives a quick overview of the process of tomography described in his book [444].

The book by Gel'fand, Gindikin, and Graev [189] is an almost breathtakingly beautiful introduction to the theoretical aspects of mathematical tomography, written by some of the pioneers in the field.

The following books give a technical and in-depth treatment of the Radon transform and tomography: Deans [124], Gel'fand and Shilov [195], Gel'fand, Graev, and Vilenkin [194] (compare, Gel'fand and Graev [192]), Ehrenpreis [149], Helgason [287, 291, 292] (compare, [277], the original edition of "The Radon Transform" [291] cited previously), Herman [296], Natterer, [444], Natterer and Wübbeling [446], Natterer and Faridani [445], Kak and Slaney [328], and Ramm and Katsevich [513].

The next set of references give either a short introduction or passing mention to the Radon transform and tomography. See *Encyclopedia of Mathematics* [612] (Vol. 9, Sto-Zyg, pp. 179–181). This is a translation of the Soviet "Mathematical Encyclopedia" and has an excellent short introduction to the Radon transform and tomography. Vilenkin et al. [611] introduce the Radon transform in chapter 8. Khavin [351] gives a one-paragraph treatment of the Radon transform, J. S. Walker [616] gives a succinct, somewhat specialized introduction to Radon transform in an appendix (pp. 380–400). Terras [597] defines the Radon transform early in the book (pp. 107–119) and has very succinct guided exercises in which the reader can develop Radon inversion, and volume two of Terras [598] has a brief mention of applications to partial differential equations. Dym and McKean [138] include a short introduction to the Fourier inversion method for the Radon transform. John in his book [325] on partial differential equations

gives a quick introduction to the definition and inversion of the Radon transform in a
few exercises.

There are some excellent survey articles on the Radon transform. Shepp and Kruskal
[556] have an elementary and fascinating introduction to both the applied and theoretical
aspects of the Radon transform. The articles by Strichartz [585] and Zalcman [625] also
are of great interest. The articles of Smith, Solmon, and Wagner [566] and Smith [562]
are very nice overviews of both the mathematical and practical aspects of tomography.
Gindikin [202] is a beautiful and concise introduction to the theory and history of the
Radon transform. This paper relates Radon's 1917 paper to modern day mathematics,
in particular, to integral geometry and tomography.

Hlawka [305–309] implements a program to apply analytical number theory to
mathematical analysis. One of his many applications is to the Radon transform.

2.9.3 Further Results

For the purposes of this section we let $C_v^\infty(\mathbb{R}^n)$ denote the space of C^∞ functions that
vanish at infinity. We provide $C_v^\infty(\mathbb{R}^n)$ with the usual Frechet space topology defined
by the sup–norm on functions and derivatives. The topological dual space to $C_v^\infty(\mathbb{R}^n)$
is called the space of ***integrable distributions***, is denoted by $\mathcal{D}'_{L^1}(\mathbb{R}^n)$, and was introduced by Schwartz [546]. This space is a proper subspace of the space of all distribu-
tions on $\mathbb{R}^n$. It contains all the common test functions (rapidly decreasing functions,
compactly supported measurable functions), all integrable functions, and compactly
supported distributions. It also contains all finite Borel measures; consequently, the
range of the Radon transform is contained in $\mathcal{D}'_{L^1}(S^{n-1} \times \mathbb{R})$ (integrable distributions
are defined on $S^{n-1} \times \mathbb{R}$ in a similar manner). However, $\mathcal{D}'_{L^1}$ does not contain all L^p
functions for $p > 1$, although it does contain some nonintegrable functions that are
integrable on almost all hyperplanes.

2.9.3.1 General Results on the Radon Transform

Hertle [298, 299] considers the Radon transform defined on the space of integrable
distributions $\mathcal{D}'_{L^1}(\mathbb{R}^n)$.

Hertle [298] shows that a continuous linear operator on certain function spaces that
behaves under rotations, dilations, and translations like the Fourier transform must be
a constant multiple of the Fourier transform. More precisely, let $\mathcal{D}(\mathbb{R}^n)$ be the space of
compactly supported C^∞ functions with the usual topology and let $\mathcal{D}'(\mathbb{R}^n)$, the space
of distributions on $\mathbb{R}^n$, be the dual space of $\mathcal{D}(\mathbb{R}^n)$. Let φ represent an arbitrary element
of $\mathcal{D}(\mathbb{R}^n)$, let $\delta > 0$ and let $a \in \mathbb{R}^n$. Let $G : \mathcal{D}(\mathbb{R}^n) \to \mathcal{D}'(\mathbb{R}^n)$ be a continuous linear
operator. Then G is a constant multiple of the Fourier transform $\mathcal{F}$, if and only if G
commutes with rotations in the obvious way, commutes with dilations in the sense
that $G(\varphi(\delta x)) = \delta^{-n}(G\varphi)(\delta^{-1}x)$, and commutes with translations in the sense that
$G(\varphi(x + a)) = e^{i\langle a, x \rangle}(G\varphi)(x)$.

In the same paper Hertle proves an analogous result for the Radon transform. The
most obvious analogy would relate to continuous linear operators from $\mathcal{D}(\mathbb{R}^n)$ to

$\mathcal{D}'\left(S^{n-1}\times\mathbb{R}\right)$ with the same behavior as the Radon transform under translations, rotations, and dilations (compare, propositions 2.41, 2.42, and 2.50). However, Hertle shows by a counterexample that there is a continuous operator from $\mathcal{D}(\mathbb{R}^n)$ to $\mathcal{D}'\left(S^{n-1}\times\mathbb{R}\right)$, which satisfies the invariance properties but which is not a constant multiple of the Radon transform. Therefore, some restriction to the range space is necessary. The appropriate range space is the space $\mathcal{D}'_{L^1}\left(S^{n-1}\times\mathbb{R}\right)$ of integrable distributions on the cylinder $S^{n-1}\times\mathbb{R}$. Hertle shows that if S is a continuous linear operator from $\mathcal{D}(\mathbb{R}^n)$ to $\mathcal{D}'_{L^1}\left(S^{n-1}\times\mathbb{R}\right)$, which satisfies the same invariance under translations, rotations, and dilations that the Radon transform does, then S is a constant multiple of the Radon transform.

Kurusa [370] presents a characterization of the Radon transform similar to Hertle's, but it avoids the condition on the range. Kurusa also has an analogous characterization of the boomerang transform B, which is defined by

$$Bf(x) = \int_{\{\theta\in S^{n-1}:\langle\theta,x\rangle\geq 0\}} f(\langle\theta,x\rangle\theta)\,d\theta.$$

The boomerang transform clearly is closely related to the backprojection operator $R^{\#}$; hence, there is a characterization of the formal adjoint of the Radon transform. The boomerang transform was introduced by Szabo [592] in connection with his study of Hilbert's fourth problem.

Madych and Nelson [413] have a similar result for the finite Radon transform on $\mathbb{R}^2$. The finite Radon transform consists of a vector of projections of the form $R_{\theta_j}f$ for a finite set of indices: $j = 1, \ldots, m$. A convenient notation is to let $\Theta = \{\theta_1, \ldots, \theta_m\}$ and to define $R_\Theta f = (R_{\theta_1}f, \ldots, R_{\theta_m}f)$. Madych and Nelson restrict attention to the situation in which the functions f are compactly supported in some closed disk. Most analytic methods of computed tomography depend on applying a linear transformation A (often a convolution) to the finite projection data $R_\Theta f$ to arrive at an approximate reconstruction of f. Madych and Nelson show that if A is such a linear, continuous operator such that $A \circ R_\Theta$ maps bounded functions to bounded functions and such that $A \circ R_\Theta$ is rotation and translation invariant, then there is a radial polynomial p of degree no greater than $2(m-1)$ such that $A \circ R_\Theta$ is a convolution operator with the polynomial kernel p:

$$A \circ R_\Theta f = f * p.$$

They develop representations of such polynomial kernels in terms of ridge functions, compare, Madych and Nelson [414–416].

It is possible to define the Radon transform on certain spaces of distributions by taking advantage of the duality proved in theorem 2.75 $\langle Rf, w\rangle = \langle f, R^{\#}w\rangle$. If u were a distribution on a space of test functions T, then one could define Ru by

$$\langle Ru, \varphi\rangle = \langle u, R^{\#}\varphi\rangle$$

provided that φ is an arbitrary function in a space of test functions with the property that $R^{\#}\varphi \in T$. For example, Helgason [291] defines the Radon transform of compactly supported distributions, whereas Gel'fand, Graev, and Vilenkin [194] and Ludwig [405] define the Radon transform of tempered distributions. Hertle [299] is able to define the Radon transform on the space of integrable distributions $\mathcal{D}'_{L^1}(\mathbb{R}^n)$, which we described previously. Hertle shows that the Radon transform is a continuous linear operator from $\mathcal{D}'_{L^1}(\mathbb{R}^n)$ to $\mathcal{D}'_{L^1}(S^{n-1} \times \mathbb{R})$. He also shows that $\mathcal{D}'_{L^1}(\mathbb{R}^n)$ is the maximal space of distributions for which the Radon transform extends uniquely. Hertle has many other interesting results in this paper. Some of them are discussed in the following subsections.

Peters [477] considers functions f that are the sum of a compactly supported C^∞ function with a linear combination of a finite number of characteristic functions of compact subsets with smooth $n-1$-dimensional boundaries. He shows that the set of discontinuities of such a function f is directly related to the set of discontinuities of certain derivatives of the backprojection of the Radon transform of f.

Hahn and Quinto [252] show how to measure the distance between Borel probability measures on $\mathbb{R}^n$ in terms of the distance between the Radon transforms of these measures. They obtain results relative to various metrics, including the Prokhorov metric and the metrics induced by various Sobolev norms.

2.9.3.2   Continuity of the Radon Transform and Its Inverse

Continuity of the inverse Radon transform is meant as sequential continuity.

The Radon transform is a bounded and, hence continuous, linear operator on $L^1(\mathbb{R}^n)$. This will be shown in chapter 3, corollary 3.25. However, the Radon transform is unbounded on $L^2(\mathbb{R}^n)$. But Hertle [299] shows that the dual Radon transform $R^\#$ is a continuous operator in the following situations:

$$R^\# : C^\infty\left(S^{n-1} \times \mathbb{R}\right) \to C^\infty\left(\mathbb{R}^n\right)$$
$$R^\# : C_v^\infty\left(S^{n-1} \times \mathbb{R}\right) \to C_v^\infty\left(\mathbb{R}^n\right).$$

An immediate consequence is that the Radon transform is a continuous operator in the following situations:

$$R : \mathcal{E}'\left(\mathbb{R}^n\right) \to \mathcal{E}'\left(S^{n-1} \times \mathbb{R}\right)$$
$$R : \mathcal{D}'_{L^1}\left(\mathbb{R}^n\right) \to \mathcal{D}'_{L^1}\left(S^{n-1} \times \mathbb{R}\right).$$

Continuity of the inverse Radon transform would be very desirable, because it would make inversion of the Radon transform a stable, or well-posed, problem. However, the inverse Radon transform is not always continuous. The continuity depends on the domain of functions chosen for the Radon transform. For example, the Radon transform is not bounded on $C^0(B)$, where B is any ball in $\mathbb{R}^n$. In fact, Hertle [299] shows that a bounded sequence of continuous functions supported on the unit ball exists, such that the Radon transform of this sequence converges to zero uniformly, but for which the sequence converges nowhere. Additionally, the sequence does not converge weakly in $L^1(\mathbb{R}^n)$. Therefore, the inverse Radon transform is not continuous on $L^1(S^{n-1} \times \mathbb{R})$.

However, if one considers the Radon transform between certain Sobolev spaces, then one can show that both the Radon transform and its inverse exist and are continuous. Define $L_\alpha^2(\mathbb{R}^n)$ to be the Sobolev space of order α consisting of all tempered distributions u defined on $\mathbb{R}^n$ with the property that $(1 + |\xi|^2)^{\frac{\alpha}{2}}\widehat{u}(\xi)$ is square integrable. A similar definition applies to $L_\alpha^2(S^{n-1} \times \mathbb{R})$, except that the Fourier transform is taken only with respect to the second variable. If W is an open subset of $\mathbb{R}^n$, then $L_\alpha^2(W)$ consists of all $u \in L_\alpha^2(\mathbb{R}^n)$, such that supp($u$) is contained in the closure of W. A similar definition applies to $L_\alpha^2(S^{n-1} \times W)$, where W is an open subset of $\mathbb{R}$.

Hertle [299] and Louis [396] proved that:

Theorem 2.83. *The Radon transform is a bicontinuous map from $L_\alpha^2(B(r)) \to L_{\alpha+\frac{n-1}{2}}^2(S^{n-1} \times [-r, r])$.*

This result was previously obtained, in the case $\alpha = 0$, by Natterer [436] and Smith, Solmon, and Wagner [566]. Similar results for Radon transforms defined on compact elliptic manifolds were obtained by Guillemin [244] and Strichartz [584].

In the case that α is a positive integer, then L_α^2 consists of square integrable functions whose weak derivatives are also square integrable. In general a function in L_α^2 is considered to be smooth of order α. Therefore, theorem 2.83 can be interpreted to mean that the Radon transform increases the smoothness of a function by $\frac{n-1}{2}$.

Finally, Hertle is able to show in [299] that, in the following cases, the Radon transform and its inverse exist and are continuous:

$$R : \mathcal{E}'\left(\mathbb{R}^n\right) \to \mathcal{E}'\left(S^{n-1} \times \mathbb{R}\right)$$

$$R : \mathcal{S}\left(\mathbb{R}^n\right) \to \mathcal{S}\left(S^{n-1} \times \mathbb{R}\right)$$

$$R : C_0^\infty\left(\mathbb{R}^n\right) \to C_0^\infty\left(S^{n-1} \times \mathbb{R}\right)$$

Note that in [299] the space $C_0^\infty(\mathbb{R}^n)$ of compactly supported, infinitely differentiable functions is denoted by $\mathcal{D}(\mathbb{R}^n)$, whereas the space $C_v^\infty(\mathbb{R}^n)$ of infinitely differentiable functions, which vanish at infinity, is denoted by $C_0^\infty(\mathbb{R}^n)$.

2.9.3.3 Inversion of the Radon Transform

The following inversion formula for the Radon transform is classical; compare, Helgason [291], Ludwig [405], also Radon [508]. We prove it in chapter 3 (see theorem 3.53).

Theorem 2.84. *Let $f \in \mathcal{S}$. Then*

1. *If n is odd, then*

$$f(x) = \frac{(-1)^{\frac{n-1}{2}}}{2^n \pi^{n-1}} \int_{S^{n-1}} \frac{\partial^{n-1} Rf}{\partial p^{n-1}} (\theta, \langle x, \theta \rangle) \, d\theta$$

and

2. If n is even, then

$$f(x) = \frac{(-1)^{\frac{n-2}{2}}}{2^n \pi^{n-1}} \int\limits_{S^{n-1}} H \frac{\partial^{n-1} Rf}{\partial p^{n-1}} (\theta, \langle x, \theta \rangle) \, d\theta$$

Here H denotes the Hilbert transform and $\frac{\partial^{n-1}}{\partial p^{n-1}}$ refers to differentiation in the second variable. Theorem 2.84 seems to apply only to smooth functions. However, Hertle [299] shows that theorem 2.84 is valid on $\mathcal{D}'_{L^1}$. He does this by observing that the inversion formula in theorem 2.84 can be written in the form $f = R^\# K Rf$, where K is the operator derived from the formulas in theorem 2.84. For example, if n is even, then $K(u) = \frac{(-1)^{\frac{n-2}{2}}}{2^n \pi^{n-1}} H \frac{\partial^{n-1}}{\partial p^{n-1}}(u)$. But $\mathcal{S} \subset C_v^\infty$ so, by duality, he can extend K to a map from $\mathcal{D}'_{L^1}$ to $\mathcal{S}'$. He then shows that for $u \in \mathcal{D}'_{L^1}(\mathbb{R}^n)$, we have $u = R^\# K Ru$, thereby extending the inversion formula to certain nonsmooth functions.

Deans [122] obtains a single inversion formula for the Radon transform that holds in both even and odd dimensions. In [123] Deans gives a Radon inversion formula for functions that are linear combinations of terms of the form $G(r) S(\theta)$, in polar coordinates, where G is a Gegenbauer polynomial and S is a spherical harmonic.

Hertle [299] proves a version of the slice-projection theorem for distributions in $\mathcal{D}'_{L^1}$. He goes on to prove that a compactly supported L^1 function f is determined by its Radon projections $R_\theta f$ for θ varying in an arbitrary open subset of S^{n-1}. A related result is that a finite measure is determined by its values on half-spaces of the form $\theta \cdot x < p$ for θ varying in an arbitrary open subset of S^{n-1}.

Inversion of the Radon transform can be accomplished via the slice-projection theorem by using inverse Fourier transformation. Practical implementations of this process discretize the data and the Fourier transform. Natterer [439] analyzes this discretization in $\mathbb{R}^n$ and provides two algorithms for choosing the finite number of directions so as to give the least error asymptotically. This least error is understood to be relative to a Sobolev norm.

2.9.3.4 More Results

Agranovsky and Quinto [9] (compare, [7]) present an application of the spherical Radon transform to approximation theory and partial differential equations. The spherical Radon transform is defined as follows:

$$R_{\text{sph}} f(x, r) = \int\limits_{S(x,r)} f$$

where $S(x, r)$ is the sphere of radius r centered at x. Hence, the spherical Radon transform integrates functions over spheres in the same way that the Radon transform integrates functions over hyperplanes. A set of injectivity for the spherical Radon transform is a set S such that $R_{\text{sph}} f(x, r) = 0, \forall x \in S, \forall r \in \mathbb{R}^+$ implies $f = 0$. The function space $\mathcal{L}(S)$ is defined as the span of translates of S of continuous radial functions on $\mathbb{R}^n$. The Coxeter system $\sum_N$ of lines in $\mathbb{R}^2$ is simply the set of lines connecting the origin to a 2Nth root of unity. In [9], Agranovsky and Quinto show:

Theorem A. *If $n = 2$, then $\mathcal{L}(S)$ is dense in $C^0(\mathbb{R}^2)$ if and only if S is not contained in any set of the form $w(\sum_N) \cup F$, where w is a rigid motion of $\mathbb{R}^2$, $\sum_N$ is a Coxeter system, and F is a finite subset of $\mathbb{R}^2$.*

They also show [9]:

Theorem B. *If $n = 2$, then the conditions in theorem A are necessary and sufficient for S to be a set of injectivity for the spherical Radon transform in $\mathbb{R}^2$.*

Theorems A and B have some very interesting consequences. Theorem B leads directly to a uniqueness result for the Darboux partial differential equation. For the heat equation $\frac{\partial u}{\partial t} = c^2 \Delta u$ on $\mathbb{R}^2 \times [0, T]$, with initial value $u(x, 0) = f(x)$, the set $Z(f)$ denotes the set where the temperature u is zero for all times t in the interval $[0, T]$. Agranovsky and Quinto show that if f is compactly supported, then either the initial temperature is identically zero or the zero temperature set $Z(f)$ is the union of a rigid motion of a Coxeter system and an isolated set of points in $\mathbb{R}^2$. Consequently for compactly supported initial distributions, it is impossible to have temperature zero all the time on a nonlinear smooth curve unless the temperature is zero everywhere and for all time.

Another consequence of theorems A and B is a similar result for the nodal sets of the membrane equation in $\mathbb{R}^2$. A third consequence is that a compactly supported continuous function f on $\mathbb{R}^2$ is uniquely determined by the values of its Riesz potentials $R_\lambda * f$ on a set $S \subset \mathbb{R}^2$, for λ in an open interval (a, b), $b \leq 2$ unless S is the union of a rigid motion of a Coxeter system and a finite subset of $\mathbb{R}^2$. See [9] for details (compare, [8]).

Of interest related to the approximation theorem A is the paper of Agranovsky, Berenstein, and Kuchment [6], which shows that the closure of the set of all continuous radial functions with centers at the points of a closed surface in $\mathbb{R}^n$ are complete in the space $L^q(\mathbb{R}^n)$, if and only if $q \geq 2n/(n+1)$.

Of interest related to theorem B are the papers of Agranovsky [4, 5] and Agranovsky, Volchkov, and Zalcman [12] who prove that a cone in $\mathbb{R}^n$ is a set of injectivity for the spherical Radon transform if and only if it is not contained in the zero set of a nontrivial homogeneous harmonic polynomial (compare, Volchkov [613, 614]).

In another paper, Agranovsky and Quinto [10] present an application of the Radon spherical transform to the geometry of stationary sets of the wave equation. We consider the following Cauchy problem for the wave equation in $\mathbb{R}^n$:

$$\frac{\partial^2 u}{\partial t^2} = \Delta u,$$
$$u(x, t) = 0 \text{ for } t > 0 \tag{2.58}$$
$$\frac{\partial}{\partial t} u(x, t) = f(x) \text{ for } t > 0.$$

It can be shown that there is a unique solution on $\mathbb{R}^n$ to this Cauchy problem, if one extends u to be zero on the half–space $t < 0$. The **stationary set of a function** $f \in C^\infty(\mathbb{R}^n)$, relative to the Cauchy problem (2.58), is defined to be the set $S(f) = \{x \in \mathbb{R}^n : u(x, t) = 0 \text{ for all } t > 0\}$, where u is the solution to the Cauchy problem (2.58). The wave equation extends directly to distributions, and Agranovsky and Quinto [10] show how to extend the definition of the stationary set to distributions. An interesting property of stationary sets is that the energy of a solution to the wave equation is constant in time on any domain Ω that is bounded by a stationary set.

Agranovsky and Quinto [10] prove that if f is a nonzero distribution supported on a finite set in $\mathbb{R}^n$, then its stationary set with respect to the wave equation (2.58) satisfies the following conditions.

1. $S(f)$ is an algebraic variety in $\mathbb{R}^n$ and is contained in the zero set of a harmonic polynomial.
2. After a suitable translation, the stationary set is of the form $S(f) = S_0 \cup V$, where V is an algebraic variety of codimension greater than 1. Furthermore, if $S_0 \neq \emptyset$, then S_0 is a harmonic cone, which is a codimension 1 real algebraic variety.

Agranovsky and Quinto also give a more specific description of the geometry of a stationary set for which the reader is referred to reference [10].

A paper of related interest is Agranovsky and Quinto [11], which gives properties of the stationary sets for the wave equation on domains related to crystallographic groups.

2.9.4 Applications of the Radon Transform

Louis has two interesting papers [400, 401], both with the title "Medical Imaging: State of the Art and Future Development." The first paper [400] is fairly elementary from the mathematical point of view but has interesting informal descriptions of the process of data collection and reconstruction of images for medical diagnosis. The second paper [401] is a detailed overview of both the process and the mathematics of various forms of medical tomography. It contains discussions of local tomography, the Radon transform, the attenuated Radon transform, the exponential Radon transform, cone beam tomography, the FDK algorithm, and Grangeat's method. It also deals with magnetic resonance imaging, ultrasound tomography, diffuse tomography, impedance tomography, and limited-view tomography. The emphasis in [401] is practical with attention being paid to discretization, stability, conditioning, and development of practical algorithms for tomography. For the reader interested in the practical side of tomography, the articles [400, 401] also provide an extensive list of references dealing with applications to medical imaging.

Another medical application of the Radon transform is to radiation therapy and radiation dose planning. Radiation therapy is used to shrink or destroy tumors. Radiation dose planning is concerned with maximizing the radiation dose at the site of the tumor while minimizing it elsewhere. This can be done by placing an array of x-ray sources of varying intensities at various locations around the body. The process of designing this array is called radiation dose planning. As we saw in chapter 1, one computes the value of the backprojection operator $R^\# g(x)$ at the point x by averaging the values of g over hyperplanes passing through x. In dimension $n = 2$ this corresponds to the idea of computing the additional effect of γ-rays passing through the point x from sources outside the body: the function $g(\theta, p)$ gives the intensity of the x-ray beam along the line defined by θ and p, and $R^\# g(x)$ gives the combined intensity of the resulting x-rays at the point x. Therefore, the backprojection operator gives a mathematical description of the process of radiation therapy. In their articles [108, 109], Cormack and Quinto apply the backprojection operator to the problem of radiation dose planning. Also see Levine, Gregerson, and Urie [388].

There are many applications of the Radon transform besides computerized tomography, which was treated in chapter 1, and other medical applications, such as those found in Louis [400, 401]. Natterer [444] and Deans [124] contain many such references which the reader may want to consult. The following applications are only a small sample of the myriad applications of the Radon transform.

Sirr and Waddle [561] report on the use of CT scanners to investigate the interior structure of violins and other bowed instruments. In one case the source of a buzz in a 1742 violin made by Carcassi in Florence was diagnosed by a CT scan. The cause of the defect was some loose dried glue in an otherwise undetectable worm track.

Michel [431] uses the Radon transform to give an infinitesimal version in dimension n of Blaschke's conjecture as to whether the canonical Riemannian metric on real projective n-space is the only one for which all geodesics are closed and of the same length.

Brédimas [69] uses the inverse of the spherical Radon transform to solve inverse problems for potentials.

Cavaretta, Micchelli, and Sharma [80] apply the Radon transform to statistics. They characterize certain extensions of univariant interpolation operators to multivariate ones in terms of the inverse Radon transform.

Peters [475] uses the Radon transform to derive laws of large numbers for certain random variables.

Richards [518] uses the Radon transform to derive some results about density functions useful in multivariate statistics.

Mayer-Wolf [429] presents an application of the discontinuous nature of the inverse Radon transform to probability theory.

Chapman [86] and Carswell and Moon [77] have applications of the Radon transform to seismology. More details can be found in chapter 5, section 5.10.

Goncharov [217, 218, 219] and Gel'fand and Goncharov [196] use the Radon transform to solve a problem arising in electron microscopy.

Maass [408, 409] studies a transform related to the Radon transform. This transform integrates functions over hyperbolas and is useful in wideband radar.

Henkin and Shananin [294, 295] use a variant of the Radon transform to study a problem in mathematical economics.

Izen [319, 320], attempts to reconstruct the index of refraction of a supersonic gas flow using a limited view three-dimensional parallel beam Radon transform.

Quinto [504, 505, 506] shows how to reconstruct a function by using the exterior Radon transform. This reconstructs a function from the knowledge of its x-ray projections outside a fixed disk or annulus. The second and third papers give a method for the nondestructive testing of rocket engine exit cones and rocket body gaskets (the title of the third paper [506] is "Computed Tomography and Rockets").

Globevnik [213] uses a support theorem for a Radon transform that integrates over circles surrounding the origin to prove several interesting facts about harmonic and analytic functions in the complex plane. Among these results is the following Morera-type theorem: if f is C^∞ near the origin and if the integral of f is zero around every circle surrounding the origin, then f is analytic. Of related interest are Globevnik [214], Globevnik and Quinto [215], and Grinberg and Quinto [240].

Roerdink [522] reviews the state of the art of CT as of 1992, including cardiac MRI.

The book [203] edited by S. G. Gindikin is a collection of applications of Radon transforms ranging from time series analysis to diffraction tomography.

Gindikin, Reeds, and Shepp [212] apply the spherical Radon transform to study dipoles of crystals.

Raymer, Beck, and McAlister [514] use tomographic methods to reconstruct the amplitude and phase structure of a quasimonochromatic wave field in a plane normal to its propagation direction by using only intensity measurements and refractive optics as input data.

Man'ko [420–422], Mancini, Man'ko, and Tombesi [419], and Alieva and Barbé [15] apply tomography to the study of quantum systems.

Berenstein [41] discusses an application of tomography to plasmas in space.

Bertero [52] describes applications of the Radon transform to medical imaging and astronomy. The applications to astronomy include image reconstruction in the Hubble telescope and in the large binocular telescope.

Desbat and Mennessier [128] study a generalized Radon transform related to Doppler stellar imaging. They show that the kernel of this type of Radon transform is nontrivial and thereby certain surface temperature distributions of stars may be invisible by this technique.

Ciotti [95] uses tomographic methods to study collisionless stellar systems.

Marzetta and Shepp [427] show that an ellipse is the only type of compact, convex subset of $\mathbb{R}^2$ for which the Radon transform depends only on the slope of the lines. They then give an application to motion detection.

Louis and Quinto [402] use local tomography (see section 3.11) to reconstruct object boundaries in shallow water by using sonar data.

Tamasan [596] uses the attenuated Radon transform to reconstruct isotropic scattering for the stationary transport of particles in a source-free scattering medium,

Cerejeiras, Schaeben, and Sommen [81] apply Radon transforms to the study of texture analysis.

Matulka and Collins [428] "rediscover" the Radon transform and apply it to reconstruct the density of a jet of gas. This has applications to aerodynamics. The input data are provided by holographic interferograms of the jet.

3

The k-Plane Transform, the
Radon–John Transform

3.1 Introduction

Let us define a k plane to be any translation of a k-dimensional subspace of $\mathbb{R}^n$. Therefore, a k plane has the form $\eta + x$, where η is a k-dimensional subspace and $x \in \mathbb{R}^n$. Note that a hyperplane is therefore an $(n-1)$ plane.

The Radon transform can be generalized so that the integration is performed on k planes instead of hyperplanes. The related transform is called a *k-dimensional Radon transform* or a *k-plane transform*. Some authors use the term *Radon–John transform*. We use the terms synonymously, and in this chapter we develop the theory of these transforms.

The main part of this chapter begins in section 3.3 with an investigation of the set of all k-dimensional linear subspaces of $\mathbb{R}^n$. This set is called the Grassmannian and is denoted by $G_{k,n}$. Grassmannians are not only sets, but they are also manifolds and measure spaces. We do not require the manifold structure, but we do need to know how to define a suitable measure on Grassmannians. This is done by introducing homogeneous spaces and Haar measure.

Once we have Grassmannians, it is easy to describe the set of all k planes and integration on k planes. This leads to the definition of the k-plane transform and its adjoint. We study the basic properties of the k-plane transform in sections 3.4 and 3.5.

An inversion formula for the k-plane transform is of great interest. We provide four main approaches to the inversion of the k-plane transform. Each approach results in essentially the same inversion formula; the differences are in which class of functions that can be inverted.

We begin with a very easy approach that yields an inversion formula for L^1 functions, satisfying a condition on the Fourier transform. This class includes rapidly decreasing functions and functions in $L^1 \cap L^2$ and hence most of the functions commonly used in practice in tomography. This inversion formula is developed in section 3.6.

More general inversion formulas are based on the Riesz potentials, which form a family of linear operators that we study in section 3.7. In section 3.7 we also study Fuglede's theorem, which demonstrates a close connection between k-plane transforms and Riesz potentials and, in fact, shows that if one could invert the Riesz potential operator, then one would have an immediate formula for the inversion of the k-plane transform.

The Riesz potential can be inverted on large classes of functions. For example, the books of Rubin [524], Samko [541], and Samko, Kilbas and Marichev [542] discuss general results on the inversion of the Riesz potential. From these results we could easily prove inversion formulas for the k-plane transform. However, the background needed would violate the prerequisites that we have established for this chapter. Therefore, in section 3.8, we develop an inversion formula for the k-plane transform based on a Fourier multiplier theorem for the Riesz potential. We are able to develop this approach using only very elementary properties of the Riesz potential along with some elementary facts about real analysis and distributions. This approach originated with the paper [564] of Smith and Keinert (for $k = 1$). The generalization to $k > 1$ was done by Keinert [348]. We have considerably simplified the proofs of these papers.

The third approach to the inversion of the k-plane transform is handled in section 3.9 by treating the k-plane transform as an unbounded operator on $L^2(\mathbb{R}^n)$.

The fourth and last approach to inversion formulas is treated in section 3.10. The inversion theorem in section 3.8 applies to a wide class of L^2 functions. In section 3.10 we describe an inversion theorem due to Rubin that applies to certain L^p functions. We describe Rubin's results without proofs because of the reasons stated earlier.

In section 3.10 we also present some properties of the k-plane transform as a bounded operator between L^p spaces.

The inversion of the k plane transform for L^p functions requires considerably more machinery than that required for L^1 functions. The reader who is only interested in inversion formulas for very well behaved functions, say compactly supported L^2 functions or rapidly decreasing functions, need only study the first inversion formula in section 3.6. However, the other inversion results are used in subsequent chapters.

As an application of the inversion formulas we derive the higher-dimensional inversion formulas that Radon stated in his 1917 paper [508]. The case of dimension $n = 2$ was already treated by other methods in chapter 2.

Roughly speaking, a Radon transform is said to be local if a function can be recovered in the neighborhood of a point by using only k planes passing close to that point. The Radon transform is local in odd dimensions but not in even dimensions. However, there are transforms related to the Radon transform that are local in even dimensions. This is studied in section 3.11 on local tomography.

Finally, in section 3.12 we investigate to what degree the k-plane transform is injective on various spaces. This leads to both uniqueness and nonuniqueness results.

The prerequisites for reading this chapter include a familiarity with the Lebesgue theory of integration and measure, the elements of group theory, the Fourier transform on $\mathbb{R}^n$, and the elementary theory of tempered distributions. Just to get to the Lebesgue theory entails having had the usual prerequisites such as advanced calculus, point-set topology, linear algebra, etc., so we will not be concerned with providing references in these areas. Ideally, the reader should also know at least the rudiments of the theory of the Haar integral and have some knowledge of Grassmann manifolds. However, we will

The Joint British Diabetes Societies have produced guidelines on the management of DKA. Treatment with fluids is the important initial therapeutic intervention, which should be followed by insulin treatment. Fluid boluses can be administered if the patient is shocked (systolic blood pressure <90 mmHg). Boluses should be 500 ml crystalloid over 15 minutes, repeated if required.

A typical fluid regime would be 1 l over 1 hour, 2 l over 4 hours, 2 l over 8 hours and then another 1 l over 6 hours with potassium supplementation as required. Although 0.9% saline ± potassium is recommended on the ward, Hartmann's is deemed a suitable alternative in critical care where potassium can be added safely. The rate of fluids should be decreased in high-risk patients (<25 years, elderly, pregnant, heart and renal failure or serious co-morbidities) to decrease the rate of complications. Once the patient's glucose levels drop to <14 mmol/l, 125 ml/hr of 10% glucose should be added.

Insulin treatment should be with a fixed-rate (not a variable-rate) IV insulin infusion (FRIII) commencing at a rate of 0.1 units/kg/hr (maximum dose 15 units/hr). The patient's long-acting insulin should also be continued.

The recommended targets for metabolic improvement are as follows:

- Reduction of the blood ketone concentration by 0.5 mmol/l/hr
- Increase in venous bicarbonate by 3 mmol/hr
- Reduction of capillary blood glucose by 3 mmol/l/hr
- Maintenance of K^+ between 4.0 and 5.0 mmol/l

If these are not being met the FRIII can be increased by 1 unit/hr.

Any precipitating cause should be identified and treated.

Factors predicting severe disease include the following:

- Ketones >6 mmol/l
- HCO_3 <5 mmol/l
- pH <7.1
- Potassium level <3.5 mmol/l on admission
- GCS <12
- Oxygen saturation <92% on air
- SBP <90 mmHg
- Heart rate >100 or <60 breaths/min
- Anion gap >16

Resolution of DKA is said to occur when the patient's pH is >7.3, and ketones are <0.3 mmol/l.

Reference

Joint British Diabetes Societies Inpatient Care Group. The Management of Diabetic Ketoacidosis in Adults. 2010. http://www.bsped.org.uk/clinical/docs/dkamanagementofdkainadultsmarch20101.pdf (accessed October 14).

C19. Which of the following non-depolarizing neuromuscular blocking drugs (NMDBs) is most likely to lead to histamine release following bolus intravenous (IV) injection?

A. Cisatracurium
B. Atracurium
C. Rocuronium
D. Pancuronium
E. Vecuronium

Answer: B

Short explanation

Atracurium and cisatracurium are benzylisoquinolinium NMDBs, which as a group are more prone to histamine release on rapid IV injection than the other listed drugs, which are all aminosteroids. Cis-atracurium is an isomer of atracurium, which is both more potent and less likely to cause histamine release.

Long explanation

NMDBs act on the post-synaptic receptors to inhibit activation by acetylcholine. In doing so, they prevent the transmission of the action potential across the neuro-muscular junction and the subsequent depolarization of the skeletal muscle. A single bolus dose of atracurium has a half-life of approximately 20 minutes, although this will vary with the patient's physiological state and the presence of other drugs. Doses may be given by infusion or repeated and titrated to effect. Whilst the drug is present at the neuromuscular junction, it will continue to exert an effect and prevent the muscle from being depolarized by a nerve impulse, rendering the patient unable to contract any skeletal muscle.

The only depolarizing NMBD in common use is suxamethonium. It also binds to the post-synaptic acetylcholine receptor and, in doing so, prevents it from being stimulated by a nerve impulse, thus rendering the muscle inactive. However, unlike the non-depolarizing drugs, in binding to the receptor, it also stimulates it, causing a brief and uncoordinated muscle contraction, seen clinically as fasciculation. Suxamethonium has a short onset time and a short half-life, thus making it useful for rapid paralysis, e.g. during a rapid sequence induction of anaesthesia. However, there are several unwanted effects, including the pain and increased potassium release from the sudden muscle contraction.

Non-depolarizing NMDBs fall into two groups: benzylisoquinoliniums (mivacurium, atracurium and tubocurarine) and aminosteroids (rocuronium, pancuronium and vecuronium). Aminosteroids tend to be responsible for less histamine release on IV injection but a greater incidence of anaphylactoid reactions.

Atracurium is a racaemic mixture of 10 isomers. One of these isomers, 1R-cis 1'R-cis atracurium, is isolated because of its improved efficacy and decreased side-effect profile and is marketed as cisatracurium. This isomer has four times the potency and a slightly longer duration of action than atracurium. As a result, a smaller dose can be given, resulting in fewer side effects, including minimal histamine release. The sudden, systemic histamine release can be associated with haemodynamic effects including peripheral vasodilation and hypotension. Avoiding these is often desirable in the unstable ICU patient population.

Reference

Appiah-Ankam J, Hunter JM. Pharmacology of neuromuscular blocking drugs. *Contin Educ Anaesth Crit Care Pain*. 2004;4(1):2–7.

C20. You are called urgently to review a patient suffering with carbon monoxide poisoning after being in a house fire.

Which of the following is LEAST likely to account for tissue hypoxia in this case?

A. Carbon monoxide's higher affinity for haemoglobin
B. Left shift of the oxyhaemoglobin dissociation curve
C. Competitive inhibition of oxygen binding with cytochrome oxidase
D. Impaired cellular utilization of oxygen
E. Increased cellular acidosis

Answer: E

Short explanation
Carbon monoxide poisoning has all of the preceding effects, but only the first four will impair oxygen delivery to the tissues. Cellular acidosis will cause a right-shift of the oxyhaemoglobin dissociation curve and therefore aid off-loading of oxygen.

Long explanation
Carbon monoxide (CO) poisoning is one of the leading causes of death after a burn injury, particularly in an enclosed space. It has a much higher affinity for haemoglobin than oxygen does, which therefore reduces oxygen carrying capacity. There is also a left-shift of the oxyhaemoglobin dissociation curve, which impairs oxygen offloading in the tissues. CO also competitively inhibits mitochondrial enzymes, which subsequently impairs aerobic metabolism. These three mechanisms all result in reduced oxygen delivery and uptake to the tissues.

The symptoms of carbon monoxide poisoning represent a spectrum related to the proportion of haemoglobin bound to CO. Patients with less than 10% COHb are unlikely to be symptomatic, but higher levels lead to headaches, difficulty concentrating and visual disturbance, with cardiovascular complications and coma at levels >50%.

The half-life of CO is roughly 6 hours, which is shortened to about 90 minutes with the administration of 100% oxygen. This can be shortened further with the administration of hyperbaric oxygen, which also has the advantage of providing an increased proportion of dissolved oxygen in the blood. Hyperbaric oxygen may be particularly useful in those with cardiovascular complications, such as acute coronary syndrome, and in those who are pregnant.

Reference
Gill P, Martin R. Smoke inhalation injury. *Contin Educ Anaesth Crit Care Pain*. 2014;15(3):143–148.

C21. You are called to the emergency department to assist with the management of a 5-year-old child who is fitting. He is a known epileptic and takes regular phenytoin. He has been fitting for about 30 minutes and has not responded to buccal midazolam given by his parents. He does not respond to an intravenous (IV) dose of lorazepam 0.1 mg/kg. His airway is maintained, and his SpO_2 is 97% on 15 l O_2 via a non-rebreathe mask.

Which of the following is the most appropriate next treatment?

A. Lorazepam 0.1 mg/kg IV bolus
B. Phenytoin 20 mg/kg IV bolus
C. Phenobarbitol 20 mg/kg IV bolus
D. Thiopentone 4 mg/kg IV bolus and intubation
E. Propofol 2 mg/kg IV bolus and intubation

Answer: C

Short explanation
NICE has published guidelines for the treatment of convulsive status epilepticus (CSE) in children. In early CSE, lorazepam can be given twice, or if midazolam has already been given in a pre-hospital setting, lorazepam can be given once. In established CSE, phenytoin should be given, but in those already on regular phenytoin, phenobarbitol should be used. In refractory CSE, general anaesthesia is required, using thiopentone.

Long explanation

The NICE guidance gives a protocol for the treatment of convulsive status epilepticus (CSE) in children. This splits the management of CSE into stages with theoretical timings from the onset of seizure activity.

Stage 1: This relates to the start of seizure activity, and priorities should be maintenance of the airway, delivering oxygen, and checking a blood glucose.

Stage 2: This is the first 5 minutes of seizure activity, and midazolam 0.5 mg/kg buccally or lorazepam 0.1 mg/kg IV (max 4 mg) should be given.

Stage 3: This is roughly 15 minutes into the seizure. IV access should have been attained, and a further dose of lorazepam 0.1 mg/kg IV given. Phenytoin should be prepared.

Stage 4: At 25 minutes into the seizure, phenytoin 20 mg/kg should be given over 20 minutes, or the same dose of phenobarbitol over 5 minutes if the patient takes regular phenytoin. Paraldehyde can also be given after the phenytoin infusion has started. The anaesthetic/intensive care team should be involved.

Stage 5: If the seizure persists to 45 minutes then general anaesthesia should be induced using thiopentone 4 mg/kg IV.

Reference

National Institute for Health and Clinical Excellence (NICE). *Appendix F: Protocols for treating convulsive status epilepticus in adults and children; children (CG137)*. London: NICE, 2012.

C22. You are about to site a central venous catheter into a neutropenic adult patient. Which of the following routes of access is associated with the highest risk of line-associated blood stream infection?

A. Basilic vein
B. External jugular vein
C. Femoral vein
D. Internal jugular vein
E. Subclavian vein

Answer: C

Short explanation

In adults, use of femoral CVCs is associated with the greatest risk of infection; subclavian lines have the lowest risk. This is related to the density of skin flora at the CVC insertion site. Femoral catheters are also associated with a greater risk for deep vein thrombosis than CVCs at other sites. In children, the risk of infection and other complications is no higher with the femoral route, making it the preferred route of access.

Long explanation

The Joint Commission published a document titled 'Preventing Central Line-Associated Bloodstream Infections: A Global Challenge, A Global Prospective' in 2012. It includes a number of recommendations, including both insertion and maintenance care bundles. A bundle is a group of factors that on their own improve care, but when applied together create an even greater improvement in care.

The CVC insertion bundle was initially trialled in Michigan adult intensive care units (ICUs) and is estimated to have saved 1500 lives over 18 months. The National Patient Safety Agency encouraged ICUs in the United Kingdom to adopt a number of bundles to match the success seen in Michigan.

The insertion of a CVC should avoid the femoral vein where possible (except in children), and full hand hygiene and barrier precautions should be implemented. An observer should be encouraged to challenge the inserter if there is a breach of aseptic technique. Chlorhexidine skin preparation is essential in the absence of allergy.

Once a CVC is sited, there are a number of additional factors that can be implemented to reduce the incidence of blood stream infections. Chlorhexidine should be applied to the site when dressings are changed which should be every 7 days for clear dressings, but more frequently if they are loose or soiled, and every 2 days for gauze dressings. Excessive changing of administration sets or caps can introduce infection, so these should not be changed more frequently than every 72 hours. An exception to this is tubing through which blood or lipids have been given, which should be changed within 24 hours. The CVC should be removed as soon as it is no longer needed.

An additional recommendation of note concerns securement of devices. They advise securing the CVC using a specialist device without the need for sutures because sutures can increase the levels of colonization, and also pose a risk of sharps injury. The use of these devices must be balanced against the risk of the CVC becoming loose or coming out.

Reference
The Joint Commission. *Preventing Central Line–Associated Bloodstream Infections: A Global Challenge, a Global Perspective.* Oak Brook, IL: Joint Commission Resources, 2012. http://www.jointcommission.org/assets/1/18/clabsi_monograph.pdf.

C23. A patient in the intensive care unit (ICU) needs to return to the operating theatre today for a second laparotomy. He underwent an emergency laparotomy 2 days ago for small bowel perforation and peritonitis.

Which of the following patient scenarios is most likely to benefit from a transfusion (packed red cells, platelets, or fresh frozen plasma)?

A. Platelets for a 50-year-old man whose platelet count has been 60×10^9/l for the past 2 days

B. Fresh frozen plasma for a 50-year-old man whose international normalized ratio (INR) has been 1.8 for the past 2 days

C. Packed red cells for a 50-year-old man whose haemoglobin has been 78 g/l for the past 2 days

D. Packed red cells for a 50-year-old man with a history of ischaemic heart disease (IHD), whose haemoglobin is 78 g/l today. He has ST-segment depression on his electrocardiogram (ECG)

E. Packed red cells for a 50-year-old man with end-stage renal failure awaiting a renal transplant whose haemoglobin has been 68 g/l for the past 2 days

Answer: D

Short explanation
The Joint United Kingdom (UK) Blood Transfusion and Tissue Transplantation Services Professional Advisory Committee guidelines advise that transfusion is unnecessary if the platelet count is above 50×10^9/l and INR <2. Transfusion of blood is inappropriate with a haemoglobin >80 g/l for haemodynamically stable patients without active bleeding. A higher transfusion threshold may be required for patients with sepsis, traumatic brain injury or acute coronary syndrome, whereas a lower threshold may be implemented in those with chronic anaemia or who are awaiting organ transplant.

Long explanation

Transfusion medicine has changed significantly in recent years, with the adverse effects of blood transfusion having increased prominence. One landmark study, the TRICC study (Transfusion Requirements in Critical Care) in 1999 demonstrated a significantly lower mortality in those patients randomized to a transfusion trigger of 70 g/l compared with those with a trigger of 100 g/l. Nearly a third of those in the lower transfusion level group did not receive a blood transfusion. This has led to a change in practice towards lower transfusion thresholds.

The *Handbook of Transfusion Medicine*, produced by the UK Blood Service, states that in the haemodynamically stable, non-bleeding patient, transfusion should be considered if the Hb is <80 g/l and is usually indicated if the Hb is <70 g/l. A single red cell unit should be transfused and the patient reassessed. A higher transfusion trigger may be beneficial in patients with some conditions. These include ischaemic stroke (>90 g/l), traumatic brain injury (90 g/l), subarachnoid haemorrhage (80–100 g/l), acute coronary syndrome (80–90 g/l) or in the early stages of severe sepsis (90–100 g/l if evidence of tissue hypoxia).

With regard to other blood products, the guidelines state that most invasive surgical procedures can be carried out safely with a platelet count $>50 \times 10^9/l$ or INR <2.0. They also note that moderate thrombocytopenia ($50–100 \times 10^9/l$) is common in critical care patients and that 'prophylactic' platelet transfusion in non-bleeding patients is not indicated.

References

Herbert P, Wells G, Blajchman MA, et al. A multicentre, randomized, controlled clinical trial of transfusion requirements in critical care. Transfusion Requirements in Critical Care Investigators, Canadian Critical Care Trials Group. *N Engl J Med.* 1999; 340(13):1056.

National Blood Service. *Guidelines for Blood Transfusion Services in the UK.* 8th edition. London: The Stationary Office, 2013.

Norfolk D, on behalf of UK Blood Services. *Handbook of Transfusion Medicine.* 5th edition. Norwich, UK: TSO Publishing, 2013.

C24. A 55-year-old male patient is about to be discharged from the intensive care unit following admission with community-acquired pneumonia. He went into atrial fibrillation (AF) on day 1 of his stay, and has remained in AF since. You decide to implement anticoagulation to reduce his stroke risk.

Which of the following is the best way to assess his stroke risk?

A. CHADS$_2$
B. CHA$_2$DS$_2$-VASc
C. HAS-BLED
D. Goldman Cardiac Risk Index
E. American Stroke Association Stroke Calculator

Answer: B

Short explanation

NICE guidance states that the CHA$_2$DS$_2$-VASc score should be used to assess stroke risk in people with AF. This should be used in combination with the HAS-BLED score, which assesses the risk of bleeding in people who are starting anticoagulation.

Long explanation

The CHA$_2$DS$_2$-VASc stroke risk score, which has replaced the CHADS$_2$ score in recent years, is used to assess stroke risk in people with AF and atrial flutter. It comprises a number of parameters that score points. The following parameters all score 1 point:

- Congestive heart failure
- Hypertension
- Age 65–74
- Diabetes
- Vascular disease
- Female sex

Two points are given to age over 74 and to previous stroke/transient ischemic attack/thromboembolism. Once the bleeding risk has been taken into account (as described subsequently), anticoagulation should normally be offered to men with a score of ≥1 and to women with a score of ≥2 (i.e. the presence of one risk factor apart from gender).

Bleeding risk must be taken into account before commencement of any anticoagulation. Assuming a risk-benefit analysis suggests a benefit to anticoagulation, then this can be with apixaban, dabigatran, rivaroxaban or a vitamin K antagonist (e.g. warfarin).

HAS-BLED, a scoring system which estimates the risk of major bleeding specifically for anticoagulation in AF, gives the following parameters one point:

- Hypertension (systolic blood pressure >160 mmHg)
- Renal dysfunction
- Liver disease
- Previous stroke
- Previous major bleeding/ongoing bleeding risk
- Unstable INR
- Age >65
- Medication affecting coagulation (e.g. nonsteroidal anti-inflammatory drugs, antiplatelets)
- Significant alcohol/drug use

With a score of 0, the risk of major bleeding in 1 year is roughly 1%; with a score of 1, it is roughly 3.5%; with a score of 3, it is roughly 6%; and with a score of 5, it is roughly 9%.

The risk of bleeding needs to be balanced against the risk of stroke, and although for most people the benefit of anticoagulation outweighs the bleeding risk, for people with an increased risk of bleeding, this may not always be the case, and careful monitoring of bleeding risk is important. A combined calculator is available online.

References

National Institute for Health and Clinical Excellence (NICE). *Atrial fibrillation: the management of atrial fibrillation (CG180)*. London: NICE, 2014.

Stroke Prevention in Atrial Fibrillation Risk Tool (SPARC). http://sparctool.com (accessed May 2015).

C25. A 26-year-old, 70-kg man with no medical problems requires a rapid sequence induction of anaesthesia (RSI). You apply 100% oxygen via a tight-fitting mask and pre-oxygenate for 3 minutes. After induction of anaesthesia, he becomes apnoeic, and you maintain his airway with manual manoeuvres.

With a tight-fitting mask supplying 100% oxygen and a patent airway, how long might this apnoeic patient's oxygen saturations be expected to stay above 90%?

A. 1 minute
B. 3 minutes
C. 8 minutes
D. 20 minutes
E. 100 minutes

Answer: E

Short explanation

With a patent airway and 100% oxygen applied, apnoeic mass-movement oxygenation will maintain oxygen saturations for up to 100 minutes. Removal of the oxygen or failing to maintain an airway will lead to much more rapid desaturation.

Long explanation

The overwhelming majority of intubations in the emergency department and intensive care unit are performed as RSIs owing to patient condition, urgency or aspiration risk. Most RSIs contain an apnoeic pause as drugs take effect. This pause can lead to significant desaturation in the critically ill, particularly if not practiced correctly. An appreciation of the physiology of apnoea is therefore important.

As blood passes through the lung, desaturated haemoglobin becomes saturated with oxygen from the alveoli and is delivered to respiring tissues. The basal oxygen consumption of the average person, and hence the rate of uptake of oxygen from the alveoli, is approximately 250 ml/min. If a patient has been fully preoxygenated, then his or her functional residual capacity acts as a reservoir of oxygen. This is the volume of the lungs at the end of passive expiration (equal to the expiratory reserve volume plus residual volume). In a 70-kg man, this is approximately 2000 ml, providing a supply of oxygen for approximately 8 minutes if the airway is closed. Without preoxygenation, the much lower partial pressure of alveolar oxygen (~13 kPa) will only contain 260 ml of oxygen, a 1-minute supply.

The uptake of oxygen from the alveoli leads to a reduction in alveolar pressure, so if the airway is held open with 100% oxygen applied, then oxygen can move down this pressure gradient (apnoeic mass-movement oxygenation). Theoretically, this mass movement could provide oxygenation indefinitely were it not for the production of CO_2 by the tissues. This is produced at a rate of approximately 200 ml/min. During normal respiration, clearance of alveolar CO_2 with every breath leads to the maintenance of a concentration gradient from the tissues to the alveoli. In apnoea, this gradient is lost and CO_2, being highly water soluble, diffuses throughout the body. As a result, the partial pressure of CO_2 in the alveoli rises slowly, at approximately 20 ml/min. As this rises, then the partial pressure of oxygen in the alveoli falls, leading to eventual desaturation after up to 100 minutes. Of course in the critically ill, many factors will reduce this time.

References

Sirian R, Wills J. Physiology of apnoea and the benefits of preoxygenation. *Contin Educ Anaesth Crit Care Pain*. 2009;9(4):105–108.

Weingart SD. Preoxygenation, reoxygenation, and delayed sequence intubation in the emergency department. *J Emerg Med*. 2011;40(6):661–667.

C26. You are called to the resuscitation room of the emergency department. A 23-year-old woman with known severe asthma arrived 2 hours earlier and is not improving despite treatment. So far, she has been treated with nebulized salbutamol and ipratropium bromide driven by oxygen and intravenous hydrocortisone.

Which of the following adjuvant drugs has the most evidence of benefit in acute severe asthma?

A   Montelukast
B. Ketamine
C. Heliox
D. Magnesium sulphate
E. Aminophylline

Answer: D

Short explanation
The only listed additional medication with evidence of benefit in acute asthma is magnesium sulphate, which has been shown to improve lung function and reduce hospital admissions. The others have been used in refractory cases but have no strong evidence of benefit.

Long explanation
Acute severe asthma is a medical emergency. There were 195 deaths attributed to asthma in the United Kingdom between February 2012 and January 2013. Most deaths occurred in patients with a history of chronically severe asthma. Risk factors for fatal or near-fatal asthma include previous ventilation, recent admissions or recurrent ED attendances and heavy use of beta-2 agonists, particularly in patients with social, emotional, mental health or learning difficulties.

Patients with acute severe asthma and hypoxaemia should receive supplemental oxygen to maintain saturations of 94 to 98%. If saturation monitoring is unavailable, then oxygen should be used.

Inhaled beta-2 agonists have evidence of rapid bronchodilator effects. In non–life-threatening asthma, inhalers may used with a spacer device; otherwise, nebulizers are preferred. Salbutamol and terbutaline are equally efficacious, but nebulized adrenaline has no advantage and is not recommended. Nebulizers should be driven with oxygen at 6 l/min and repeated as required. Continuous nebulization may be required in life-threatening cases. There is little evidence to support the use of intravenous beta-2 agonists. The addition of ipratropium bromide nebulizers is known to improve bronchodilation and reduce length of illness and should be used in severe asthma.

Corticosteroids are known to reduce mortality and hospital admissions. Oral prednisolone is as effective as intravenous hydrocortisone, but there may be difficulty in its administration in life-threatening cases. Doses of 40 to 50 mg prednisolone or 100 mg hydrocortisone QDS are required.

Magnesium sulphate has some evidence to support its use in acute severe asthma, with a large study suggesting that lung function is improved and intubation rates reduced. However, the use of repeated boluses of IV magnesium sulphate is not supported by evidence of benefit or even safety and is not recommended.

Other therapies are mentioned in the British Thoracic Society asthma guidelines as having been attempted in refractory acute severe asthma but with no evidence of benefit. These include aminophylline, which may benefit a subgroup of the sickest patients; montelukast; antibiotics; heliox; intravenous fluids; and nebulized furosemide. None of these therapies are recommended but are known to have been tried and/or studied. Other therapies with case reports in the literature include ketamine, inhaled anaesthetic agents and recombinant human DNase.

References

British Thoracic Society, Scottish Intercollegiate Guidelines Network. British guideline on the management of asthma. A national clinical guideline. *Thorax*. 2014;69:i1.

Royal College of Physicians. *Why Asthma Still Kills: The National Review of Asthma Deaths (NRAD) Confidential Enquiry report.* London: Royal College of Physicians, 2014.

C27. A 78-year-old woman presents to the emergency department with severe abdominal pain, nausea and diarrhoea. She has a past medical history of hypertension and atrial fibrillation and smokes 20 cigarettes per day. On examination, she is tachycardic and febrile with cool peripheries. Chest and abdominal radiographs are reported as normal. Initial blood results show the following:

Haemoglobin 119 g/l; white blood cell count 26.9×10^9/l; platelets 286×10^9/l
Na^+ 139 mmol/l; K^+ 4.7 mmol/l; urea 11.3 mmol/l; creatinine 124 μmol/l
pH 7.18; pO_2 12.0 kPa; pCO_2 4.1 kPa; base excess −11.5; lactate 7.2 mmol/l

What is the most likely diagnosis?

A. Diverticulitis
B. Gallstone ileus
C. Cholecystitis
D. Mesenteric ischaemia
E. Colonic obstruction

Answer: D

Short explanation

This patient has risk factors for the development of mesenteric ischaemia (age >70; hypertension, smoking, atrial fibrillation), along with classical clinical features and a severe metabolic acidosis. Diverticulitis and cholecystitis rarely cause significant lactic acidosis. Neither obstruction nor ileus is likely in the presence of normal plain abdominal radiographs.

Long explanation

Acute mesenteric ischaemia is a rare but life-threatening disease. The annual incidence is approximately 1 to 2 per 1000 patient years. Mortality rates range from 50 to 100%, depending on case series.

Mesenteric ischaemia is most commonly caused by arterial occlusion, either thrombotic or embolic. Other causes include mesenteric venous thrombosis and hypoperfusion, either due to systemic hypotension or non-occlusive mesenteric ischaemia (NOMI).

Risk factors for arterial mesenteric ischaemia include increased age (>70), atrial fibrillation, cardiovascular disease (ischemic heart disease, stroke) and previous arterial emboli. Risk factors for thrombotic venous mesenteric ischaemia include previous deep vein thrombosis/pulmonary embolism, portal hypertension, pancreatitis or pancreatic cancer and hypercoagulable states.

Presentation of mesenteric ischaemia is usually with severe abdominal pain out of keeping with examination findings, nausea, vomiting and diarrhoea, often with bloody stool. Laboratory tests may be non-specific, although a raised white cell count and lactic acidosis resistant to fluid therapy are hallmarks of the disease.

Computed tomography (CT) is the investigation of choice for mesenteric ischaemia, although patients with peritonism, signs of perforation or haemodynamic instability may be taken straight to theatre for exploratory laparotomy. CT with contrast is necessary to effectively rule out mesenteric ischaemia, although the risks of intravenous contrast should be considered. Magnetic resonance imaging is a possible alternative but is clearly a logistically challenging investigation in patients with potential for haemodynamic instability. Findings consistent with mesenteric ischaemia include thrombosis (arterial or venous), intramural gas, gas in the portal system or biliary tree, focal lack of bowel-wall enhancement and visceral infarcts.

Treatment is traditionally 'damage-control surgery': an emergency laparotomy with washout, resection of the affected bowel segment and mesentery, although revascularization is an increasingly common option for the treatment of mesenteric arterial occlusion without peritonitis or perforation. This can be an embolectomy, local thrombolysis or stenting and performed either as an open or endovascular procedure.

References

Acosta S, Björck M. Modern treatment of acute mesenteric ischaemia. *Br J Surg*. 2014;101:100–108.

Jones J, Cudnik MT, Stockton S, et al. The diagnosis of acute mesenteric ischemia: a systematic review and meta-analysis. *Ann Emerg Med*. 2012;60:S23.

Kougias P, Lau D, El Sayed HF, et al. Determinants of mortality and treatment outcome following surgical interventions for acute mesenteric ischemia. *J Vasc Surg*. 2007;46:467–474.

C28. A 43-year-old man is admitted to the intensive care unit (ICU) with severe respiratory failure. He is known to have HIV but has not yet started highly active antiretroviral therapy (HAART). He is intubated and ventilated with 60% oxygen, achieving a PaO$_2$ of 8.4 kPa. A chest radiograph demonstrates bilateral diffuse perihilar infiltrates.

Which initial treatment regimen is most appropriate?

A. Trimethoprim and sulphamethoxazole
B. Trimethoprim, sulphamethoxazole and prednisolone
C. Trimethoprim, sulphamethoxazole, prednisolone and HAART
D. HAART only
E. Trimethoprim, sulphamethoxazole and HAART

Answer: B

Short explanation

This patient is likely to have pneumocystis pneumonia (PCP), which is the commonest cause of respiratory failure in patients with HIV. First-line therapy is trimethoprim and sulphamethoxazole. In patients with severe hypoxia (PaO$_2$ <9.2 kPa on room air), corticosteroids are of benefit. Starting HAART during an acute illness is risky, owing to the increased risk posed by immune reconstitution inflammatory syndrome.

Long explanation

With advances in antiretroviral therapy, HIV has become a chronic disease in the developed world, and patients with HIV are increasingly admitted to ICU with diseases unrelated to HIV. However, complications of HIV are still common, particularly in patients with undiagnosed or untreated HIV. The commonest reason for ICU admission in HIV patients (35–40% of cases) remains respiratory failure. Sepsis is an increasing reason for ICU admission (15–20% of cases), with other reasons being neurological, cardiac or gastrointestinal disease, trauma and post-surgical care. By far the most common cause of respiratory failure in HIV is pneumocystis pneumonia (PCP), although asthma, bacterial pneumonia and chronic obstructive pulmonary disease are increasingly seen.

PCP is an opportunistic alveolar infection caused by an atypical fungus, *Pneumocystis jirovecii*. It is thought to lie dormant in the lung tissue of up to 20% of healthy individuals, only causing significant infection in people with immunocompromise. It is commonly associated with HIV infection, and indeed it is the commonest AIDS-defining illness, with infection usually only seen in patients with a CD4 count of <200 cells/μl. However, patients with non-HIV immunosuppression are also at risk of PCP, particularly recipients of transplants (solid organ and haematopoietic stem cell) and patients on cancer chemotherapy or long-term corticosteroids.

Clinical features include dyspnoea, cough and fever of gradual onset. Chest radiographs may be normal in up to one-quarter of patients but typically show bilateral perihilar interstitial infiltrates that become more diffuse and widespread as the disease progresses. Patients with PCP are at increased risk of pneumothorax, particularly those requiring invasive positive pressure ventilation.

First-line therapy of PCP is trimethoprim/sulphamethoxazole. Patients who do not tolerate this may be treated with other agents such as dapsone, clindamycin/primaquine or pentamidine. In patients with significant hypoxia, defined as a PaO_2 of <70 mmHg (9.2 kPa) on air or an A-a gradient of ≥35 mmHg (4.6 kPa), corticosteroids have been shown to significantly reduce mortality in PCP. Current recommendations are prednisolone 40 mg twice daily for 5 days.

The question of when to start HAART is controversial. Although HAART undoubtedly improves the outcomes of patients with HIV, starting it during acute illness is risky. This is due to immune reconstitution inflammatory syndrome, in which the newly reactivated host immune system causes a dramatic increase in inflammation at sites of previous or current infection. In severe PCP, this can be fatal. Current guidance is to start HAART within 2 weeks of the diagnosis of PCP.

References

Monnet X, Vidal-Petiot E, Osman D, et al. Critical care management and outcome of severe *Pneumocystis* pneumonia in patients with and without HIV infection. *Crit Care*. 2008;12(1):R28.

Saccente M. Intensive care of patients with HIV infection. *N Engl J Med*. 2006;355:173–181.

Thomas CF, Limper AH. *Pneumocystis* pneumonia. *N Engl J Med*. 2004;350:2487–2498.

C29. You are called to the emergency department as part of the major trauma team. A 27-year-old pedestrian has been hit by a car. He has the following findings on primary survey:

Airway: Intact

Breathing: Respiratory rate 24/min; SpO$_2$ 97% on 15 l oxygen. Bilateral air entry, no added sounds.

Circulation: Heart rate 105 bpm; blood pressure 105/85 mmHg. Cool peripheries. No external haemorrhage. Unstable pelvis.

Disability: Glasgow Coma Score 15/15; pupils equal and reactive; in pain.

Initial trauma radiographs are taken. Chest radiograph shows no abnormality; pelvic radiograph demonstrates an iliac wing fracture.

What is the best way to assess the extent of internal bleeding from his pelvic fracture?

A. Abdominal ultrasound (focused assessment with sonography in trauma [FAST] scan)
B. Diagnostic peritoneal lavage
C. Emergency 'damage-control' laparotomy
D. Computed tomography (CT) with contrast
E. Further pelvic radiographs including Judet views

Answer: D

Short explanation

This patient has a significant pelvic fracture but is haemodynamically stable. A CT with contrast is the most sensitive test for internal haemorrhage because it can show both the volume of intraperitoneal blood and ongoing bleeding during the arterial phase of the scan. FAST scan is a useful test but is neither as specific nor as sensitive as a CT scan. Peritoneal tap has a lower false-negative rate than peritoneal lavage, but neither are now routinely performed. If internal haemorrhage cannot be otherwise excluded, a laparotomy may be necessary.

Long explanation

Pelvic fracture is usually the result of blunt trauma and often associated with other significant injuries. There are different ways to classify pelvic fractures. The Tile classification is based on the stability of the pelvic ring and the displacements possible due to the injury. Pelvic fractures are either stable, partially stable (allowing displacement in one direction) or completely unstable. The Young-Burgess classification is based on radiological findings and the mechanism of injury. Patients with a mechanically unstable pelvic fracture are not necessarily haemodynamically unstable, although the greater the damage and/or displacement, the greater the chance of a life-threatening haemorrhage.

In patients with possible intra-abdominal or internal pelvic haemorrhage, the choice of investigation depends chiefly on their haemodynamic stability. Abdominopelvic CT with intravenous contrast is the investigation of choice but may not always be possible because of the patient's haemodynamic condition. CT is valuable for determining the presence of other visceral injuries, haemorrhage, exact fracture delineation and determination of the requirement for laparotomy or emergency angio-embolization.

In unstable patients with pelvic fractures, CT may not be possible and so other investigations should be performed to assess for the presence of intra-abdominal haemorrhage. Ultrasound (FAST scan) is a useful tool with an excellent positive predictive value for intraperitoneal fluid; however, its false-negative rate is unacceptably high, and so FAST cannot be used to safely exclude internal haemorrhage.

In haemodynamically unstable patients with a negative or equivocal FAST scan, diagnostic peritoneal tap (DPT) and lavage (DPL) are recommended by the Eastern Association of Trauma Surgeons and in the Advanced Trauma Life Support guidelines. In the United Kingdom, neither of these investigations are commonly performed. There is a lower false-positive rate with DPT than DPL, so this is the test that should be performed. In certain situations, proceeding straight to damage-control laparotomy may be the necessary course of action.

References

Cullinane DC, Schiller HJ, Zielinski MD, et al. Eastern Association for the Surgery of Trauma Practice Management guidelines for hemorrhage in pelvic fracture – update and systematic review. *J Trauma Inj Infect Crit Care*. 2011;71(6):1850–1868.

Tile M. Acute pelvic fractures: I. causation and classification. *J Am Acad Orthop Surg*. 1996;4(3):143–151.

C30. A 68-year-old woman is on intensive care. She develops signs of severe sepsis, and blood cultures from her haemofiltration catheter and peripheral samples have grown *Candida* (exact species and sensitivities awaited). Other urine and sputum cultures have not yielded positive results.

Which of the following is the LEAST important to instigate?

A. Perform an echocardiograph (echo)
B. Treat with intravenous Caspofungin
C. Get an ophthalmology review
D. Perform a renal tract ultrasound
E. Remove and replace the haemofiltration catheter

Answer: D

Short explanation

Invasive candidaemia is likely to be the source of this patient's sepsis. As a potential source, the haemofiltration catheter should be removed and treatment commenced with an appropriate antifungal agent such as Caspofungin. An ophthalmology review looking for endophthalmitis and an echocardiogram looking for fungal balls should be performed in all patients with candidaemia. A renal tract ultrasound would identify the presence of fungal balls in the kidney however the absence of candiuria makes this unlikely.

Long explanation

The majority of invasive fungal infections result from infections with fungi of the *Candida* species, most commonly *Candida albicans*. Differentiation between colonization with *Candida* and invasive disease can be difficult. This patient is likely to have invasive disease because she has associated symptoms of severe sepsis and *Candida* grown simultaneously from several sources. Her critical care admission, recent ventilation, broad-spectrum antibiotics for pneumonia, central venous line presence and renal replacement therapy are all risk factors for invasive disease.

The Infectious Diseases Society of America (IDSA) produced guidelines in 2009 regarding the management of patients with invasive candidiasis and mucosal candidiasis. It recommends treatment for confirmed or suspected candidaemia with fluconazole or an echinocandin (Caspofungin, micafungin, anidulafungin) in non-neutropenic patients. ISDA prefers echinocandins to treat those with a moderate to severe degree of illness, which goes along with the European Society of Clinical Microbiology and Infectious Diseases (ESCMID) guidelines. The IDSA guidelines also recommend echinocandins as treatment for candidaemia in neutropenic patients.

The guidelines strongly recommend removing intravenous catheters in patients who are not neutropenic. There is less evidence of benefit to removing lines in patients who are neutropenic (these patients are more likely to have an alternative source causing candidaemia; however, removal should be considered on an individual patient basis).

All patients with candidaemia should be screened for spread to distant organs. An ophthalmology review should be sought to screen for endophthalmitis, which can cause irreversible visual loss in all patients, even if they are asymptomatic. The ESCMID guidelines recommend performing an echocardiogram to rule out endocarditis in these patients.

Imaging of the kidney should be performed to look for candida pyelonephritis, fungal balls or abscesses in patients with candiduria. Ultrasound should be performed first line; however, computed tomography or magnetic resonance imaging can also be of diagnostic value. The presence of fungal balls is rare in adult patients but is a far more common finding in infants and neonates on intensive care units who present with candiduria.

References

Cornely O.A, Bassetti M, Calandra T, et al. ESCMID guideline for the diagnosis and management of Candida diseases 2012: non-neutropenic adult patients. *Clin Microbiol Infect*. 2012;18(Supp 7):19–37.

Kauffman CA, Fisher JF, Sobel JD, Newman CA. Candida urinary tract infections – diagnosis. *Clin Infect Dis*. 2011;52(Suppl 6):S452–S456.

Pappas PG, Kauffman CA, Andes D, et al. Clinical practice guidelines for the management candidiasis: 2009 Update by the Infectious Diseases Society of America. *Clin Infect Dis*. 2009;48(5):503–535.

Exam D: Questions

D1. A 28-year-old man has been admitted to intensive care with severe respiratory failure. He is sedated and ventilated. A viral throat swab confirms the presence of influenza A, and a diagnosis of viral pneumonia is made.

Which of the following is the most appropriate antiviral agent?

A. Aciclovir
B. Ritonavir
C. Zanamivir
D. Ganciclovir
E. Oseltamivir

D2. A 33-year-old patient has collapsed in the hospital canteen, suffering from suspected anaphylaxis following ingestion of some peanuts. He has acute airway swelling and stridor. His blood pressure is 73/40 mmHg, and heart rate is 110 bpm.

Which is the most appropriate dose and route of adrenaline?

A. 10 ml 1:10,000 intravenous (IV)
B. 1 ml 1:1000 IV
C. 0.5 ml 1:1000 IV
D. 0.5 ml 1:1000 intramuscular (IM)
E. 1 ml 1:1000 IM

D3. Which of the following treatments in the management of adult traumatic brain injury is MOST likely to be beneficial?

A. Avoidance of systolic blood pressure <90 mmHg
B. Avoidance of SpO_2 <90%
C. Prophylactic hypothermia
D. Maintenance of a cerebral perfusion pressure (CPP) above 70 mmHg
E. Graduated compression stockings or pneumatic compression stockings

D4. A 44-year-old man presents to hospital with fever, myalgia, night sweats and shortness of breath. He is found to have a pan-systolic murmur, splenomegaly and microscopic haematuria. A transthoracic echocardiogram demonstrates an oscillating mass associated with the mitral valve.

What is the most likely causative organism?

A. Pseudomonas aeruginosa
B. *Staphylococcus aureus*
C. *Streptococcus bovis*
D. *Escherichia coli*
E. *Coxiella burnetti*

D5. A 68-year-old man has been on intensive care for 4 days with acute respiratory distress syndrome (ARDS) and multiorgan failure after an emergency laparotomy for large bowel perforation with faecal peritonitis. He is requiring 80% oxygen, inotropic and vasoconstrictor support and renal replacement therapy. He weighs 90 kg and is 175 cm tall.

Which of the following is the most appropriate ventilator setting:

A. Tidal volume (V_T) = 490 ml; positive end expiratory pressure (PEEP) = 14 cmH_2O
B. V_T = 540 ml; PEEP = 10 cmH_2O
C. V_T = 420 ml; PEEP = 6 cmH_2O
D. V_T = 540 ml; PEEP = 6 cmH_2O
E. V_T = 420 ml; PEEP = 14 cmH_2O

D6. A 45-year-old man was admitted to the ICU 10 days earlier after complete spinal cord injury at T2. He has undergone surgical fixation of his spinal fracture, and he is medically stable. He has had a tracheostomy sited, and you would like to commence respiratory weaning.

Which of the following is most likely to impair the weaning process?

A. Pulmonary compliance of 50 ml/cm H_2O
B. FiO_2 of 0.35
C. 8-mm internal diameter tracheostomy
D. Vital capacity of 750 ml
E. High opioid requirements for analgesia

D7. Drug A is 98% protein bound and is metabolized in the liver with a high hepatic extraction ratio. Which of the following changes that may occur in a patient on ICU is most likely to alter the circulating levels of the drug?

A. Reduced hepatic blood flow due to low cardiac output
B. Fall in serum albumin concentration
C. Co-administration of a hepatic enzyme inducer
D. Acute kidney injury
E. Ileus

135

D8. A patient with urinary sepsis has received high-volume fluid resuscitation with normal saline but remains hypotensive and tachypneic on the ward. He is transferred to the intensive care unit, and a noradrenaline infusion is started. Cardiac output studies after commencement of noradrenaline suggest a high cardiac index with appropriate filling and vasoconstriction.

Arterial blood gas (ABG) on air:

pH 7.29, pCO_2 3.1 kPa, pO_2 12 kPa, HCO_3 16 mmol/l, base excess −16, lactate 1.2 mmol/l, Na^+ 144 mmol/l, K^+ 3.6 mmol/l, Cl^- 124 mmol/l, venous saturation 76%

What would be the MOST appropriate next intervention?

A. Fluid bolus with Hartmann's
B. Commence a dobutamine infusion
C. 100 ml 8.4% sodium bicarbonate
D. Invasively ventilate to restore normocapnia
E. Increase noradrenaline infusion rate

D9. A new blood test is developed to identify patients who develop ventilator-associated pneumonia. The test is described by the company representative as having a high positive predictive value (PPV). Which of the following statements is most likely to be accurate regarding the value of this test?

A. The area under the receiver operating curve (AUROC) is likely to be close to 0.5.
B. The negative predictive value (NPV) will probably also be high.
C. The PPV will be similar when tested in two different populations.
D. The PPV can be calculated by dividing the number of true positives by the number of false-positive results.
E. The PPV can be calculated by dividing the number of true positives by the sum of all positive results.

D10. You are looking after a patient with the following findings. With reference to the Child-Pugh scoring system, which is associated with the worst 1-year survival?

A. Bilirubin 45 μmol/l
B. Serum albumin 26 g/l
C. International normalized ratio (INR) 2.1
D. Mild ascites
E. Grade II hepatic encephalopathy

D11. A 45-year-old smoker is admitted in cardiogenic shock after an acute myocardial infarction. He underwent percutaneous coronary intervention with the insertion of two stents and had an intra-aortic balloon pump (IABP) inserted within 2 hours of admission.

Which of the following is LEAST accurate regarding IABPs?

A. Using the arterial trace, balloon inflation should be timed with the dicrotic notch.
B. The tip of the balloon should lie 2 to 3 cm distal to the origin of the left subclavian artery.
C. Insertion of an IABP has good evidence of reduced mortality in these patients.
D. An increase should be seen in this patient's cardiac output (CO).
E. His augmented diastolic pressure should be higher than his systolic pressure.

D12. An 18-year-old woman is admitted to the intensive care unit with acute liver failure after a paracetamol overdose.

Which of the following criteria is LEAST useful in predicting her need for a liver transplant?

A. Post-resuscitation arterial pH 7.28
B. Grade III encephalopathy
C. Creatinine 320 µmol/l
D. Prothrombin time 108 sec
E. Bilirubin 320 µmol/l

D13. A 30-year-old trauma patient is transferred to the intensive care unit after 4 hours of emergency surgery to remove a ruptured spleen following a road traffic accident. He has received 10 units of packed red cells, 1 unit of platelets and 8 units of fresh frozen plasma during the resuscitation. He has also received two doses of tranexamic acid 1 g.

To guide further therapy, you decide to run a rotational thrombo elastometry (ROTEM) test. Which of the following is most likely to explain the following results?

Measure	Normal range	EXTEM result	FIBTEM result
Clotting time	40–80 seconds	69 seconds	185 seconds
Clot formation time	34–160 seconds	263 seconds	—
a-angle	55–78 degrees	58 degrees	—
Maximum clot formation	50–72 mm	38 mm	3 mm

A. The tranexamic acid given in theatre
B. A low platelet count
C. Acidosis
D. A low fibrinogen level
E. Disseminated intravascular coagulation

D14. Concerning oxygen toxicity, which of the following potential effects of oxygen is LEAST likely to occur after prolonged exposure to 100% oxygen at atmospheric pressure?

A. Increased peripheral vascular resistance
B. Decreased intracranial pressure following severe head injury
C. Nausea, dizziness and headache
D. Diffuse alveolar damage, similar to that seen in acute respiratory disease syndrome
E. Acute tracheobronchitis

D15. A 45-year-old patient presents to hospital with an acute kidney injury. His potassium level is elevated at 7.2 mmol/l and is refractory to medical management. He has a background of alcoholic cirrhosis and has abnormal liver function tests but normal clotting. He is being admitted to the intensive care unit for continuous renal replacement therapy.

Which is the BEST anticoagulant regimen to prescribe?

A. Unfractionated heparin
B. No anticoagulation regimen
C. Predilation fluids
D. Citrate
E. Epoprostenol

D16. You are asked about analgesia for a 35-year-old former intravenous (IV) drug user who is being observed in the ICU with partial thickness burns to both lower limbs after a house fire. She takes regular high-dose buprenorphine. She has so far had 30 mg IV morphine in the past hour and is still in severe pain.

What is the best description for the action of buprenorphine in this scenario?

A. Non-competitive partial agonist
B. Inverse agonist
C. Competitive antagonist
D. Non-competitive antagonist
E. Irreversible antagonist

D17. You are called urgently by the nurses on intensive care. A patient, who had a percutaneous tracheostomy earlier that day to aid ventilator weaning, has suddenly desaturated. As you attend, you notice the tracheostomy appears to be several centimetres out from where it should sit. The patient is still breathing, and you administer oxygen both via the mouth and via the tracheostomy.

What is the most appropriate next step?

A. Remove the inner tube and attempt to pass a suction catheter down the tracheostomy
B. Advance the tracheostomy tube
C. Remove the tracheostomy tube and oxygenate via the mouth
D. Remove the tracheostomy tube and insert a size 6.0 endotracheal tube through the tracheostomy hole.
E. Call the resuscitation team

D18. You are asked by the nurses to review a slim (body mass index 18) 45-year-old man who was admitted to the intensive care unit 2 hours ago from the operating theatre after a negative laparotomy. The indication for surgery was abdominal pain, severe diarrhoea, tachycardia, fever and agitation. He now has a heart rate of 160, a blood pressure of 90/55, a temperature of 39.3°C and oxygen saturations of 98% in air. His creatine kinase is normal.

Which of the following is the most likely diagnosis?

A. Thyroid storm
B. Malignant hyperpyrexia
C. Phaeochromocytoma
D. Neuroleptic malignant syndrome (NMS)
E. Uncontrolled pain

D19. You are called to the emergency department to review a patient in respiratory failure. Which of the following patients would be most likely to benefit from noninvasive ventilation (NIV)?

A. A 45-year-old patient with an acute exacerbation of chronic obstructive pulmonary disease (COPD) with a pH of 7.31, $PaCO_2$ of 7.9, PaO_2 9.1, who has a left-sided pneumothorax
B. A 56-year-old patient with an acute exacerbation of COPD with a pH of 7.29, $PaCO_2$ of 7.1, PaO_2 7.9, who is currently vomiting
C. A 61-year-old patient with an acute exacerbation of COPD with a pH of 7.35, $PaCo_2$ of 8.5, PaO_2 8.5
D. A 48-year-old patient with an acute exacerbation of COPD with a pH of 7.25, $PaCO_2$ of 7.5, PaO_2 15 on arrival in hospital
E. A 46-year-old patient with an acute exacerbation of COPD with a pH of 7.27, $PaCO_2$ of 8.0, PaO_2 8.5, who underwent a lobectomy 2 weeks ago

D20. You have just intubated a 42-year-old woman with a history of alcohol excess, who attended the emergency department with a massive upper gastrointestinal bleed. An endoscopy team is on the way, but she is continuing to bleed profusely. You decide to insert a Sengstaken-Blakemore tube. You insert the tube to 50 cm.
What is the next step?

A. Confirm gastric placement, inject 250 ml into the gastric balloon, apply traction to the tube with a 500-ml bag of fluid

B. Inject 250 ml into the gastric balloon, apply traction to the tube with a 1000 ml bag of fluid

C. Confirm gastric placement, inject 250 ml into the oesophageal balloon, inject 250 ml into the gastric balloon, apply traction to the tube with a 500-ml bag of fluid

D. Confirm gastric placement, inject 250 ml into the oesophageal balloon, inject 250 ml into the gastric balloon, apply traction to the tube with a 1000-ml bag of fluid

E. Confirm gastric placement, inject 250 ml into the gastric balloon, apply traction to the tube with a 1000 ml bag of fluid, inject 250 ml into the oesophageal balloon

D21. You are called by an anaesthetist who is asking for a post-operative intensive care unit bed for a 57-year-old woman who is booked for a hemicolectomy for a tumour. She had a myocardial infarction 6 months previously and has renal impairment and chronic obstructive pulmonary disease.
Which of the following is the best way to calculate her predicted mortality from this operation?

A. APACHE III SCORE
B. P-POSSUM Score
C. Goldman Risk Index
D. American Society of Anaesthesiologists (ASA) physical status classification
E. New York Heart Association (NHYA) Classification

D22. You are called to help with the management of a 50-year-old man with septic shock. His blood pressure is 80/40.
With reference to the Surviving Sepsis Campaign guidelines, which of the following is the best initial fluid bolus?

A. 2 units of packed red cells
B. 500 ml 4.5% human albumin solution
C. 500 ml hydroxyethyl starch solution (e.g. Hespan, Voluven)
D. 500 ml Hartmann's solution
E. 2000 ml Hartmann's solution

D23. A 23-year-old man is admitted to the intensive care unit after a road traffic accident. He has a traumatic brain injury and an intracranial pressure (ICP) monitor in situ.
Which of following pressure monitors is the most appropriate?

A. Subarachnoid
B. Subdural
C. Epidural
D. Parenchymal
E. Ventricular

D24. A 40-year-old woman in the intensive care unit (ICU) is intubated and venti-lated with pneumonia and severe acute respiratory distress syndrome (ARDS). She has been started on a propofol infusion at 100 mg/hr and alfentanil at 2 mg/hr and appears settled. Because of increasing hypoxaemia, a cis-atracurium infusion has been started at 37.5 mg/hr. The patient requires a chest drain insertion on the ICU, and you are concerned about awareness during the procedure.

Which is the most appropriate course of action?

A. Double the rate of propofol infusion
B. Add a midazolam infusion at 5 mg/hr
C. Check a 'train of four' neuromuscular response before starting
D. Give a 100-mg bolus of propofol before the procedure
E. Attach a bispectral index (BIS) monitor and titrate sedation to maintain a score of 40–60 throughout

D25. A 36-year-old woman presents to hospital with acute abdominal pain and has an elevated lipase level of four times the normal range.

What is the MOST likely underlying cause of her current illness?

A. Hypertriglyceridemia
B. Hypercalcaemia
C. Gallstones
D. Excess alcohol intake
E. Cytomegalovirus

D26. You are called to the emergency department as part of the trauma team. A 27-year-old woman was struck by a van whilst crossing a road near to the hospital. She has chest wall and lower limb injuries. The pre-hospital team has secured intravenous access, intubated her and brought her by ambulance to the emergency department. Her primary survey findings are as follows:

Airway: Oral ETT at 21 cm; normal $ETCO_2$ trace; attached to ventilator
Breathing: SpO_2 94% (FiO_2 1.0); reduced breath sounds and hyperresonance on the left
Circulation: Heart rate 140; blood pressure 74/44; capillary refill time >5 s; no external bleeding; soft abdomen; stable pelvis
Disability: Pupils 3 mm, equal and reactive; sedated and paralyzed
Exposure: Bilateral closed midshaft femoral fractures

What is the first step in her management?

A. Thoracostomy followed by intercostal chest drain
B. 1 litre bolus of crystalloid
C. Splinting of femoral fractures
D. Chest radiograph
E. Infusion of 4 units O negative packed red cells

D27. A 49-year-old woman with known liver cirrhosis is admitted with grade III hepatic encephalopathy. An ascitic tap is performed, which demonstrates 350 neutrophils/µl, no red blood cells and no organisms visible on Gram stain.

Which initial treatment is the most likely to lead to a sustained improvement in her encephalopathy?

A. Piperacillin and tazobactam
B. Flumazenil
C. Lactulose
D. Rifaximin
E. L-ornithine-L-aspartate (LOLA)

D28. A 73-year-old male patient is ventilated in the intensive care unit (ICU) for post-operative pneumonia after laparotomy for small bowel obstruction. By day 5, his respiratory support has been weaned, but when the propofol sedation is reduced he is too agitated to allow extubation. The CAM-ICU screening tool demonstrates he is suffering from delirium.

Which of the following is the LEAST appropriate drug to help manage his delirium?

A. Dexmedetomidine infusion
B. Haloperidol as required
C. Regular quetiapine
D. Lorazepam as required
E. Increased dose propofol infusion

D29. A 64-year-old male smoker with a history of mild emphysema is admitted with breathlessness. Chest X-ray demonstrates a large pneumothorax. A 12-F intercostal drain is inserted, and his symptoms improve. At 48 hours, there is still evidence of a persistent pneumothorax on chest X-ray, so high-volume/low-pressure suction is applied. Twenty-four hours later, a persistent pneumothorax remains.

What would be the next MOST appropriate treatment strategy?

A. Referral to thoracic surgery to consider chemical pleurodesis
B. Increase the size of the chest drain to 24-F
C. Convert the suction to a high-pressure system
D. Referral to thoracic surgery to consider surgical pleurectomy
E. Perform a medical pleurodesis with talc

D30. A 55-year-old male patient has suffered an ST-elevation myocardial infarction (STEMI). He underwent primary percutaneous coronary intervention (PCI), having a bare metal stent inserted into his circumflex artery. He was previously well and has no increased bleeding risk.

What would be the MOST appropriate antiplatelet regime for this patient to continue post-PCI on discharge?

A. Aspirin and ticagrelor
B. Aspirin and clopidogrel
C. Aspirin alone
D. Aspirin and tirofiban
E. Clopidogrel alone

Exam D: Answers

D1. A 28-year-old man has been admitted to intensive care with severe respiratory failure. He is sedated and ventilated. A viral throat swab confirms the presence of influenza A, and a diagnosis of viral pneumonia is made.

Which of the following is the most appropriate antiviral agent?

A. Aciclovir
B. Ritonavir
C. Zanamivir
D. Ganciclovir
E. Oseltamivir

Answer: E

Short explanation

Aciclovir is effective against herpes simplex and varicella zoster viruses. Ritonavir is an antiretroviral used to treat HIV infection. Ganciclovir is used in cytomegalovirus infections. Zanamivir and oseltamivir are active against influenza A. NICE recommends the use of oseltamivir as first-line antiviral in such cases.

Long explanation

Influenza viruses are single-strand RNA viruses in the Orthomyxoviridae family of viruses. There are three types of influenza virus: A, B and C. Influenza A virus has been responsible for many outbreaks of pandemic influenza, particularly in 1918, 1968 and 2009. There are many subclasses of influenza A virus, based on the expression of different types of proteins, haemagglutinin and neuraminidase, on the surface. This gives rise to subtype nomenclature such as H1N1 or H7N5. Currently, only three subtypes are circulating in humans, H1N1, H1N2 and H3N2.

Three drugs are available on the United Kingdom with activity against influenza A; neuraminidase inhibitors (oseltamivir and zanamivir) and amantadine. Amantadine is not recommended for prophylaxis or treatment of influenza. Of the neuraminidase inhibitors, NICE recommends that either oseltamivir or zanamivir may be used to treat influenza, but the decision should be based on which of the two is the cheaper option. Zanamivir costs twice as much as oseltamivir.

References

Joint Formulary Committee. *British National Formulary*. 69th ed. London: BMJ Group and Pharmaceutical Press; 2015.

National Institute of Clinical Excellence (NICE). (2009). *Amantadine, oseltamivir and zanamivir for the treatment of influenza [TA168]*. London: National Institute for Health and Care Excellence.

Centre for Disease Control. Types of influenza viruses. Available online http://www.cdc.gov/flu/about/viruses/types.htm (accessed 25 February 2015).

D2. A 33-year-old patient has collapsed in the hospital canteen, suffering from suspected anaphylaxis following ingestion of some peanuts. He has acute airway swelling and stridor. His blood pressure is 73/40 mmHg, and heart rate is 110 bpm.

Which is the most appropriate dose and route of adrenaline?

A. 10 ml 1:10,000 intravenous (IV)
B. 1 ml 1:1000 IV
C. 0.5 ml 1:1000 IV
D. 0.5 ml 1:1000 intramuscular (IM)
E. 1 ml 1:1000 IM

Answer: D

Short explanation

The adult dose of adrenaline (epinephrine) is 0.5 ml of 1:1000 (500 µg) IM repeated after 5 minutes if no better. Adrenaline may also be given intravenously (in 50–100 µg aliquots guided by response) by trained experienced specialists in appropriate settings – that is, with established IV access and continuous electrocardiogram and blood pressure monitoring.

Long explanation

Life-threatening anaphylaxis is recognized by the presence of acute, life-threatening airway, breathing or circulatory compromise in the presence of a known or likely trigger, or history suggestive of allergy. Emergency management consists of calling for help, laying the patient flat and raising his or her legs (so long as this doesn't worsen airway or breathing difficulties), removing the trigger and administering adrenaline, followed by second-line therapies such as antihistamines and steroids. The adult dose of adrenaline (epinephrine) is 0.5 ml of 1:1000 (500 µg) IM repeated after 5 minutes if no better. Patients may carry their own EpiPen, which usually contains a reduced dose of 300 µg (or 150 µg in the EpiPen Jr). It is appropriate to use this if available whilst waiting for the emergency team to arrive.

Adrenaline may instead be given intravenously (in 50–100 µg aliquots guided by response) by trained experienced specialists in an appropriate settings. It is dangerous to administer IV adrenaline in this setting without proper monitoring and if not trained to do so. This usually requires the patient to be in a high dependency area of the hospital.

A mast cell tryptase level should be taken as sent as soon as practicably possible within the first hour and again at 1 to 2 hours with a further baseline sample more than 24 hours after the event. Referral should be made to the local allergy clinic for identification and testing of the trigger and a full and detailed record of all drugs given including exposure to latex, chlorhexidine etc included in the referral. It is also important to consider completing a 'yellow' form if a drug reaction is considered likely and alerting the patient and their general practitioner to the reaction.

Reference

Working Group of the Resuscitation Council (UK). *Emergency treatment of anaphylactic reaction. Guidelines for healthcare providers*. Resuscitation Council UK, 2008 http://www.resus.org.uk/pages/reaction.htm (accessed April 2015).

D3. Which of the following treatments in the management of adult traumatic brain injury is MOST likely to be beneficial?

A. Avoidance of systolic blood pressure <90 mmHg
B. Avoidance of SpO_2 <90%
C. Prophylactic hypothermia
D. Maintenance of a cerebral perfusion pressure (CPP) >70 mmHg
E. Graduated compression stockings or pneumatic compression stockings

Answer: A

Short explanation

The Brain Trauma Foundation's (BTF) international guidelines are based on three levels of recommendation; 1 being the strongest and 3 the weakest. Of the preceding options, only option A (avoidance of hypotension) is a level 2 recommendation; B, C, and E are level 3, whereas maintenance of a CPP >70 mmHg has a level 2 recommendation to avoid it (CPP of 50–70 mmHg is the target).

Long explanation

The Brain Trauma Foundation's guidelines, last published in 2007, offer a thorough review of the evidence available at that time. The levels of recommendation match the levels of evidence, where level 1 equates to a good quality randomized controlled trial (RCT), and level 3 to a poor RCT, a case series, or moderate/poor case-control or cohort studies.

With reference to blood pressure control, hypotension (systolic BP <90 mmHg) should be avoided (level 2 recommendation). Oxygen levels should also be monitored, either invasively or noninvasively, and low levels (SpO_2 <90%, or PaO_2 <60 mmHg) avoided (level 3).

Intracranial pressure (ICP) should be treated when above 20 mmHg (level 2 recommendation). With regard to cerebral perfusion pressure (CPP), both low (CPP <50 mmHg, level 3) and high (CPP >70 mmHg, level 2) values have worse outcomes and should be avoided. The target range is therefore a CPP of 50–70 mmHg.

Mannitol can be used in the treatment of raised ICP, but the wide dose range in this level 2 recommendation, of 0.25 to 1 g/kg, reflects concern over the profound diuresis it generates. Some authors would advise initial treatment at the low end of this dose range, then repeating doses as required. Evidence published since the BTF guidelines suggests that hypertonic saline is as effective as mannitol with fewer potential side effects.

Sedation with barbiturates to achieve burst suppression on electroencephalogram monitoring is not recommended on a prophylactic basis but can be used as a rescue measure when maximal medical and surgical treatment has been unsuccessful at controlling raised ICP (level 2).

Prophylactic anticonvulsant use remains controversial, as there is no evidence that early post-traumatic seizures affect outcomes. Despite this, there is a level 2 recommendation that anticonvulsants are given to reduce the incidence of these seizures occurring, due to their effect on cerebral oxygen demand.

Hyperventilating a patient to a $PaCO_2$ of 3.3 kPa or less is associated with worse outcomes if done on a prophylactic basis and should therefore not be done (level 2). If, however, this is a temporary rescue measure to control raised ICP, it can be utilized (level 3) so long as it is not in the first 24 hours after injury (level 3).

With regard to prophylactic hypothermia, the evidence is rather inconclusive when studies are pooled together, but there is no evidence of a decrease in mortality and therefore cannot be recommended as prophylaxis (as opposed to rescue). Steroid use should be avoided because it is associated with increased mortality (level 1).

Patients suffering from traumatic brain injury are at high risk of thromboembolism and therefore should have both mechanical and pharmacological prophylaxis unless contraindicated (both level 3). Pharmacological prophylaxis has the potential to worsen any haemorrhagic lesion, and administration will need to be a multidisciplinary team decision.

Reference

Brain Trauma Foundation. Guidelines for the management of severe traumatic brain injury. *J Neurotrauma*. 2007, Vol 24, S1-S106. http://www.braintrauma.org/pdf/protected/Guidelines_Management_2007w_bookmarks.pdf. (Accessed May 2015).

D4. A 44-year-old man presents to hospital with fever, myalgia, night sweats and shortness of breath. He is found to have a pan-systolic murmur, splenomegaly and microscopic haematuria. A transthoracic echocardiogram demonstrates an oscillating mass associated with the mitral valve.

What is the most likely causative organism?

A. *Pseudomonas aeruginosa*
B. *Staphylococcus aureus*
C. *Streptococcus bovis*
D. *Escherichia coli*
E. *Coxiella burnetti*

Answer: B

Short explanation
This patient has infective endocarditis. The commonest causative organism is *Staphylococcus aureus*, which is isolated in more than 40% of cases in which an organism is identified. Streptococcal species are responsible for more than one-third of cases, but only 10% are caused by *Streptococcus bovis*. The other organisms are rarer causes of endocarditis.

Long explanation
Infective endocarditis (IE) is inflammation of the endocardium, predominantly caused by bacterial infection. The commonest site of infection are the valves, although chordae tendineae or any other structure may be affected. The likelihood of infection of these normally sterile tissues is greatly increased by damage, for example, by rheumatic fever. Risk factors for developing IE include pre-existing cardiac lesions, such as prosthetic valves, rheumatic fever or intracardiac devices, and risk factors for bacteraemia, such as indwelling intravascular catheters, intravenous drug use or chronic infections.

IE typically presents with fever, malaise, myalgia and fatigue. Many patients develop night sweats, shortness of breath or symptoms due to septic emboli, such as stroke. Examination findings include new regurgitant cardiac murmurs and/or signs of cardiac failure with vascular embolic or immunological phenomena. Vascular lesions include petechiae, splinter haemorrhages, conjunctival haemorrhage and Janeway lesions (macular lesions on palms and soles). Immunological phenomena include Osler nodes (painful nodules in fingers and toes), Roth spots (retinal haemorrhagic lesions) and glomerulonephritis.

Echocardiography is key in diagnosing IE. Detection of a vegetation associated with a valve or other structure is diagnostic, with a specificity of almost 100% even with transthoracic echocardiography (TTE). A negative TTE is not sufficient to rule out IE, having a sensitivity of 30 to 60%, and so transoesophageal echocardiography should be performed in patients in whom IE is suspected. Blood cultures should be taken before commencing antibiotics, with at least three separate sets separated by 30 minutes. It is, however, important to avoid unnecessary delays in starting antibiotics in patients with sepsis. Empirical antibiotics should be started immediately after cultures – within 60 minutes. For native valves, co-amoxiclav with gentamicin is recommended (vancomycin, gentamicin and ciprofloxacin for penicillin allergic patients). Patients with new (<12 months) prosthetic valves should be treated with vancomycin, gentamicin and rifampicin.

The commonest causative organisms, where one has been identified, are *Staphylococcus* (>50%, 80% of which are *S. aureus*), *Streptococcus, Enterococcus,* and Gramnegative rods (*E. coli* and *P. aeruginosa* most commonly). Nonbacterial causes (e.g. fungi, viruses) make up around 2% of all cases of IE.

References
Bor DH, Woolhandler S, Nardin R, et al. Infective endocarditis in the U.S., 1998–2009: a nationwide study. *PLoS One*. 2013;8(3):1–8.

Habib G, Hoen B, Tornos P, et al. Guidelines on the prevention, diagnosis, and treatment of infective endocarditis (new version 2009). *Eur Heart J*. 2009;30:2369–2413.

Hoen B, Duval X. Infective endocarditis. *N Engl J Med*. 2013;368:1425–1433.

D5. A 68-year-old man has been on intensive care for 4 days with acute respiratory distress syndrome (ARDS) and multiorgan failure after an emergency laparotomy for large bowel perforation with faecal peritonitis. He is requiring 80% oxygen, inotropic and vasoconstrictor support and renal replacement therapy. He weighs 90 kg and is 175 cm tall.

Which of the following is the most appropriate ventilator setting?

A. Tidal volume (V_T) = 490 ml; positive end expiratory pressure (PEEP)= 14 cmH$_2$O
B. V_T = 540 ml; PEEP = 10 cmH$_2$O
C. V_T = 420 ml; PEEP = 6 cmH$_2$O
D. V_T = 540 ml; PEEP = 6 cmH$_2$O
E. V_T = 420 ml; PEEP = 14 cmH$_2$O

Answer: E

Short explanation
Patient with ARDS should be ventilated according to the ARDSnet mechanical ventilation protocol. Tidal volumes should be 6 ml/kg predicted body weight (PBW) based on his height. His PBW is 70 kg, so a tidal volume of 420 ml is appropriate. Secondly, PEEP should increase with FiO$_2$; ARDSnet PEEP tables suggest PEEP levels of 14 to 22 cmH$_2$O for an FiO$_2$ of 0.8, although this should be tailored to individual patients.

Long explanation
ARDS is an acute inflammatory process of the lung leading to reduced gas exchange and hypoxaemia. It can be caused by diseases of the lung such as pneumonia (direct ARDS) or systemic disease such as sepsis (indirect ARDS). Up to 20% of all patients in a typical intensive care unit fit the diagnostic criteria for ARDS. Mortality estimates for ARDS vary from study to study but fall between 20 and 40%.

In 2000, the ARDS network published a landmark study that demonstrated a clear mortality benefit for patients with ARDS who were ventilated with tidal volumes (V_T) of 6 ml/kg predicted body weight (PBW) compared with patients ventilated with V_T of 12 ml/kg PBW, as had been common practice until then. This finding has been found to be robust and is now the standard of practice around the world. Low tidal volume ventilation should also be adopted in patients without ARDS because it has been shown to reduce the risk of developing a ventilator-associated lung injury.

The V_T required is based on the predicted body weight (PBW) for which there are many formulae. The formulae, using the gender and height of the patient, from the ARDSnet trial is as follows:

Males: PBW (kg) = 50 + 0.91 (height (cm) − 152.4);
Females: PBW (kg) = 45.5 + 0.91 (height (cm) − 152.4).

Positive end-expiratory pressure is crucial in minimizing airway collapse and atelectrauma. Evidence suggests that compared with low PEEP, high PEEP improves oxygenation and may reduce mortality in ARDS. PEEP can be set either using pressure-volume curves (set at the lower inflection point of the static pressure-volume curve) or using a PEEP/FiO_2 table. The PEEP table published by the ARDS network specifies two levels of PEEP for a given FiO_2; in this case with $FiO_2 = 0.8$, a PEEP of 14 cmH_2O (low) or 22 cmH_2O (high) is recommended. When increasing PEEP, patients must be closely monitored for signs of cardiovascular compromise or increased airway pressures; plateau pressure should be kept below 30 cmH_2O to reduce the risk of barotrauma.

References
The Acute Respiratory Distress Syndrome Network. Ventilation with lower tidal volumes as compared with traditional tidal volumes for acute lung injury and the acute respiratory distress syndrome. *N Engl J Med*. 2000;342:1301–1308.

ARDSnet ventilator protocol. http://www.ardsnet.org/files/ventilator_protocol_2008-07.pdf (accessed February 2016).

De Beer JM, Gould T. Principles of artificial ventilation. *Anaesth Intensive Care Med*. 2007;14:83–93.

Santa Cruz R, Rojas J, Nervi R, et al. High versus low positive end-expiratory pressure (PEEP) levels for mechanically ventilated adult patients with acute lung injury and acute respiratory distress syndrome [review]. *Cochrane Database Syst Rev*. 2013;(6).

D6. A 45-year-old man was admitted to the ICU 10 days earlier after complete spinal cord injury at T2. He has undergone surgical fixation of his spinal fracture, and he is medically stable. He has had a tracheostomy sited, and you would like to commence respiratory weaning.

Which of the following is most likely to impair the weaning process?

A. Pulmonary compliance of 50 ml/cm H_2O
B. FiO_2 of 0.35
C. 8-mm internal diameter tracheostomy
D. Vital capacity of 750 ml
E. High opioid requirements for analgesia

Answer: E

Short explanation
Patients who are agitated or who have pain are less likely to successfully wean. Tracheostomies often aid weaning, and the ideal diameter is 8 mm. Weaning should commence once the vital capacity is 150 ml and is more likely to succeed at higher volumes. An FiO_2 of ≤0.4 and a pulmonary compliance of ≥50 ml/cmH_2O are prerequisites for weaning success.

Long explanation

Spinal cord injury affects 12 to 16 people per million population in the United Kingdom and is most commonly caused by trauma. Patients are often young, and the effect on quality of life is devastating. Spinal cord injuries affecting the thoracic vertebrae may cause some degree of respiratory embarrassment depending on the level affected. These injuries are often complicated by associated trauma to ribs and the lungs. Spinal cord injuries in the cervical region will have a greater effect on respiratory mechanics, with those above C4/5 likely to need immediate ventilator support due to diaphragmatic involvement. Those patients who require ventilation are best served by receiving an early tracheostomy as rapid weaning is unlikely to be effective. A surgical tracheostomy is often preferred when there is an unstable cervical spine injury. Alongside the mechanical effects of spinal cord trauma, respiratory compromise is compounded by increased secretions and bronchospasm due to autonomic dysfunction.

Over the initial few days to weeks, muscle spasticity increases thoracic wall tone and reduces chest wall compliance. Some degree of respiratory effort is required to begin a weaning programme, but simply triggering a ventilator does not equate to useful function.

Spinal Cord Injury UK provides guidance for those caring for patients with spinal cord injury. Weaning from respiratory support in this group may take months to years. Progress is slow and complicated by other injuries, recurrent sepsis and other complications of paralysis such as reduced nutritional intake and catabolism of muscle mass. Dedicated weaning plans with an engaged patient and multidisciplinary team are vital.

Weaning involves the steady increase of ventilator-free breathing (VFB). Duration of VFB is based on vital capacity (VC) and is recommended to commence once the VC exceeds 150 ml. There should be 1 to 2 hours on the ventilator between weaning episodes to reduce fatigue, and this may be aided by high tidal volumes to reduce atelectasis. Other adjuncts to weaning such as salbutamol nebulizers, remaining supine, not sitting and improved nutrition via gastrostomy are recommended. Weaning should only commence if the FiO_2 is <40%, the PEEP is around 5 cmH_2O, the patient is awake, co-operative, not in pain but not overly sedated, has no ongoing infection and has a pulmonary compliance of ≥50 ml/cmH_2O.

References

Intensive Care Society Weaning Guidelines 2007. http://www.ics.ac.uk/ics-home page/guidelines-and-standards (accessed April 2015).

Respiratory Information for Spinal Cord Injury UK weaning guidelines. http://www .risci.org.uk/NSCISB%20RISCI%20final.doc (accessed April 2015).

D7. Drug A is 98% protein bound and is metabolized in the liver with a high hepatic extraction ratio.

Which of the following changes that may occur in a patient on ICU is most likely to alter the circulating levels of the drug?

A. Reduced hepatic blood flow due to low cardiac output
B. Fall in serum albumin concentration
C. Co-administration of a hepatic enzyme inducer
D. Acute kidney injury
E. Ileus

Answer: A

Short explanation

Drugs with high hepatic extraction ratios have a clearance that is dependent on liver blood flow. Protein binding and hepatic enzyme activity will predominantly affect drugs with a lower hepatic extraction ratios. Acute kidney injury will only affect drugs metabolized and/or excreted by the kidneys. Ileus will only affect drug concentration of enteric drugs.

Long explanation

Drugs with high hepatic extraction ratios have a clearance that is dependent on liver blood flow. For drugs with a low extraction ratio and high protein binding, clearance is dependent on protein binding, and displacement by other drugs will alter clearance. For drugs with a low extraction ratio and low protein binding, clearance is independent of flow and more dependent on liver enzyme function.

The extraction ratio (ER) of a drug by a specific organ relates to the fraction of the drug removed for each pass of that organ. For example, if the drug levels in the blood flowing from the liver are half those of the blood supplying the liver, the hepatic ER would be 50%. ER is dependant on the rate of drug delivery (blood flow), availability of uptake (amount of free drug) and enzyme activity. Drugs with a high rate of uptake and high enzyme capacity such as propofol are only limited by blood flow. Drugs with a limited enzyme capacity and high protein binding (such as phenytoin) vary greatly with changes in enzyme function (inducers/inhibitors) and changes in protein binding but not with blood flow. Drugs with limited enzyme capacity but low protein binding are less affected by changes in protein binding but still limited by enzyme capacity.

Subtle changes in free drug levels will matter more for those drugs with a narrow therapeutic window. Some drugs are inactivated by liver enzymes, whereas other (prodrugs) may be activated. Therefore the same reduction in liver blood flow may increase the levels of active drug for some medicines and simultaneously reduce it for others.

Reference

Peck TE, Hill SA, Williams M. *Pharmacology for Anaesthesia and Intensive Care.* 3rd edition. Cambridge: Cambridge University Press, 2008, p 11.

D8.   A patient with urinary sepsis has received high-volume fluid resuscitation with normal saline but remains hypotensive and tachypnoeic on the ward. He is transferred to the intensive care unit, and a noradrenaline infusion is started. Cardiac output studies post commencement of noradrenaline suggest a high cardiac index with appropriate filling and vasoconstriction.

Arterial blood gas (ABG) on air:

pH 7.29, pCO_2 3.1 kPa, pO_2 12 kPa, HCO_3 16 mmol/l, base excess −16, lactate 1.2 mmol/l, Na^+ 144 mmol/l, K^+ 3.6 mmol/l, Cl^- 124 mmol/l, venous saturation 76%

What would be the MOST appropriate next intervention?

A. Fluid bolus with Hartmann's
B. Commence a dobutamine infusion
C. 100 ml 8.4% sodium bicarbonate
D. Invasively ventilate to restore normocapnia
E. Increase noradrenaline infusion rate

Answer: C

Short explanation

This patient's ABG demonstrates a normal-anion gap acidosis likely secondary to high-volume fluid resuscitation with normal saline on the ward. The hypocapnia is a compensatory response, therefore treatment to restore normocapnia should not be implemented. The cardiac output studies do not suggest treatment with a fluid bolus or inotropes would be of benefit. Sodium bicarbonate would be appropriate treatment to reverse the hyperchloraemic acidosis.

Long explanation

This patient has a normal anion gap metabolic acidosis (often referred to as a hyper-chloraemic metabolic acidosis). It occurs due to an increase in chloride relative to strong cations (especially sodium) or a loss of cations with chloride retention. An infusion of sodium bicarbonate will increase the patient's sodium level in relation to chloride thus treating this patient's metabolic acidosis.

The use of bicarbonate in the treatment of metabolic acidosis has been controversial with concerns regarding the worsening of intracellular acidosis and causing a rise in $PaCO_2$. When metabolic acidosis is secondary to bicarbonate loss (renal/gut) replacement of bicarbonate would be an appropriate treatment modality. The mainstay of management for metabolic acidosis should be treatment of the underlying disease process. It has been suggested that, in conjunction with this, treatment with sodium bicarbonate would be appropriate for patients whose pH is <7.1.

Causes of normal anion gap metabolic acidosis include the following:

- Renal causes
  - Renal tubular acidosis (type I, II and IV): intrinsic renal disease where there is an inability to appropriately acidify the urine.
  - Drugs that result in reduced ability of the kidney's to acidify the urine
    - Carbonic anhydrase inhibitor: acetazolamide
    - Aldosterone antagonists: spironolactone
- Extrarenal causes
  - Iatrogenic acids
    - Total parenteral nutrition
    - Chloride rich fluids such as 0.9% sodium chloride
  - Alkaline gastrointestinal losses:
    - Diarrhoea
    - Small bowel fistula
    - Pancreatic fistulae
  - Ureteric diversion: ileal conduit
    - Urea is degraded by bacteria in the gut and the resultant metabolites (NH_3 and H^+)$^+$ are then absorbed causing an acidosis

References

McGee WT, Headley JM, Frazier JA. Quick guide to cardiopulmonary care. Edwards Critical Care Education. http://ht.edwards.com/scin/edwards/site collectionimages/products/pressuremonitoring/ar11206-quickguide3rded.pdf (accessed March 2015).

Morris CG, Low J. Metabolic acidosis in the critically ill: part 2. *Causes Treatment Anaesth*. 2008;63(4):396–411.

D9. A new blood test is developed to identify patients who develop ventilator-associated pneumonia. The test is described by the company representative as having a high positive predictive value (PPV).

Which of the following statements is most likely to be accurate regarding the value of this test?

A. The area under the receiver operating curve (AUROC) is likely to be close to 0.5.
B. The negative predictive value (NPV) will probably also be high.
C. The PPV will be similar when tested in two different populations.
D. The PPV can be calculated by dividing the number of true positives by the number of false positive results.
E. The PPV can be calculated by dividing the number of true positives by the sum of all positive results.

Answer: E

Short explanation
An AUROC of 0.5 would signify a useless test. Almost always, when a test has a high PPV, it has a low NPV, and vice versa. PPVs are only valid for the population in which they are tested because the population demographics will affect the likelihood of different outcomes. Answer E is the most accurate statement.

Long explanation
Receiver operating characteristic (ROC) curves are plots of true positives versus false positives. They were developed and named by radio operators, looking for a way of quantifying the accuracy of tests developed to identify planes on radar in the Second World War. The area under this curve has a value from 0 to 1 with the perfect test having a value of 1 and a value of 0.5 representing a test with a 50:50 chance of giving a true positive or a false positive – that is, a useless test.

The PPV and NPV of a test are useful to clinicians. They answer the following questions: How likely is it the patient has the disease, given that the test is positive (PPV) or negative (NPV)? For example, given a positive D-dimer result, what is the chance the patient has venous thromboembolism? In this example, the answer might be 'not very high' because this test notoriously has a high false-positive rate. However, it also has a low false-negative rate, so for the reverse situation – given a negative D-dimer, how likely is it the patient doesn't have a blood clot – the answer is *very* likely'. Therefore the NPV of the test is good but the PPV is poor.

It is also clear that the PPV and NPV of a test will change, depending on the population. For example, if you perform D-dimers on a population of post-operative surgical patients, it is likely that the results will largely come back positive, yet many of these are likely to be false positives, and the true rates of blood clots in this group is low. However, in a population of fit and well patients presenting with calf swelling to their general practitioner, the false positive rate will likely be lower. In other words, the PPV of a D-dimer test for detecting a blood clot is higher in a population that has a lower false-positive rate and also higher in a population that has a higher true-positive rate – that is, a population in whom the disease is more prevalent.

Reference
Lalkhen AG, McCluskey A. Clinical tests: sensitivity and specificity. *Contin Educ Anaesthesia, Crit Care Pain.* 2008;8(6):221–223. http://bjarev.oxfordjournals.org/cgi/doi/10.1093/bjaceaccp/mkn041 (accessed April 2015).

D10. You are looking after a patient with the following findings. With reference to the Child-Pugh scoring system, which is associated with the worst 1-year survival?

A. Bilirubin 45 µmol/l
B. Serum albumin 26 g/l
C. International normalized ratio (INR) 2.1
D. Mild ascites
E. Grade II hepatic encephalopathy

Answer: B

Short explanation
The Child-Pugh score is used to assess the prognosis of chronic liver disease, mainly cirrhosis. It uses five parameters, which each score 1, 2 or 3. The total score is used to classify the patient into Child-Pugh class A to C. All the parameters in the question feature in the Child-Pugh score, but the albumin level of 26 g/l is the only one that scores 3 points; the others are 2 points each.

Long explanation
The Child-Pugh score was originally used to predict mortality during surgery but is now used to determine the prognosis of chronic liver disease. The score includes five clinical measures of liver disease. Each is scored 1 to 3, with 3 indicating the most severe derangement.

Parameter	1 point	2 points	3 points
Ascites	None	Mild	Moderate/severe
Albumin (g/l)	>35	28–35	<28
Bilirubin (µmol/l)	<34	34–50	>50
Hepatic encephalopathy	None	Grade 1–2	Grade 3–4
INR	<1.7	1.7–2.3	>2.3

The total score corresponds to a Child-Pugh class: 5–6 points are considered class A, 7–9 points class B, and 10 or more points are class C. Without stimulus or surgical intervention, the 1-year survival ranges from about 100% in class A to roughly 45% in class C.

Recognizing the severity of liver disease has important implications on prognostication, as well as on day-to-day management in the intensive care unit. Liver failure has effects both locally and throughout the organ systems. Some of the key functions of the liver include synthesis, metabolism and storage, and therefore impairment of these can lead to reduced production of key proteins such as clotting factors and albumin, delayed metabolism of many drugs and reduced storage of glycogen, leading to hypoglycaemia.

Systemically there are numerous effects, including reduction in the systemic vascular resistance, activation of the renin-angiotensin system and retention of sodium and water. The onset of ascites can cause a restrictive ventilation defect, whereas capillary dilation can worsen shunt. Portal hypertension, caused by hepatic scaring and reduced nitric oxide production, will lead to varices, which can cause profound bleeding. Leading causes of death amongst this patient group include sepsis, haemorrhage and renal failure.

References
Durand F, Valla D. Assessment of the prognosis of cirrhosis: Child–Pugh versus MELD. *J Hepatol*. 2005;42;1:S100–107.

Vaja R, McNicol L, Sisley I. Anaesthesia for patients with liver disease. *Contin Educ Anaesth Crit Care Pain.* 2010;10(1):15–19.

D11. A 45-year-old smoker is admitted in cardiogenic shock after an acute myocardial infarction. He underwent percutaneous coronary intervention with the insertion of two stents and had an intra-aortic balloon pump (IABP) inserted within 2 hours of admission.

Which of the following is TRUE regarding IABPs?

A. Using the arterial trace, balloon inflation should be timed with the dicrotic notch.
B. The tip of the balloon should lie 2 to 3 cm distal to the origin of the left subclavian artery.
C. Insertion of an IABP has good evidence of reduced mortality in these patients.
D. An increase should be seen in this patient's cardiac output (CO).
E. His augmented diastolic pressure should be higher than his systolic pressure.

Answer: C

Short explanation

If the balloon pump is correctly placed, distal to the origin of the left subclavian artery and timed to inflate during diastole (with the dicrotic notch on the arterial waveform), the patient's CO should increase and his augmented diastolic pressure should be higher than the systolic pressure. Although IABPs have traditionally been used to treat cardiogenic shock associated with acute myocardial infarctions, recent evidence has demonstrated no decrease in 30-day or 12-month mortality.

Long explanation

An IABP is inserted into a femoral artery and should be positioned with the balloon tip about 2 to 3 cm distal to the origin of the left subclavian artery. The balloon is inflated with helium and when inflated should partially occlude the descending aorta, not exceeding 80 to 90% of its diameter. Inflation and deflation of the balloon should be co-ordinated with the cardiac cycle. This can be timed either electrically using the electrocardiogram (ECG) waveform as a trigger or mechanically using the arterial pressure trace. Inflation should be timed to occur at the onset of diastole, which coincides with the middle of T wave on the ECG or the dicrotic notch on the arterial pressure waveform. Deflation should occur at the onset of systole before the aortic valve opens. The triggers are the R wave on the ECG or just before the upstroke of systole on the pressure trace.

Balloon inflation during diastole increases diastolic blood pressure, with the aim of improving coronary perfusion. Deflation occurs at the onset of systole resulting in a reduced afterload, reduced left ventricular end diastolic pressure (LV-EDP) and LV wall tension, thus improving CO. Coronary perfusion should increase because of increased diastolic pressure and a reduction in LV wall tension. The augmented diastolic pressure should ideally be higher than systolic pressure if timing of inflation and deflation are correct.

The insertion of an IABP has traditionally been used to treat patients in cardiogenic shock after an acute myocardial infarction. The IABP-SHOCK II trial randomized 600 patients with cardiogenic shock in association with an acute myocardial infarction who were planned to receive early revascularization therapy. They demonstrated no significant decrease in 30-day mortality rates in those patients randomized to receive treatment with an IABP rather than no IABP. A follow-up trial also demonstrated that insertion of an IABP in these patients did not reduce mortality at 12 months.

References

Krishna M, Zacharowski K. Principles of intra-aortic balloon pump counterpulsation. *Contin Educ Anaesth Crit Care Pain*. 2009;9:24–28.

Thiele H, Zeymer U, Neumann F-J et al. Intraaortic balloon support for myocardial infarction with cardiogenic shock. *N Engl J Med*. 2012;367(14):1287–1296.

Thiele H, Zeymer U, Neumann F-J et al. Intra-aortic balloon counterpulsation in acute myocardial infarction complicated by cardiogenic shock (IABP-SHOCK II): final 12 month results of a randomised, open-label trial. *Lancet* 2013;382(9905):1638–1645.

D12. A 18-year-old female is admitted to the intensive care unit with acute liver failure after a paracetamol overdose.

Which of the following criteria is LEAST useful in predicting her need for a liver transplant?

A. Post-resuscitation arterial pH 7.28
B. Grade III encephalopathy
C. Creatinine 320 μmol/l
D. Prothrombin time 108 sec
E. Bilirubin 320 μmol/l

Answer: E

Short explanation

All of the preceding are values in the King's College Hospital Criteria that are used to identify patients with acute liver failure who need a liver transplant. Indications after paracetamol overdose include an arterial pH <7.3 following adequate fluid resuscitation or a prothrombin time >100 seconds with a creatinine of >300 μmol/l and grade III or IV encephalopathy. A bilirubin of >300 μmol/l is a criteria used in patients with acute liver failure from aetiologies other than paracetamol.

Long explanation

Acute liver failure is characterized by a triad of jaundice, coagulopathy and encephalopathy in patients with previously normal liver function. It is classified according to the time from the onset of jaundice to the development of encephalopathy:

- Hyperacute disease: <7 days
- Acute disease: 1–4 weeks
- Subacute disease: 4–12 weeks (some definitions quote up to 6 months)

In the United Kingdom, paracetamol overdose is the commonest cause, accounting for up to 70% of all cases of acute liver failure, and presents as hyperacute disease. Despite the greatest degree of coagulopathy, cerebral oedema and encephalopathy being seen in this cohort of patients, they also have the one of the best prognoses; transplant-free survival has now risen to ≥50%.

Introduced in 1989, the King's College Hospital Criteria are still used to predict the need for liver transplant in patients with acute liver failure.

King's College Hospital Criteria for liver transplantation:

Patients presenting with acute liver failure following paracetamol overdose:

- Arterial pH <7.3 after adequate fluid resuscitation
OR

- All of the following:
  - Prothrombin time >100 seconds (international normalized ratio [INR] >6.5)
  - Creatinine ≥300 μmol/l
  - Grade III–IV encephalopathy

Patients presenting with acute liver failure from non-paracetamol overdose aetiologies

- Prothrombin time >100 seconds (INR >6.5)

OR

- Any three criteria from the following:
  - Age <10 or >40 years
  - Prothrombin time >50 seconds
  - Bilirubin ≥300 μmol/l
  - Onset of encephalopathy >7 days after the development of jaundice
  - Unfavourable disease aetiology: non-A hepatitis, non-B hepatitis, halothane hepatitis, drug-induced liver failure/idiosyncratic drug reaction

The King's Criteria have been modified with the addition of lactate measurement. A lactate level >3 mmol/l after adequate fluid resuscitation is an indication for liver transplant in patients post-paracetamol overdose.

After liver transplant, 1-year survival is approximately 80%.

References

Bernal W, Donaldson N, Wyncoll, D, Wendon J. Blood lactate as an early predictor of outcome in paracetamol induced acute liver failure: a cohort study. *Lancet* 2002;359(9306):558–563.

Goldberg E, Chopra S, Brown RS, Travis AC. UpToDate: acute liver failure in adults: management and prognosis. http://www.uptodate.com/contents/acute-liver-failure-in-adults-management-and-prognosis?source=search'result&search=kings+criteria+liver+transplant&selectedTitle=1%7E150 (accessed September 14).

Maclure P, Salman B. Management of acute liver failure in critical care. Anaesthesia tutorial of the week 251. 2012. http://www.aagbi.org/education/educational-resources/tutorial-week/my-events/tutorial?page=3 (accessed September 2014).

Willars C. Update in intensive care medicine: acute liver failure. Initial management, supportive treatment and who to transplant. *Curr Opin Crit Care.* 2014;20(2):202–209.

D13. A 30-year-old trauma patient is transferred to the intensive care unit after 4 hours of emergency surgery to remove a ruptured spleen following a road traffic accident. He has received 10 units of packed red cells, 1 unit of platelets and 8 units of fresh frozen plasma during the resuscitation. He has also received two doses of tranexamic acid 1 g.

To guide further therapy, you decide to run a rotational thromboelastometry (ROTEM) test. Which of the following is most likely to explain the following results?

Measure	Normal range	EXTEM result	FIBTEM result
Clotting time	40–80 seconds	69 seconds	185 seconds
Clot formation time	34–160 seconds	263 seconds	—
a-angle	55–78 degrees	58 degrees	—
Maximum clot formation	50–72 mm	38 mm	3 mm

A. The tranexamic acid given in theatre
B. A low platelet count
C. Acidosis
D. A low fibrinogen level
E. Disseminated intravascular coagulation

Answer: D

Short explanation

Tranexamic acid inhibits fibrinolysis. Both thrombocytopenia and hypofibrinogenaemia may extend the clot formation time and decrease maximal clot firmness. However, the FIBTEM test eliminates the influence of platelets by irreversible inactivation with cytochalasin D and therefore further reveals fibrinogen problems. Answers A, C and E would show different patterns.

Long explanation

The use of point-of-care coagulation testing is increasingly common in settings from cardiac surgery to emergency departments, general surgery and obstetrics. There are guidelines recommending the use of either thromboelastography (TEG) or ROTEM in trauma to guide resuscitation. Both TEG and ROTEM provide real-time markers of clot formation, strength and degradation. They differ in some of the terminology and in the test mechanism.

TEG uses a rotating cup containing a sample of blood activated by kaolin. A wire hanging in the sample is attached to a strain gauge, and as the blood clots, the wire is perturbed more and more by the movements of the cup. This is displayed as the TEG trace. In the ROTEM system, the cup is attached to the strain gauge, and the wires act as a stirring device, sitting in the sample and moving the cup more and more as the clot forms.

The time taken for the clot to start forming is the clotting time. The rate at which the clot starts to form is displayed as the a-angle, and the increasing portion of the curve is the clot formation time (CFT). The maximal clot firmness (MCF) and maximal lysis are the subsequent widest and narrowest points of the curve. In this example, both the clotting time and a-angle are normal. The CFT is prolonged and the MCF is decreased. This matches a pattern of hypofibrinogenaemia and the recommendation might be to administer cryoprecipitate or fibrinogen concentrate.

In both systems additional assays can be run using other activators such as tissue factor (EXTEM) and with added enzymes such as heparinase to eliminate the effect of any heparin given to the patient (HEPTEM). Platelet deactivators can be added to remove platelet function from the clot formation and therefore study the effectiveness of the fibrinogen within the sample (FIBTEM).

References

Gonzalez E, Pieracci FM, Moore EE, Kashuk JL. Coagulation abnormalities in the trauma patient: the role of thromboelastography. *Sem Thromb Hemost*. 2010; 2617.725 7071.

ROTEM product website. http://www.rotem.de/en/methodology/result interpretation (accessed April 2015).

D14. Concerning oxygen toxicity, which of the following potential effects of oxygen is LEAST likely to occur after prolonged exposure to 100% oxygen at atmospheric pressure?

A. Increased peripheral vascular resistance
B. Decreased intracranial pressure following severe head injury
C. Nausea, dizziness and headache
D. Diffuse alveolar damage, similar to that seen in acute respiratory disease syndrome
E. Acute tracheobronchitis

Answer: C

Short explanation

The central nervous system effects of hyperoxia include mild neurological signs or generalized seizures. However, the effects have only been demonstrated at high pressures (>1.8 atm underwater and >2.8 atm in a hyperbaric chamber). Other effects may been seen at normal pressures after different degrees of exposure. Acute tracheobronchitis is the earliest manifestation, followed by sternal pain, cough and thickened secretions and, later, diffuse alveolar damage.

Long explanation

It is important to recognize oxygen as a drug with potential harmful side effects. In the majority of patients, oxygen therapy is relatively benign and harmless. However, in certain groups of patients, this is not the case, and care should be used when administering oxygen. The risks of oxygen to the neonate are well documented, and neonatal resuscitation is commonly performed using air rather than oxygen due to the risks of developing retinopathy of prematurity.

Oxygen has been researched in animal models for a range of clinical uses, including its effects on haemodynamic control, increasing peripheral vascular resistance. Oxygen causes a reduction in inflammatory processes and suppression of inflammatory mediators including inducing nitric oxide synthase. It also has antimicrobial effects, particularly on anaerobic organisms, and can lead to enhanced tissue repair. However, none of these benefits have been clinically demonstrated in humans at non-hyperbaric pressures.

The potential toxic effects of oxygen are better documented and associated with both duration of use and partial pressure, with some toxic effects only seen during hyperbaric therapy. The first organ system to manifest symptoms is the respiratory system with acute tracheobronchitis, followed by sternal pain, cough and thickened secretions. Diffuse alveolar damage may occur after 48 hours' exposure to more than 50% oxygen. The clinical picture in diffuse alveolar damage is similar to ARDS, and there may be significant overlap. The neurological sequelae only occur at hyperbaric pressures and include seizures and coma.

Oxygen may be toxic through the production of free radicals and reactive oxygen species, but our understanding of these effects is still limited. In chronic lung disease patients with an increased pCO_2, and hypoxic respiratory drive, even low levels of oxygen (>30%) can lead to hypoventilation, profound hypercapnia and loss of consciousness. Considerable caution is required when prescribing oxygen to chronic obstructive pulmonary disease patients.

The benefits of hyperbaric oxygen therapy are described in specific circumstances such as carbon monoxide poisoning, air embolism, decompression sickness and *Clostridium* myonecrosis. However, hyperbaric therapy comes with its own challenges and risks.

Reference
Bitterman H. Bench-to-bedside review: oxygen as a drug. *Crit Care Med.* 2009 13:205.

D15. A 45-year-old patient presents to hospital with an acute kidney injury. His potassium level is elevated at 7.2 mmol/l and is refractory to medical management. He has a background of alcoholic cirrhosis and has abnormal liver function tests but normal clotting. He is being admitted to the intensive care unit for continuous renal replacement therapy.

Which is the BEST anticoagulant regimen to prescribe?

A. Unfractionated heparin
B. No anticoagulation regimen
C. Predilution fluids
D. Citrate
E. Epoprostenol

Answer: A

Short explanation
The KDIGO guidelines recommend the use of anticoagulation in patients on continuous renal replacement therapy (CRRT) unless they have an increased bleeding risk, pre-existing impaired coagulation or are receiving systemic anticoagulation. First-line anticoagulation would be citrate unless contraindicated by liver disease or shock and unfractionated heparin is second-line treatment. Predilution fluids alone are not adequate and epoprostenol is not recommended.

Long explanation
During continuous renal replacement therapy (CRRT), contact of blood with the extracorporeal filter circuit results in activation of platelets and the coagulation cascade. Although not mandatory, anticoagulation is usually used to prevent the filter from clotting and to maintain the permeability of the filter membrane thus prolonging the life span of the filter.

Methods to anticoagulate the CRRT circuit:

- Systemic anticoagulation
- Unfractionated heparin
- Low molecular weight heparins
- Direct thrombin inhibitors: lepirudin, argatroban, bivaluridin
- Anti-thrombin dependent factor Xa inhibitors: danaparoid, fondaparinux
- Other anticoagulants: epoprostenol
- Regional circuit anticoagulation
- Citrate

KDIGO guidelines include recommendations regarding the use of anticoagulation during renal replacement therapy. Anticoagulation (systemic or regional) is recommended for all patients providing they do not have an increased bleeding risk, pre-existing impaired coagulation or are receiving systemic anticoagulation.

Citrate is recommended as first line unless contraindications such as severe liver disease or shock are present because of concerns about citrate accumulation and toxicity. Additionally, citrate anticoagulation should only be used in units with an established protocol. This method allows selective anticoagulation of the circuit overcoming the risks associated with systemic anticoagulation and can be used in patients with an increased risk of bleeding.

Second line modalities include unfractionated heparin or low molecular weight heparins (LMWH). Alternatives to heparins in patients with immune mediated heparin induced thrombocytopenia (HIT) include direct thrombin inhibitors and antithrombin dependent factor Xa inhibitors. KDIGO recommend argatroban as the preferable treatment in this circumstance unless patients have severe liver disease.

Alternative anticoagulants such as prostaglandins, which act as platelet inhibitors (e.g. epoprostenol) have also been used. Studies have demonstrated limited efficacy when used alone, and they have a side-effect profile including inducing systemic hypotension and so have not been recommended by KDIGO.

KDIGO recommend additional non-anticoagulant strategies:

- Good vascular access allowing high blood flows to be reliably achieved
- Reducing blood viscosity through the filter
  - Predilution fluids
  - Treatments that involve diffusion as opposed to ultrafiltration
- Reducing blood-air contact in the bubble trap
- Ensuring a prompt response to the filter alarms resulting in rapid correction of any suboptimal filter conditions which may have arisen

Reference
Kidney Disease: Improving Global Outcomes. The KDIGO clinical practice guideline for acute kidney injury. http://www.kdigo.org/clinical_practice_guidelines/pdf/KDIGO%20AKI%20Guideline.pdf (accessed July 2014).

D16. You are asked about analgesia for a 35-year-old former intravenous drug user who is being observed in the ICU with partial thickness burns to both lower limbs after a house fire. She takes regular high-dose buprenorphine. She has so far had 30mg IV morphine in the last hour, and is still in severe pain.

What is the best description for the action of buprenorphine in this scenario?

A. Non-competitive partial agonist
B. Inverse agonist
C. Competitive antagonist
D. Non-competitive antagonist
E. Irreversible antagonist

Answer: C

Short explanation
On its own, buprenorphine works as a partial agonist at opioid receptors. However, in the presence of morphine, a full agonist, it competes with this full agonist for receptor sites, and therefore works as a competitive antagonist.

Long explanation
A partial agonist is a substance that produces a sub-maximal effect in response to its binding to the receptor. A full agonist will generate a full response when it binds to the receptor. With regard to the opioid receptor, a partial agonist will still have affinity

for the receptor, but it will produce a reduced response (termed the intrinsic activity). No matter how much the dose of a partial agonist is increased, it will still be unable to elicit a maximal response. On its own, this is the effect that buprenorphine has at the mu receptor.

In the presence of a full agonist, partial agonists can begin to act as competitive antagonists. In this example, morphine is a full agonist at mu opioid receptors, and on its own would produce a maximal response (high intrinsic activity). When buprenorphine is present in addition to morphine, there will be competition for the binding sites. Where buprenorphine binds instead of morphine, there will be a sub-maximal response, and hence it is functioning as a competitive antagonist. When this happens, a higher dose of full agonist will be necessary to skew the competition in morphine's favour and obtain the full response.

References

Cross M, Plunkett E. Affinity, efficacy and potency: section 4: pharmacodynamics. In *Physics, Pharmacology and Physiology for Anaesthetists, Key Concepts for the FRCA*. Cambridge: Cambridge University Press, 2008.

Drug Action, Chapter 3. In Peck T, Hill S Williams M. *Pharmacology for Anaesthesia and Intensive Care*. 3rd edition. Cambridge: Cambridge University Press, 2008.

D17. You are called urgently by the nurses on intensive care. A patient, who had a percutaneous tracheostomy earlier that day to aid ventilator weaning, has suddenly desaturated. As you attend, you notice the tracheostomy appears to be several centimetres out from where it should sit. The patient is still breathing, and you administer oxygen both via the mouth and via the tracheostomy.

What is the most appropriate next step?

A. Remove the inner tube and attempt to pass a suction catheter down the tracheostomy
B. Advance the tracheostomy tube
C. Remove the tracheostomy tube and oxygenate via the mouth
D. Remove the tracheostomy tube and insert a size 6.0 endotracheal tube through the tracheostomy hole.
E. Call the resuscitation team

Answer: A

Short explanation

The National Tracheostomy Safety project algorithm for emergency tracheostomy management with a patent upper airway states that in a patient who is breathing, the inner tube of the tracheostomy should be removed, and a suction catheter passed down the tracheostomy tube. If this is not possible and the patient is not improving, the tracheostomy tube should be removed.

Long explanation

The national confidential enquiry into patient outcomes and death (NCEPOD) in 2014 reviewed the care received by patients who underwent a tracheostomy. One of the principle recommendations of this report was that all staff must be able to manage complications of tracheostomy use, including obstruction or displacement, in accordance with the algorithms produced by the National Tracheostomy Safety Project. The Intensive Care Society guidelines follow these algorithms.

As soon as a displaced or obstructed tracheostomy is suspected, expert help should be summoned and the airway assessed by looking, listening and feeling for signs of breathing. Capnography is a valuable aid in this. If the patient is breathing, high-flow oxygen should be applied to both the tracheostomy site and the face. The resuscitation team should be called for any patients who are not breathing and cardiopulmonary resuscitation started if there is no pulse.

The patency of the tracheostomy should be verified as a matter of urgency. This should be before any attempts to bag ventilate because positive pressure applied to a displaced tracheostomy can quickly result in catastrophic surgical emphysema. Any speaking valves and inner tubes should be removed and a suction catheter passed down the tracheostomy to assess patency. If unsuccessful, the tracheostomy cuff should be deflated and breathing reassessed. Patent tracheostomies may be cautiously used, but should be thoroughly assessed for partial displacement or partial obstruction. If the patient continues to deteriorate, the tracheostomy should be removed.

If the patient is not breathing after removal of the tracheostomy, bag ventilation should be attempted. Initial attempts should focus on the oral airway, with standard mask ventilation with adjuncts or supraglottic devices. The tracheostomy site may need to be covered to allow effective ventilation. Alternatively, the tracheostomy may be used for mask ventilation, using a paediatric mask. If intubation is necessary, again, the oral route should be attempted first. If this proves impossible, then the tracheostomy site may be intubated, using either an oral tube or a tracheostomy tube. Fibre-optic bronchoscopes and other airway assistance devices are invaluable in such circumstances.

References

McGrath BA, Bates L, Atkinson D, Moore JA. Multidisciplinary guidelines for the management of tracheostomy and laryngectomy airway emergencies. *Anaesthesia*. 2012;67(9):1025–1041.

Wilkinson K, Martin I, Freeth H, et al. and the National Confidential Enquiry into Patient Outcome and Death. *On the Right Trach? A review of the care received by patients who underwent a tracheostomy*. 2014. http://www.ncepod.org.uk/2014 report1/downloads/On%20the%20Right%20Trach FullReport.pdf (accessed January 2016).

D18. You are asked by the nurses to review a slim (body mass index 18) 45-year-old man who was admitted to the intensive care unit 2 hours ago from the operating theatre after a negative laparotomy. The indication for surgery was abdominal pain, severe diarrhoea, tachycardia, fever and agitation. He now has a heart rate of 160, a blood pressure of 90/55, a temperature of 39.3°C and oxygen saturations of 98% in air. His creatine kinase (CK) is normal.

Which of the following is the most likely diagnosis?

A. Thyroid storm
B. Malignant hyperpyrexia
C. Phaeochromocytoma
D. Neuroleptic malignant syndrome (NMS)
E. Uncontrolled pain

Answer: A

Short explanation

Thyroid storm is the most likely cause of his current state and of his presentation. NMS, phaeochromocytoma and pain would be expected to cause hypertension, not hypotension. A normal CK and lack of rigidity or desaturation would make malignant hyperpyrexia unlikely.

Long explanation

Thyroid storm is a medical emergency that occurs in patients with pre-existing hyperthyroidism after a trigger. The most common trigger is a stress response, usually due to an infective or surgical insult. This leads to additional release of thyroid hormones, and a hypermetabolic state ensues. Typically, it presents with pyrexia, tachycardia and hypotension, although hypertension can sometimes occur. Cognitive disturbance is also common, along with gastrointestinal disturbance and tachyarrhythmias, particularly atrial fibrillation.

Investigations will demonstrate a low thyroid-stimulating hormone level and high levels of T3 and T4, but diagnosis is usually clinical. Thyroid storm occurs in patients with a background of hyperthyroidism. The symptoms of this are many but include weight loss, an intolerance to heat, agitation, loose stools, palpitations and loss of libido.

Hyperthyroidism is usually treated with carbimazole (which reduces T3/T4 production), propylthiouracil (same action as carbimazole, but also reduces peripheral conversion of T4 to T3) and propranolol (to control the sympathetic effects). These agents may all be used in the management of thyroid storm, with beta-blockade (propranolol or alternative) given intravenously. General supportive measures, optimization of electrolytes and treatment of any underlying cause are all required. Hydration and active cooling are particularly relevant in hypermetabolic states such as this, and patients may be profoundly hyperpyrexial in thyroid storm. Dantrolene also has some evidence to support its use in this scenario at a dose of 1 mg/kg.

It is worth noting that a significant proportion of T3 and T4 are carried bound to plasma proteins, and so administration of any medications that compete for these binding sites (for example, aspirin) will cause a rise in free T3 and T4 and a worsening of the clinical state.

References

Farling P. Thyroid disease. *Br J Anaesth.* 2000;85(1):15–28.

Migneco A, Ojetti V, Testa A, et al. Management of thyrotoxic crisis. *Eur Rev Med Pharmacol Sci.* 2005;9(1):69–74.

D19. You are called to the emergency department to review a patient in respiratory failure. Which of the following patients would be most likely to benefit from non-invasive ventilation (NIV)?

A. A 45-year-old patient with an acute exacerbation of chronic obstructive pulmonary disease (COPD) with a pH of 7.31, $PaCO_2$ of 7.9, PaO_2 9.1, who has a left-sided pneumothorax

B. A 56-year-old patient with an acute exacerbation of COPD with a pH of 7.29, $PaCO_2$ of 7.1, PaO_2 7.9, who is currently vomiting

C. A 61-year-old patient with an acute exacerbation of COPD with a pH of 7.35, $PaCO_2$ of 8.5, PaO_2 8.5

D. A 48-year-old patient with an acute exacerbation of COPD with a pH of 7.25, $PaCO_2$ of 7.5, PaO_2 15 on arrival in hospital

E. A 46-year-old patient with an acute exacerbation of COPD with a pH of 7.27, $PaCO_2$ of 8.0, PaO_2 8.5, who underwent a lobectomy 2 weeks ago

Answer: E

Short explanation

An undrained pneumothorax and vomiting are both listed as exclusions to NIV in the British Thoracic Society (BTS) guidelines. These state that NIV should be considered in all COPD patients with respiratory acidosis despite 1 hour of full medical management. Lobectomy is not listed as an exclusion to NIV, although it seems sensible that caution should be used.

Long explanation

The BTS, in association with the Intensive Care Society, published guidelines in 2008 on the use of NIV in the management of patients with COPD admitted to hospital with acute type II respiratory failure (T2RF). NIV has been demonstrated to halve mortality in patients with T2RF due to COPD (with a number needed to treat of ten.

Medical management must be initiated first, with oxygen (titrated for target SpO_2 of 88–92%), beta-2 agonists, ipratropium and steroids all given within the first hour. If the patient remains in respiratory acidosis, with a pH <7.35 and $PaCO_2$ >6 kPa, they should be offered NIV. Before initiation of NIV, the patient must have a decision made about his or her suitability for escalation to intensive care, intubation and ventilation if NIV fails.

Patients who are conscious, consenting and able to protect their airway, without excessive secretions, should be offered NIV. NIV should not be initiated in those who are comatose or moribund or those without a reasonable chance of meaningful recovery. The exception to this is patients for whom tracheal intubation is inappropriate and who are unconscious due to hypercapnia.

Patients who have copious respiratory secretions or are vomiting should not receive NIV. Other contraindications include pneumothorax (if not drained), upper gastrointestinal surgery, and facial trauma and burns.

Reference

Royal College of Physicians, British Thoracic Society, Intensive Care Society. *Chronic obstructive pulmonary disease: non-invasive ventilation with bi-phasic positive airways pressure in the management of patients with acute type 2 respiratory failure.* Concise Guidance to Good Practice series, No. 11. London: Royal College of Physicians, 2008.

D20. You have just intubated a 42-year-old woman with a history of alcohol excess, who attended the emergency department with a massive upper gastrointestinal bleed. An endoscopy team is on the way, but she is continuing to bleed profusely. You decide to insert a Sengstaken-Blakemore tube. You insert the tube to 50 cm. What is the next step?

A. Confirm gastric placement, inject 250 ml into the gastric balloon, apply traction to the tube with a 500-ml bag of fluid

B. Inject 250 ml into the gastric balloon, apply traction to the tube with a 1000-ml bag of fluid

C. Confirm gastric placement, inject 250 ml into the oesophageal balloon, inject 250 ml into the gastric balloon, apply traction to the tube with a 500-ml bag of fluid

D. Confirm gastric placement, inject 250 ml into the oesophageal balloon, inject 250 ml into the gastric balloon, apply traction to the tube with a 1000-ml bag of fluid

E. Confirm gastric placement, inject 250 ml into the gastric balloon, apply traction to the tube with a 1000-ml bag of fluid, inject 250 ml into the oesophageal balloon

Answer: A

Short explanation

Gastric placement needs to be confirmed before the gastric balloon is fully inflated. Once the gastric balloon is inflated and traction has been applied, bleeding often stops. Only if bleeding continues should the oesophageal balloon be inflated, and only then with careful pressure monitoring and with minimal pressure for short periods of time.

Long explanation

A Sengstaken-Blakemore tube (SBT) is used as a temporary measure to control bleeding from gastric and/or oesophageal varices. It usually has four ports – gastric balloon, gastric aspirate, oesophageal balloon and oesophageal aspirate. The gastric balloon should not be inflated unless there is reasonable certainty that it is in the stomach. The oesophagogastric junction is usually situated at 38 to 40 cm from the incisors, and so the tube will usually need to be inserted beyond this to ensure the gastric balloon is in the stomach. If there is doubt as to its position, the gastric balloon can be inflated with 50 ml of air or contrast and a chest X-ray taken to ensure the balloon is below the diaphragm.

Once position has been confirmed, the gastric balloon is generally inflated with 250 ml of air or saline to ensure traction can be applied and pressure maintained on the varices. Once the balloon is adequately inflated, metal clamps are applied to prevent leakage of air. The gastric port should be aspirated to measure ongoing blood loss. A small amount of traction is then applied to the SBT to ensure adequate tamponade of the varices. This may be achieved by suspending a 500-ml (or 1000-ml) bag of fluid from the SBT. The larger bag may have a higher risk of causing tissue necrosis or pressure sores of the mouth.

Inflating the oesophageal balloon carries a much higher risk to the patient because there is an increased danger of oesophageal necrosis occurring. The oesophageal balloon is inflated by air until a manometer reads 35 to 40 mmHg, the pressure is then reduced to the lowest pressure required to prevent bleeding. Oesophageal balloons should be deflated every few hours to assess for cessation of bleeding; gastric balloons are generally left inflated for 12 to 24 hours. The longer the balloons are left inflated, and traction applied, the higher the risk of necrosis. There is also increased risk of rebleeding when the balloons are deflated.

References

Allen P, Booth J. Insertion of a Sengstaken-Blakemore tube in critical care. In Waldmann C, Soni N, Rhodes A. *Oxford Desk Reference*. Oxford: Oxford University Press, 2008, Chaper 4, p 74–75.

Kupfer Y, Cappell M, Tessler S. Acute gastrointestinal bleeding in the intensive care unit. The intensivist's perspective. *Gastroenterol Clin North Am*. 2000; 29(2):275–307.

D21. You are called by an anaesthetist who is asking for a post-operative intensive care bed for a 57-year-old woman who is booked for a hemicolectomy for a tumour. She had a myocardial infarction 6 months previously, has renal impairment and chronic obstructive pulmonary disease.

Which of the following is the best way to calculate her predicted mortality from this operation?

A. APACHE III SCORE
B. P-POSSUM Score
C. Goldman Risk Index
D. American Society of Anaesthesiologists (ASA) physical status classification
E. New York Heart Association (NHYA) Classification

Answer: B

Short explanation

The P POSSUM score will predicted morbidity and mortality for the intended operation. The NHYA is a classification of functional status of heart disease but is not a predictor of mortality. The Goldman Risk index is used to predict the risk of cardiac complications from non-cardiac surgery, and the ASA, although simple, does not predict the risk of a procedure. APACHE score is calculated within the first 24 hours of intensive care unit (ICU) admission.

Long explanation

The NHYA Functional Classification provides a simple way of classifying heart disease (originally cardiac failure), which is useful for preoperative assessment. It places patients into one of four categories, based on limitations to physical activity:

1. No symptoms and no limitation in ordinary physical activity
2. Mild symptoms and slight limitation during ordinary activity
3. Marked limitation in activity due to symptoms, even during less-than-ordinary activity; comfortable only at rest
4. Severe limitations; experience symptoms even while at rest

The Goldman Cardiac Risk Index is a predictor of cardiac risk in non-cardiac surgery. Points are allocated across a number of different parameters, with higher scoring patients having a higher risk of death. It was created in 1977 and has since been superseded with Lee's revised index (1999), which has simplified it to six parameters, the presence of two or more of which are associated with a significant increase in myocardial infarction or death.

The ASA classification is a surgical risk assessment tool introduced in 1963:

1. Healthy
2. Mild systemic disease
3. Severe systemic disease
4. Severe systemic disease that is a constant threat to life
5. Moribund; not expected to survive without the operation

There are several limitations of the ASA system, such as the exclusion of the presenting condition, but its main advantage is its simplicity.

The APACHE III system was designed to estimate disease severity and prognosis after ICU admission with high accuracy. The score is calculated within the first 24 hours of ICU admission, but given the copyright on the APACHE III (and subsequent APACHE IV) statistical analysis, most units still use APACHE II.

P-POSSUM lies somewhere between the APACHE and ASA scoring. Although it has 12 physiological and six operative parameters to calculate an estimated risk for an operation, it can be completed preoperatively with just the physiological parameters and still predict outcome well.

References

American Heart Association. Classes of Heart Failure. http://www.heart.org/HEARTORG/Conditions/HeartFailure/AboutHeartFailure/Classes-of-Heart-Failure_UCM_306328_Article.jsp (accessed March 2015).

Bouch D, Thompson J. Severity scoring systems in the critically ill. *Contin Educ Anaesth Crit Care Pain.* 2008;8(5):181–185.

Goldman L, Caldera DL, Nussbaum SR, et al. Multifactorial index of cardiac risk in noncardiac surgical procedures. *N Engl J Med.* 1977;297:845–850.

D22. You are called to help with the management of a 50-year-old man with septic shock. His blood pressure is 80/40.

With reference to the Surviving Sepsis Campaign guidelines, which of the following is the best initial fluid bolus?

A. 2 units of packed red cells
B. 500 ml 4.5% human albumin solution
C. 500 ml hydroxyethyl starch solution (e.g. Hespan, Voluven)
D. 500 ml Hartmann's solution
E. 2000 ml Hartmann's solution

Answer: D

Short explanation
The guidelines recommend crystalloids as the initial fluid of choice in the initial management of severe sepsis and septic shock. Large volumes may be required, but fluids should be administered as boluses with reassessment. Albumin is recommended for use in patients requiring substantial volumes of crystalloid, but not initially. Colloids are not recommended initially, and hydroxyethyl starch has recommendations against its use.

Long explanation
The surviving sepsis campaign guidelines, updated in 2012, offer key recommendations on all aspects of the management of patients with sepsis. With regard to fluid therapy, the recommendations are as follows:

– Initial fluid resuscitation in severe sepsis and septic shock should be with crystalloids.
– Hydroxyethyl starches should NOT be used.
– Albumin should be used in patients with sepsis who require large volumes of crystalloids.
– In patients with septic shock, initial fluid resuscitation should be at least 30 ml/kg (e.g. 2000 ml for a 70-kg patient), although significantly more may be required in some cases.
– Fluid administration should take the form of rapid boluses followed by reassessment. This should continue until there is no further improvement in haemodynamic parameters. Sepsis is an evolving condition, however, and fluid boluses may need to be reinstated subsequently.

With regard to blood product administration, the guidelines recommend a haemoglobin target of 70 to 90 g/l and a transfusion threshold of 70 g/l, in the absence of conditions such as myocardial ischaemia or acute haemorrhage. This recommendation is contrary to early goal-directed therapy (EGDT) protocols (e.g. Rivers et al. 2001), which advised that in the early resuscitation of patients with septic shock, transfusion may be indicated in patients with low central venous SvO_2, to achieve a target haematocrit of 30%. Recent studies have questioned the value of the EGDT approach to early management of sepsis.

References
Dellinger R, Levy M, Rhodes A, et al. Surviving Sepsis Campaign Guidelines Committee. Surviving sepsis campaign: international guidelines for management of severe sepsis and septic shock: 2012. *Crit Care Med*. 2013; 41(2): 580–637.
Rivers E, Nguyen B, Havstad S, et al. Early Goal-Direct Therapy Collaborative Group: early goal-directed therapy in the treatment of severe sepsis and septic shock. *N Engl J Med*. 2001;345:1368–1377.

D23. A 23-year-old man is admitted to the intensive care unit after a road traffic accident. He has a traumatic brain injury and an intracranial pressure (ICP) monitor in situ.

Which of following pressure monitors is the most appropriate?

A. Subarachnoid
B. Subdural
C. Epidural
D. Parenchymal
E. Ventricular

Answer: E

Short explanation

The ventricular catheter connected to an external strain gauge, which can be recalibrated in situ, is the most reliable and accurate way to measure ICP. Options A through C are all less accurate, whereas option D is initially accurate, but once monitoring is commenced, it cannot be recalibrated.

Long explanation

The Brain Trauma Foundation guidelines for the management of severe traumatic brain injury (TBI) advise that ICP should be monitored in all patients who have a acceptable chance of survival, a Glasgow Coma Score (GCS) <9, and computed tomographic (CT) evidence of TBI. It should also be monitored in patients with a normal CT head scan, if they have **more** than one of the following risk factors: age >40, motor score <4 or hypotension (SBP <90 mmHg).

In patients with less severe TBI (i.e. GCS ≥9), the risks of ICP monitor insertion may outweigh the benefits of the device, although the intention of monitor insertion is to minimize any secondary injury through control of ICP and maintenance of cerebral perfusion pressure. The risks of ICP monitoring can be caused by the insertion, such as haemorrhage; caused by its presence inside the skull, such as infection; or caused by the line itself, such as malfunction or obstruction. These risks have to be balanced against the benefits of ICP monitor insertion.

When ICP monitoring is indicated, an appropriate device must be used. Although ventricular monitoring is the reference against which other devices are compared, all devices need to have a number of specifications, in accordance with the Association for the Advancement of Medical Instrumentation. All devices should monitor pressure across a wide range, from at least 0 to 100 mmHg, as well as having less than 10% error from 20 to 100 mmHg, and be accurate to ±2 mmHg below this. Most new devices far exceed this minimum threshold, but the ventricular site still provides the most accurate readings, at low cost, with minimal complications.

Reference

Guidelines for the Management of Severe Traumatic Brain Injury. 3rd edition. A joint project of the Brain Trauma Foundation, and American Associate of Neurological Surgeons, and Congress of Neurological Surgeons. New York, NY: Brain Trauma Foundation, 2007. https://www.braintrauma.org/uploads/06/06/Guidelines_Management_2007w_bookmarks_2.pdf (accessed January 2016).

D24. A 40-year-old woman in the intensive care unit (ICU) is intubated and venti-lated with pneumonia and severe acute respiratory distress syndrome (ARDS). She has been started on a propofol infusion at 100 mg/hr and alfentanil at 2 mg/hr and appears settled. Due to increasing hypoxaemia, a cis-atracurium infusion has been started at 37.5 mg/hr. The patient requires a chest drain insertion on the ICU, and you are concerned about awareness during the procedure.

Which is the most appropriate course of action?

A. Double the rate of propofol infusion
B. Add a midazolam infusion at 5 mg/hr
C. Check a 'train of four' neuromuscular response before starting
D. Give a 100 mg bolus of propofol before the procedure
E. Attach a bispectral index (BIS) and titrate sedation to maintain a score of 40 to 60 throughout

Answer: E

Short explanation

A patient undergoing a procedure in the ICU requires additional sedation, especially when receiving neuromuscular blocking agents. Answers A and B may or may not avoid awareness, and answer D is likely to wear off before the procedure is finished. Answer C may help you decide whether the patient is paralyzed but will not help avoid awareness.

Long explanation

There is a clear difference between the sedation aims in the ICU compared with the aims of sedation or a general anaesthetic for surgery. In theatres or for a procedure, the aim of using anaesthetic drugs should be to render the patient amnesic of events and unaware of noxious stimuli. The aims of sedation in the ICU are often counter to this and are to provide the minimum amount of sedation required to facilitate the pro-vision of therapies, including tolerating an endotracheal tube and a ventilator, often whilst remaining both aware and rousable. It is therefore important to remember that when performing a painful procedure on a patient in ICU who appears sedated, there may well be a requirement for increasing both analgesia and sedation before starting. It is also important to remember that patients in ICU may be weak and delirious and therefore unable to respond in a normal manner to a painful stimulus, despite being aware.

Doubling an infusion rate or adding a new infusion without a bolus will take approximately five half-lives of the drug before achieving the targeted steady state. This would take too long to provide anaesthesia before starting a procedure. Midazo-lam has a half-life of 1.5 to 2.5 hours, and propofol has a half-life of 40 minutes. Both drugs are therefore commonly given as a loading bolus, followed by an infusion. Once at steady state, a usual infusion rate for maintenance of midazolam sedation is 1 to 7 mg/hr, and propofol is given usually at a rate of 20 to 200 mg/hr for a 70-kg adult. High infusion rates of propofol for a prolonged period risks triggering propofol infusion syndrome.

Conversely, a single bolus of propofol will redistribute rapidly and wear off before the procedure is complete. When injected intravenously in a bolus dose of 2 mg/kg, the onset of anaesthesia is rapidly achieved within 10 to 30 seconds. However, due to rapid redistribution into fat-soluble tissues and protein binding, the plasma levels fall rapidly and the offset time from the bolus dose is 3 to 10 minutes. Most surgical procedures therefore require either an infusion of propofol to follow or another form of anaesthesia maintenance. Propofol fits a three compartment model for distribu-tion among plasma, highly fat-soluble compartments and less soluble components. Therefore, calculating estimates of plasma concentration is extremely complex. For this reason, models have been developed and programmed into infusion pumps that can be set to automatically vary the infusion rate to achieve a targeted plasma or effector site level.

Reference

Barr J, Fraser GL, Puntillo K, et al. Clinical practice guidelines for the management of pain, agitation, and delirium in adult patients in the intensive care unit. *Crit Care Med.* 2013;41(1):263–306.

D25. A 36-year-old woman presents to hospital with acute abdominal pain and has an elevated lipase level of four times the normal range.

What is the MOST likely underlying cause of her current illness?

A. Hypertriglyceridemia
B. Hypercalcaemia
C. Gallstones
D. Excess alcohol intake
E. Cytomegalovirus

Answer: C

Short explanation

This patient has acute pancreatitis. All of the preceding are causes of acute pancreatitis, but gallstones and alcohol are the main causes. In developed countries, gallstones account for approximately 40% of cases, whereas alcohol accounts for approximately 35% of cases.

Long explanation

Causes:

- Biliary or gallstone pancreatitis (~40%). It occurs when small gallstones pass into the bile duct and get lodged at the sphincter of Oddi. This obstructs both biliary and pancreatic ducts and therefore the outflow of pancreatic secretions. The gallstone may have dislodged and hence be undetectable by the time of diagnosis.
- Alcohol is the second commonest overall cause of acute pancreatitis (~35–40%), more common in men than women and the commonest cause of chronic pancreatitis. Acute alcoholic pancreatitis has the highest mortality.
- Iatrogenic pancreatitis:
  - Post-endoscopic retrograde cholangiopancreatography (ERCP)
  - Post-surgical
- Abdominal trauma: this occurs more commonly with penetrating than blunt trauma
- Drug-induced pancreatitis; drugs thought to be responsible include the following:
  - anticonvulsant: valproic acid
  - antihypertensive: methyldopa
  - antimicrobial: tetracycline, sulfonamides, metronidazole, nitrofurantoin, isoniazid,
  - corticosteroids
  - chemotherapy agents: asparaginase, cisplatin
  - diuretics: furosemide, thiazides
  - immunosuppressants: azathioprine, 6-mercaptopurine
  - non-steroidal anti-inflammatories: salicylate, piroxicam
  - other: oestrogens, octreotide, statins

- Infection; this tends to be more common in children than adults, and the disease tends to be milder than that due to binary disease or alcohol-induced pancreatitis; causes include the following:
  - Viral infections – mumps virus, Coxsackie virus, cytomegalovirus (CMV), hepatitis virus, Epstein-Barr virus (EBV), echovirus, varicella-zoster virus (VZV), measles virus, rubella virus
  - Bacterial infections – Mycoplasma pneumoniae, Salmonella, Campylobacter, Mycobacterium tuberculosis, Legionella
  - Fungal infections – *Aspergillus*
  - Parasitic infections – Toxoplasma, Cryptosporidium
- Inherited pancreatitis: genetic mutations are thought to increase the risk of patients developing pancreatitis
- Hypercalcaemia, which can be caused by
  - Hyperparathyroidism
  - Excessive doses of vitamin D
  - Total parenteral nutrition
  - Malignancy (bony metastases or multiple myeloma)
  - Sarcoidosis
- Hypertriglyceridemia (typically >500 mg/dl): Triglyceride levels can be artificially elevated during an episode of acute pancreatitis, so levels need to be repeated following the acute episode
- Autoimmune pancreatitis
- Toxins:
  - Organophosphate insecticides
  - Scorpion and snake bites
- Hypothermia
- Malignancy:
  - Pancreatic ductal carcinoma
  - Ampullary carcinoma
  - Islet cell tumour
  - Sarcoma
  - Lymphoma
  - Cholangiocarcinoma
- Vascular disease
  - Vasculitis
  - Ischaemia
- Idiopathic

References

VanWoerkom R, Adler DG. Acute pancreatitis: review and clinical update. *Hosp Phys.* 2009;45(1):9–19.

Wang G-J, Gao C-F, Wei D, et al. Acute pancreatitis: etiology and common pathogenesis. *World J Gastroenterol.* 2009;15(12):1427–1430.

D26. You are called to the Emergency Department as part of the Trauma team. A 27-year-old woman was struck by a van whilst crossing a road near to the hospital. She has chest wall and lower limb injuries. The prehospital team has secured intravenous access, intubated her and brought her by ambulance to the emergency department. Her primary survey findings are as follows:

Airway: Oral ETT at 21 cm; normal $ETCO_2$ trace; attached to ventilator

Breathing: SpO$_2$ 94% (FiO$_2$ 1.0); reduced breath sounds and hyperresonance on the left

Circulation: Heart rate 140; blood pressure 74/44; capillary refill time >5 s; no external bleeding; soft abdomen; stable pelvis

Disability: Pupils 3 mm, equal and reactive; sedated and paralyzed

Exposure: Bilateral closed midshaft femoral fractures

What is the first step in her management?

A. Thoracostomy followed by intercostal chest drain
B. 1-l bolus of crystalloid
C. Splinting of femoral fractures
D. Chest radiograph
E. Infusion of 4 units O negative packed red cells

Answer: A

Short explanation
This woman has signs of a tension pneumothorax. This is life threatening and should be treated immediately by thoracostomy followed by insertion of a chest drain. Once treated, the other management options become appropriate.

Long explanation
Tension pneumothorax is an uncommon but rapidly life-threatening complication of thoracic trauma, occurring in 1 to 3% of trauma cases. A tension pneumothorax develops when air enters the pleural space but is unable to leave. In spontaneously ventilating patients, this is thought to occur through an internal pleural defect that acts as a one-way flap valve, leading to a gradual accumulation of intrapleural air, which eventually causes decompensation. In patients receiving assisted ventilation, high intrathoracic pressures lead to rapidly progressive pneumothoraces which progress to tension pneumothorax and cardiorespiratory collapse.

Classically, tension pneumothorax is a clinical diagnosis that should be easily identifiable without recourse to imaging. However, signs may in fact be difficult to discern, particularly in the context of multiply injured trauma patients in a noisy, stressful environment. The classical features are anxiety, dyspnoea, tachycardia, hypotension with reduced breath sounds and hyperresonant percussion on the same side of the chest.

A recently published meta-analysis has quantified the presenting features and management of tension pneumothorax. In spontaneously ventilating patients, chest pain, dyspnoea and shortness of breath were common symptoms but seen in only 30 to 50% of patients. The classical signs were not as common as would be expected: tachypnoea (46%), hypotension (16%), tachycardia (43%), hyperresonance (27%) and decreased breath sounds (58%). Diagnosis is even more difficult in artificially ventilated patients: hypotension (66%), tachycardia (31%), hyperresonance (8%) and decreased breath sounds (45%).

Imaging has classically been discouraged because any delay may prove fatal. However, in the meta-analysis, more than 50% of patients with tension pneumothorax had a chest radiograph before treatment. It has been suggested that ultrasound may help with diagnosis because it is readily available and may be more sensitive than chest radiography.

Tension pneumothorax should be treated as a life-threatening emergency. After thoracostomy, an intercostal chest drain should be sited as quickly as possible to decompress the pleural cavity. If the necessary equipment is unavailable, then insertion of a large-bore needle or cannula into the second intercostal space in the mid-clavicular line can be undertaken. It is important to be aware that standard intravenous cannulae are too short to reach the pleural space in many patients. Also, cannulae are easily kinked or dislodged; this is simply an interim measure before definitive chest drain insertion can occur.

References

American College of Surgeons (ACS) Committee on Trauma. *Advanced Trauma Life Support Student Course Manual.* 9th edition. Chicago, IL: ACS, 2012

Roberts DJ, Leigh-Smith S, Faris PD, et al. Clinical presentation of patients with tension pneumothorax, a systematic review. *Ann Surg.* 2015;261(6):1068–1078.

D27. A 49-year-old woman with known liver cirrhosis is admitted with grade III hepatic encephalopathy. An ascitic tap is performed, which demonstrates 350 neutrophils/µl, no red blood cells and no organisms visible on Gram stain.

Which initial treatment is the most likely to lead to a sustained improvement in her encephalopathy?

A. Piperacillin and tazobactam
B. Flumazenil
C. Lactulose
D. Rifaximin
E. L-ornithine-L-aspartate (LOLA)

Answer: A

Short explanation

Treatment of the underlying cause of decompensation is the key in management of decompensated hepatic encephalopathy. The presence of >250 neutrophils/µl indicates a diagnosis of spontaneous bacterial peritonitis. Flumazenil can lead to a transient improvement in encephalopathy. The other treatments lead to an improvement in encephalopathy by reducing ammonium levels in the blood.

Long explanation

Hepatic encephalopathy is a reversible impairment of brain function associated with liver failure. The severity of hepatic encephalopathy may be graded according to the West-Haven criteria, according to conscious level, cognition and findings on neurological examination (see table). A patient with normal consciousness and cognition but subtle neurological examination findings may have minimal hepatic encephalopathy (MHE), which is thought to be a precursor to overt hepatic encephalopathy.

Severity	Level of consciousness	Cognition/behaviour	Neurological examination
Grade 0	Normal	Normal	May be impaired
Grade 1	Mild confusion	Reduced attention	Mild asterixis/tremor
Grade 2	Lethargy	Disorientation, inappropriate behaviour	Asterixis, slurred speech
Grade 3	Somnolent but rouseable	Bizarre behaviour	Rigidity, clonus, hyperreflexia
Grade 4	Comatose	Comatose	Abnormal posturing

The pathogenesis of hepatic encephalopathy is not fully understood, but is thought to stem from a failure of hepatic clearance of ammonia from enteric sources. This ammonia enters the systemic circulation and interferes with brain function by causing neuroinflammation, increasing neuronal glutamine levels and interfering with neurotransmitters such as GABA, dopamine, serotonin and acetylcholine. Neurotoxicity, cerebral oedema and impaired neurotransmission result, leading to the clinical manifestations outlined above.

Management of grade 3 hepatic encephalopathy includes intensive care unit admission with organ support as necessary and attempts to reduce the ammonia levels in the blood. However, the most important part of management is to find and treat any precipitating factor, such as spontaneous bacterial peritonitis (SBP), gastrointestinal bleeding, electrolyte disturbance or sedative medications.

SBP is a common complication of cirrhosis and should be suspected in any patient with known cirrhosis and abdominal pain or signs of systemic inflammation. Diagnostic paracentesis should be performed immediately. The presence of >250 neutrophils/μl is diagnostic of SBP and empirical broad-spectrum antibiotics should be started before culture results are available. Up to 60% of cases of SBP have negative microbiological microscopy and culture results.

Ammonia levels may be reduced by the use of lactulose and rifaximin, which reduce ammonia production in the gut by increasing transit time and reducing bacterial numbers. Strategies aimed at removing ammonia from the blood include LOLA, which increases the extrahepatic metabolism of ammonia to glutamine. However, this treatment is not widely available and has little evidence of benefit.

References
Bajaj JS. Review article: the modern management of hepatic encephalopathy. *Aliment Pharmacol Ther*. 2010;31:537–547.

Felipo V. Hepatic encephalopathy: effects of liver failure on brain function. *Nat Rev Neurosci*. 2013;14(12):851–858.

Gines P, Angeli P, Lenz K, et al. Clinical practice guidelines: EASL clinical practice guidelines on the management of ascites, spontaneous bacterial peritonitis, and hepatorenal syndrome in cirrhosis. *J Hepatol*. 2010;53:397–417.

D28. A 73-year-old male patient is ventilated in the intensive care unit (ICU) for post-operative pneumonia after laparotomy for small bowel obstruction. By day 5, his respiratory support has been weaned, but when the propofol sedation is reduced he is too agitated to allow extubation. The CAM-ICU screening tool demonstrates he is suffering from delirium.

Which of the following is the LEAST appropriate drug to help manage his delirium?

A. Dexmedetomidine infusion
B. Haloperidol as required
C. Regular quetiapine
D. Lorazepam as required
E. Increased dose propofol infusion

Answer: D

Short explanation

All the preceding treatment strategies can be used to manage delirium in adult ICU patients if the patient requires sedation to prevent them from harming him- or herself. Benzodiazepines are associated with worsening of delirium and should be avoided if possible unless the patient is suffering from alcohol or benzodiazepine withdrawal.

Long explanation

Delirium in ICU patients is associated with worse patient outcomes. This includes increased mortality, prolonged duration of mechanical ventilation, increased ICU and hospital length of stay and increased cognitive impairment post-ICU. Patients are also more likely to self-extubate and pull out invasive lines and devices. Delirium is also associated with increased health care costs. Prevention, early identification of its presence and prompt treatment of ICU delirium is therefore important.

Primary prevention interventions – non-pharmacological:

- Repeated reorientation of patients; orientating stimuli include 24-hour clocks, glasses, hearing aids
- Provision of cognitively stimulating activities (several times/day)
- Minimize sleep disturbance: optimization of the patients' environments to reduce light and noise stimulation at night, cluster patient care activities to the daytime and decrease patient stimulation at night
- Early mobilization and physiotherapy with range-of-movement exercises
- Timely removal of catheters/lines and physical restraints
- Early correction of dehydration and any physiological derrangements

Primary prevention interventions – pharmacological:

- Stop unnecessary medications that exacerbate delirium (e.g. sedatives, opiates, anticholinergics)
- Daily sedation holds or sedation management plans for all ventilated patients unless contraindicated
- Pain management protocol
- There is some evidence that prophylactic melatonin or ramelteon (a melatonin agonist) may have a protective effect and reduce the incidence of the development of delirium, although this was not in a critical care setting

The treatment of delirium involves implementation of primary prevention interventions as well the following strategies.

Non-pharmacological:

- Reassurance
- Bedside sitters to keep the patient safe without the need for pharmacological interventions

Pharmacological:

- Ensure adequate analgesia including with opiates before other sedating agents are used
- Haloperidol to treat symptoms although evidence does not suggest it reduces their duration
- The duration of delirium may be reduced by treatment with atypical antipsychotics (e.g. quetiapine, risperidone, olanzapine)
- Recent guidelines from the Society of Critical Care Medicine (SCCM) do not recommend the use of antipsychotics in patients at significant risk for torsades de pointes
- Sedation with dexmedetomidine; there is some evidence to suggest that it is associated with less delirium than treatment with benzodiazepines
- Benzodiazepines should be avoided if possible; however, they have a role in the management of patients with alcohol or benzodiazepine withdrawal
- The SCCM guidelines do not recommend the use of rivastigmine (cholinesterase inhibitor) in the treatment or prevention of delirium

References
Barr J, Fraser GL, Puntillo K, et al. Clinical practice guidelines for the management of pain, agitation, and delirium in adult patients in the intensive care unit. *Crit Care Med*. 2013;41(1):263–306.
Hatta K, Kishi Y, Wada K, et al. Preventive effects of ramelteon on delirium. A randomised placebo-controlled trial. *JAMA Psychiatry*. 2014;71(4):397–403.
Reade MC, Finfer S. Sedation and delirium in the intensive care unit. *N Engl J Med*. 2014;370(5):444–454.

D29. A 64-year-old male smoker with a history of mild emphysema is admitted with breathlessness. Chest X-ray demonstrates a large pneumothorax. A 12-F intercostal drain is inserted, and his symptoms improve. At 48 hours, there is still evidence of a persistent pneumothorax on chest X-ray, so high-volume/low-pressure suction is applied. Twenty-four hours later, a persistent pneumothorax remains.
What would be the next MOST appropriate treatment strategy?

A. Referral to thoracic surgery to consider chemical pleurodesis
B. Increase the size of the chest drain to 24-F
C. Convert the suction to a high-pressure system
D. Referral to thoracic surgery to consider surgical pleurectomy
E. Perform a medical pleurodesis with talc

Answer: D

Short explanation
This patient has a secondary pneumothorax which persists after 72 hours of drainage. High-pressure suction and large-bore chest drains are not recommended by the British Thoracic Society (BTS) for the management of pneumothoraces. Surgical pleurectomy is recommended in preference to surgical or medical pleurodesis because of the lower associated recurrence rate.

Long explanation

Secondary spontaneous pneumothoraces occur in patients with pre-existing lung disease. Pneumothoraces can be asymptomatic and clinical assessment is an unreliable indicator of size although severe symptoms and signs of respiratory distress suggest the presence of a tension pneumothorax. Diagnosis of a pneumothorax is made using inspiratory chest X-rays or computed tomography (CT) in more complex cases. In addition, ultrasound is increasingly used to diagnose pneumothorax. Pneumothorax size is assessed as the size of the rim between the lung margin and the chest wall at the level of the hilum; >2 cm is stated as the cutoff between large and small pneumothoraces. CT scans can more accurately assess this as well as identify the presence of bullous lung disease.

The BTS recommend the following for the management of secondary pneumothoraces:

- Patients require admission to hospital.
- Oxygen therapy should be administered.
- If the pneumothorax is small and the patient is asymptomatic, aspiration may be attempted.
- A small-bore (8- to 14-F) chest drain should be inserted if aspiration has been unsuccessful. Large chest tubes (20- to 24-F) are not proven to be advantageous over smaller ones and are not recommended.
- All patients should be referred to a respiratory physician within 24 hours of admission.
- The routine use of suction is not recommended because of the risk of re-expansion pulmonary oedema if commenced initially. There are also concerns that it might increase the risk of persistent air leak occurring. If used, high-volume/low-pressure suction should be chosen.
- Referral to thoracic surgeons should be sought if there is a persistent air leak or failure of the lung to re-expand after 48 hours. A review should be sought within 5 days of pneumothorax onset for ongoing drain management including the consideration of the use of suction as well as the need for surgical interventions.
  - Surgical options are the preferred treatment options if the preceding have failed to resolve the pneumothorax because they are associated with a lower recurrence rate than medical pleurodesis:
    - Pleurectomy via open thoracotomy (recurrence rate 1%)
    - Pleurectomy and pleural abrasion via video-assisted thoracoscopic surgery (VATS) (recurrence rate 5%)
    - Surgical chemical pleurodesis (less commonly used following the advent of VATS procedures)
  - Alternative options (for those unfit for surgery):
    - Medical pleurodesis
    - Insertion of a long-term Heimlich valve

Reference

MacDuff A, Arnold A, Harvey J. Management of spontaneous pneumothorax: British Thoracic Society Pleural Disease Guideline 2010. *Thorax*. 2010;65(Suppl 2):ii18–ii31.

D30. A 55-year-old male patient has suffered an ST-elevation myocardial infarction (STEMI). He underwent primary percutaneous coronary intervention (PCI), having a bare metal stent inserted into his circumflex artery. He was previously well and has no increased bleeding risk.

What would be the MOST appropriate antiplatelet regime for this patient to continue post-PCI on discharge?

A. Aspirin and ticagrelor
B. Aspirin and clopidogrel
C. Aspirin alone
D. Aspirin and tirofiban
E. Clopidogrel alone

Answer: A

Short explanation
This patient has had a stent inserted and so requires dual antiplatelet therapy on discharge. Tirofiban is an intravenous drug and thus is no use in the long term. The National Institute of Clinical Excellence (NICE) and the European Society of Cardiology (ESC) recommend the use of ticagrelor or prasugrel in combination with aspirin in preference to clopidogrel.

Long explanation
- Antiplatelet agents:
  - Cyclooxygenase-1 (COX-1) inhibitor
    - Aspirin irreversibly inhibits COX-1 in platelets resulting in decreased thromboxane A_2 production, which is required for platelet aggregation.
  - $P2Y_{12}$ receptor antagonists (oral drugs)
    - Adenosine diphosphate (ADP) $P2Y_{12}$ receptor antagonists
      - These are thienopyridine derivatives and are prodrugs. The active metabolites irreversibly bind the ADP $P2Y_{12}$ receptor, preventing it from activating intracellular processes that result in increased platelet aggregation.
        - First-generation thienopyridine: ticlopidine (this has clinically been replaced by newer drugs)
        - Second-generation thienopyridine: clopidogrel
        - Third-generation thienopyridine: prasugrel
    - Other $P2Y_{12}$ receptor antagonists
      - Ticagrelor is a cyclopentyl-triazolo-pyrimidine analogue that reversibly binds the $P2Y_{12}$ receptor directly. Its active metabolite is responsible for a proportion of its action.
    - Both prasugrel and ticagrelor have been demonstrated to be more potent and effective at inhibiting the $P2Y_{12}$ receptor and producing a more reliable clinical effect on platelet aggregation than clopidogrel.
- Glycoprotein IIb/IIIa (GP IIb/IIIa) antagonists (intravenous drugs)
  - Activation of the GP IIb/IIIa receptor by fibrinogen stimulates the final common pathway in platelet aggregation; by blocking this, they have a powerful antiplatelet action
  - Abciximab
  - Eptifibate
  - Tirofiban
- Phosphodiesterase inhibitor
  - Dipyridamole, a pyridopyrimidine derivative that inhibits platelet aggreagation by inhibiting adenosine uptake

- Others drugs that exert an antiplatelet effect
  - Epoprostenol, a prostaglandin I_2 (prostacyclin) analogue that works by increasing cAMP and thus decrease intracellular calcium
  - Dextran

The ESC guidelines recommend the use of ticagrelor or prasugrel in combination with aspirin in preference to clopidogrel. This should be continued for up to 12 months post-PCI. They also recommend an additional intravenous anticoagulant around the time of the procedure for patients undergoing PCI. Prasugrel should not be used in patients with a history of previous cerebrovascular accident or transient ischaemic attack, age ≥75 years or weight <60 kg and with moderate-severe hepatic dysfunction. Ticagrelor is contraindicated in a patients with a history of haemorrhagic stroke, active bleeding or moderate-severe hepatic dysfunction.

References
Capodanno D, Ferreiro JL, Angiolillo DJ. Antiplatelet therapy: new pharmacological agents and changing paradigms. *J Thromb Haemost*. 2013;11(Suppl 1):316–312.

The Task Force on the Management of ST-Segment Elevation Acute Myocardial Infarction of the European Society of Cardiology (ESC). ESC Guidelines for the management of acute myocardial infarction in patients presenting with ST-segment elevation. *Eur Heart J*. 2012;33:2569–2619.

Exam E: Questions

E1. A 24-year-old woman with normal lungs is breathing 60% oxygen via a face mask. On arterial blood gas analysis, her $PaCO_2$ is 4.5 kPa.

Which of the following is the most likely value of the partial pressure of oxygen in her alveoli?

A. 42 kPa
B. 51 kPa
C. 54 kPa
D. 60 kPa
E. 80 kPa

E2. A 64-year-old male patient is admitted intubated and ventilated after an out-of-hospital cardiac arrest. Unfortunately, he has suffered a significant hypoxic brain injury, and withdrawal of life-sustaining therapy is being considered.

When making ongoing treatment/withdrawal decisions for this patient, which of the following is the BEST method?

A. Treat/withdraw treatment after consultation with an independent mental capacity advisor (IMCA)
B. Treat/withdraw treatment with consent from the patient's lasting power of attorney (LPA)
C. Treat/withdraw treatment after consultation with the patient's family to establish the patient's wishes
D. Treat/withdraw treatment according to patient's wishes using a valid advanced directive
E. Treat/withdraw treatment according to the team's opinion of the patient's best interests

E3. A patient with a history of cirrhotic liver disease is referred to critical care for organ support.

Which of the following features suggests the worst prognosis in terms of 1-year mortality?

A. Child-Pugh score 6
B. MELD score 8
C. Spontaneous bacterial peritonitis (SBP)
D. Presence of oesophageal varices
E. Hepatic encephalopathy

E4. A 48-year-old man has just been admitted to the intensive care unit (ICU) with bacterial pneumonia and multiorgan failure. He is significantly hypoxaemic and acidaemic, and so a decision is made to intubate him. A rapid sequence induction of anaesthesia (RSI) is performed, but intubation proves impossible over three attempts. It is not possible to ventilate him via face mask, nor via supraglottic airway. A surgical airway is required. You have a scalpel, a tracheal dilator and a size 6 tracheostomy tube.

How should tracheal access be achieved?

A. Send for percutaneous tracheostomy kit
B. Horizontal skin incision at laryngeal prominence; blunt dissection with dilator; vertical incision through cricothyroid membrane
C. Stab incision through skin at level of tracheal rings 1–2; enlarge incision with dilator
D. Vertical skin incision from below laryngeal prominence; blunt dissection with dilator; horizontal incision through cricothyroid membrane
E. Stab incision through skin and cricothyroid membrane; enlarge incision with dilator

E5. Which of the following intensive care unit (ICU) scoring systems has the highest sensitivity and specificity as measured by the best (i.e. closest to 100%) area under the receiver-operating curve?

A. ICNARC 2007
B. SAPS II
C. APACHE III
D. APACHE IV
E. $MPMII_0$

E6. A 24-year-old female patient with a background of mild asthma is admitted to the intensive care unit with Guillain-Barré syndrome. She is complaining of finding it difficult to breathe.

Which of the following measurements is it BEST guide to monitor this patient's respiratory function and identify her need for intubation and ventilation?

A. Peak flow measurement
B. Patient's self report of symptoms
C. Pulse oximetry (SpO_2)
D. Arterial blood gases (ABGs)
E. Measurement of vital capacity

E7. A 64-year-old patient has been confirmed as brain-stem dead after a massive intracerebral bleed, and the decision has been made to proceed to organ retrieval. The transplant team are en route. He is not on any vasoactive drugs and is euvolaemic. His heart rate is 74 bpm, blood pressure is 74/48 mmHg, cardiac index is 3.1 l/min/m^2 and his systemic vascular resistance index is 1014 dynes.sec/cm^5/m^2.

Which of the following is the BEST treatment for his hypotension?

A. Noradrenaline
B. Vasopressin
C. Adrenaline
D. Dobutamine
E. Tri-iodothyronine

E8. A 40-year-old woman presents to the emergency department with a 3-week history of lethargy, generalized weakness and pain on movement with polyuria and polydipsia. She has upper abdominal pain and nausea. She appears mildly confused. Past medical history includes a breast lump removed 5 years ago, but she is otherwise well. Her electrocardiogram (ECG) demonstrates a shortened QT interval.

What is the most important treatment to initiate first for this condition?

A. Intravenous (IV) calcium gluconate
B. IV saline with furosemide
C. IV magnesium replacement
D. IV potassium replacement
E. Insulin

E9. A patient suffered a subarachnoid haemorrhage (SAH) 6 days earlier, and the aneurysm has been successfully secured by interventional radiology. Her Glasgow Coma Score (GCS) has dropped acutely from 13 to 10 and does not improve over the next hour. Transcranial Doppler (TCD) ultrasonography is performed and the ratio of flow velocity in the right middle cerebral artery compared with the right internal carotid artery is 3.7.

What is the MOST likely cause of her reduction in GCS?

A. Hydrocephalus
B. Seizure
C. Cerebral oedema
D. Vasospasm
E. Rebleeding

E10. A 35-year-old is admitted with a cough productive of green sputum. Initial observations demonstrate he is pyrexial at 39.5°C and has a heart rate of 120, a blood pressure of 76/36, a respiratory rate of 28 and oxygen saturations of 88% in air. Investigations reveal a lactate of 4, a C-reactive protein of 289, a white cell count (WCC) of 18×10^9/l, a platelet count of 98×10^9/l and chest X-ray demonstrated left lower lobe consolidation.

Which is the BEST term to describe his condition?

A. Hypovolaemic shock
B. Sepsis
C. Systemic Inflammatory Response Syndrome
D. Septic Shock
E. Severe Sepsis

E11. A 43-year-old is admitted to the intensive care unit at midnight requiring intubation and ventilation for acute respiratory distress syndrome (ARDS) secondary to acute severe pancreatitis.

Following stabilization, what feeding regime should be commenced initially for this patient?

A. Oral nutrition
B. Nil by mouth
C. Nasogastric feed
D. Nasojejunal feed
E. Total parenteral nutrition

E12. Research has been carried out into a new drug, measuring the effect on heart rate before and after administration of the drug in a population of healthy volunteers. Which of the following statistical tests would be most appropriate?

A. Analysis of variance (ANOVA)
B. Wilcoxon signed rank
C. Friedman
D. Kruskal-Wallis
E. Fisher's exact test

E13. Which of the following methods of oxygen analysis is most appropriate for measuring partial pressure of oxygen in an arterial blood gas (ABG) sample?

A. A Severinghaus electrode
B. A mass spectrometer
C. A paramagnetic analyzer
D. A Clark electrode
E. A system based on the ratio of light absorption in oxygenated and deoxygenated haemoglobin

E14. A 66-year-old woman has been brought in to hospital after several days of vomiting and diarrhoea. Her vital signs are as follows: pulse rate 95; blood pressure 110/50; respiratory rate 24; capillary refill time 4 seconds; cool skin; core temperature 36.3°C. An intravenous cannula is sited and routine blood tests sent.
Which IV fluid prescription is appropriate in this scenario?

A. 1 l 0.18% NaCl with 4% glucose and 27 mmol/l potassium over 12 hours
B. 500 ml 0.9% NaCl over 15 minutes
C. 3 l Hartmann's solution (compound sodium lactate) over 12 hours
D. 500 ml 6% tetrastarch (hydroxyethyl starch 130/0.4) over 30 minutes
E. 500 ml 4.5% human albumin solution (HAS)

E15. Which of the following components of the APACHE IV scoring system has the greatest impact on the final score?

A. Acute physiological parameters
B. Acute diagnosis
C. Chronic health conditions
D. Age
E. Admission source and previous length of stay

E16. A 73-year-old patient who is anticoagulated for a mechanical aortic valve has just been admitted to the intensive care unit with severe bilateral pneumonia and has required intubation.
Which is the MOST appropriate stress ulcer regime for this patient to be commenced on admission?

A. Omeprazole (enterally)
B. Ranitidine (intravenously)
C. Enteral feed
D. Sucralfate (enterally)
E. None required

E17. You have just intubated a patient in the intensive care unit for airway protection due to a low Glasgow Coma Score. The patient is haemodynamically stable.

In conjunction with clinical assessment, which of the following methods is the best way to confirm placement of the endotracheal tube in the airway?

A. Colometric end tidal carbon dioxide detection
B. Chest X-ray
C. Visualisation of tube passing through the cords
D. Oesophageal detection device
E. Continuous end tidal capnography

E18. In which condition is treatment with magnesium LEAST useful?

A. Pre-eclampsia
B. Epilepsy
C. Refeeding syndrome
D. Asthma
E. Ventricular tachyarrhythmias

E19. Which of the following is LEAST likely to potentiate (prolong) the effect of a bolus of cis-atracurium?

A. Acidosis
B. Low plasma cholinesterase levels due to liver disease
C. Cyclosporin
D. Gentamicin
E. Verapamil

E20. A 64-year-old patient is in the intensive care unit undergoing haemofiltration for acute renal failure. She is receiving a blood transfusion as her haemoglobin is 67 g/l.

Which of the following complications is the MOST likely to occur from the blood transfusion?

A. Transfusion-related acute lung injury (TRALI)
B. Transfusion-associated circulatory overload (TACO)
C. Acute transfusion reaction (ATR)
D. Transfusion-transmitted infection (TTI)
E. Haemolytic transfusion reaction (HTR)

E21. A 56-year-old man presents after an out-of-hospital cardiac arrest. He suffered an unwitnessed cardiac arrest. His initial rhythm was asystole, and return of spontaneous circulation (ROSC) was achieved after a further 34 minutes. His Glasgow Coma Score (GCS) is 3 in the emergency department (ED). He is admitted to the intensive care unit for targeted temperature management.

Which of the following is MOST useful in prognosticating poor neurological outcome in this patient?

A. Electroencephalogram (EEG) with an unreactive baseline at 72 hours post-rewarming
B. Absent bilateral N20 somatosensory evoked potentials (SSEPs) 24 hours post-rewarming
C. Absent corneal and pupillary reflexes, bilateral N20 SSEPs and GCS 3 at 72 hours post-arrest
D. Clinical history, fixed dilated pupils and a GCS motor score of 1 on admission to ED
E. Loss of grey-white matter differentiation on admission computed tomography (CT)

E22. You are asked by the nurses in the intensive care unit to review a 50-year-old man who was admitted earlier in the day. He has septic shock secondary to a community acquired pneumonia and is intubated and ventilated. He has just developed a tachyarrhythmia, with a heart rate of 175 and a blood pressure of 70/30. The electrocardiogram (ECG) shows fast atrial fibrillation (AF).

Which of the following is the most appropriate initial management?

A. Amiodarone IV
B. Digoxin IV
C. Beta-blockade IV
D. Synchronized DC shock
E. Non-synchronized DC shock

E23. With regard to oxygen delivery to the tissues, assuming cardiac output remains the same, which of the following patients will have the greatest oxygen delivery?

A. SaO_2 100%, Hb 100 g/l, PaO_2 12 kPa
B. SaO_2 100%, Hb 150 g/l, PaO_2 12 kPa
C. SaO_2 100%, Hb 100 g/l, PaO_2 85 kPa
D. SaO_2 80%, Hb 150 g/l, PaO_2 12 kPa
E. SaO_2 80%, Hb 150 g/l, PaO_2 85 kPa

E24. You are called urgently to help the emergency department with a 20-year-old gentleman who has taken a mixture of recreational drugs including MDMA (ecstasy), cocaine, LSD and alcohol. He was markedly agitated in the community and has already been intubated. He has a temperature of 39.5°C, is tachycardic and flushed, and had a seizure before intubation. You diagnose likely serotonin syndrome.

Which of the following is the best way to manage him?

A. Supportive treatment only
B. Dantrolene 2.5 mg/kg
C. Benzodiazepine infusion
D. Sedation with propofol and fentanyl
E. Bicarbonate infusion

E25. All of the following patients have sustained a head injury. Which of them is LEAST likely to require a computed tomography (CT) head within 1 hour?

A. A 40-year-old with a Glasgow Coma Score (GCS) of 142 hours after falling down a flight of stairs
B. A 30-year-old with a GCS of 12 on arrival to the emergency department after falling down a flight of stairs
C. A 35-year-old with a GCS of 15 who has vomited twice after falling down a flight of stairs
D. A 30-year-old with a GCS of 15 who had a brief generalized seizure that self-terminated after falling down a flight of stairs
E. A 40-year-old with a GCS of 15 who takes warfarin for atrial fibrillation who fell down a flight of stairs

E26. You are called to the emergency department to assist with the management of a 25-year-old man who is fitting. He is a known epileptic and takes regular levetiracetam. He has been fitting for 25 minutes and has not responded to diazepam 10 mg rectally given pre-hospital, nor to two 4-mg intravenous (IV) doses of lorazepam. His airway is not currently at risk, and his SpO_2 is 97% on 15 l oxygen via a non-rebreathe mask.

Which of the following is the most appropriate next treatment?

A. Lorazepam 0.1 mg/kg IV bolus
B. Phenytoin 18 mg/kg IV bolus
C. Thiamine 250 mg IV bolus
D. Thiopentone 4 mg/kg IV bolus and intubation
E. Glucose 50% 50 ml IV bolus

E27. You have been asked to transfer an intubated 64-year-old with a sub-arachnoid haemorrhage to the regional neuro intensive care unit for decompressive surgery. His Glasgow Coma Score was 4/15 before intubation (M2, V1, E1). He is sedated and ventilated.

Which of the following is the LEAST important to monitor on the transfer?

A. Central venous pressure (CVP)
B. Invasive blood pressure (BP)
C. End-tidal CO_2 (ETCO$_2$)
D. Temperature
E. Airway pressure

E28. You are administering prophylactic heparin to an intensive care patient as thromboprophylaxis.

Which of the following complications of heparin is associated with the highest mortality?

A. Haemorrhage
B. Type I thrombocytopenia (non-immune mediated)
C. Type II thrombocytopenia (immune mediated)
D. Arrhythmias
E. Osteoporosis

E29. You are called to the emergency department as part of the major trauma team. A 33-year-old woman has been run over by a bus and has an unstable pelvic fracture. She has the following findings on primary survey:

Airway: Intact

Breathing: Respiratory rate 30/min; SpO$_2$ 97% on 15 l oxygen; bilateral air entry, no added sounds

Circulation: Heart rate: 135; blood pressure 75/45; cold peripheries; no external haemorrhage; deformed pelvis

Disability: GCS 15/15; pupils equal and reactive; in pain

Whole body computed tomography (CT) scan shows a isolated, 'open-book' pelvic injury. She has had intravenous access established and oxygen given but no other treatment.

What is the most appropriate next step in her management?

A. Interventional radiology for angio-embolization

B. Immediate transfer to theatre for pelvic fixation

C. Attempted stabilization of pelvic fracture using an external binder

D. Rapid transfusion of 1000 ml crystalloid before blood products

E. Further plain radiographs of her pelvis including Judet views

E30. You have just inserted a radial arterial line into a patient, but when you connect it to the monitor, the display shows a hyper-resonant trace:

Which of the following is the most likely reason for this?

A. Using a long arterial cannula

B. Using a wide arterial cannula

C. Using a stiff cannula

D. Clot formation in the cannula

E. Air bubble in the catheter tubing

Exam E: Answers

E1. A 24-year-old woman with normal lungs is breathing 60% oxygen via a face mask. On arterial blood gas analysis, her $PaCO_2$ is 4.5 kPa.

Which of the following is the most likely value of the partial pressure of oxygen in her alveoli?

A. 42 kPa
B. 51 kPa
C. 54 kPa
D. 60 kPa
E. 80 kPa

Answer: B

Short explanation

The alveolar gas equation is required to estimate the partial pressure of oxygen in the alveoli, as follows:

$$p_AO_2 = FiO_2 (P_{atm} - P_{H_2O}) - p_aCO_2/RQ$$
$$= 0.6 (101 - 6.25) - 4.5/0.8$$
$$= 51.225 \text{ kPa}$$

Long explanation

Calculation of the partial pressure of oxygen in the alveoli (p_AO_2) is important for the estimation of the alveolar-arterial (A-a) gradient. The alveoli contain a mixture of gases: oxygen, nitrogen, carbon dioxide and water vapour. Determining the partial pressures of each of these is necessary to accurately estimate the p_AO_2. As the p_ACO_2 is as difficult to measure as the p_AO_2, it is assumed to be equal to the p_aCO_2 as CO_2 is highly soluble and diffuses quickly. The partial pressure of water is equal to the saturated vapour pressure at atmospheric pressure and 37°C, as inspired air is quickly saturated with water vapour as is travels through the upper airways.

The equation is as follows:

$$p_AO_2 = FiO_2 (P_{atm} - P_{H_2O}) - p_aCO_2/RQ$$
$$= 0.6 (101 - 6.25) - 4.5/0.8$$
$$= 51.225 \text{kPa}$$

p_AO_2 partial pressure of oxygen in the alveoli
FiO_2 fraction of inspired oxygen
P_{atm} atmospheric pressure
P_{H_2O} saturated vapour pressure of water at atmospheric pressure and body temperature
p_aCO_2 partial pressure of CO_2 in arterial blood
RQ   respiratory quotient (usually 0.8 for mixed diet)

Reference
Cruikshank S, Hirschauer N. The alveolar gas equation. *Contin Educ Anaesthesia, Crit Care Pain*. 2004;4(1):24–27.

E2. A 64-year-old male patient is admitted intubated and ventilated after an out-of-hospital cardiac arrest. Unfortunately, he has suffered a significant hypoxic brain injury and withdrawal of life-sustaining therapy is being considered.

When making ongoing treatment/withdrawal decisions for this patient, which of the following is the BEST method?

A. Treat/withdraw treatment after consultation with an independent mental capacity advisor (IMCA)
B. Treat/withdraw treatment with consent from the patient's lasting power of attorney (LPA)
C. Treat/withdraw treatment after consultation with the patient's family to establish the patient's wishes
D. Treat/withdraw treatment according to patient's wishes using a valid advanced directive
E. Treat/withdraw treatment according to the team's opinion of the patient's best interests

Answer: D

Short explanation
A valid advance decision by the patient that applies to the particular circumstances experienced by the patient during that specific situation is a legal document that clinicians must follow. All other options are opinions of third parties of what they think the patient would wish for.

Long explanation
The Mental Capacity Act of 2005 states that all patients must be assumed to have capacity, and it is the clinician's responsibility to prove whether patients lack it. A two-stage capacity test should be used. The first stage is to identify an impairment or disturbance (temporary or permanent) of the patient's mind or brain. If this is present, the second stage is to identify whether this affects his or her ability to make the specific decision required at that point in time. To have capacity, patients must be able to take in the information given to them, retain it, weigh it upto make a decision and to communicate that decision back.

Capacity is decision dependent, and a lack can be temporary or permanent. All practical steps need to be made to aid patients to achieve capacity.

All decisions regarding treatment and care for patients lacking capacity should be made using the principle of the patient's best interests. Only emergency treatment (the least restrictive and invasive option available) can be provided without identifying patient's likely wishes for ongoing care. This information can be obtained from a patient's pre-existing advanced directive, which needs to be valid and cover the particular circumstances that the patient is currently experiencing. Another person may have previously been granted the ability to consent for patients if they lack capacity. This can be a lasting power of attorney (LPA), registered with the Office of the Public Guardian, who has been appointed by the patient themselves or a deputy appointed by the Court of Protection. Decisions regarding life-sustaining treatment can only be made by an LPA if this has been formally documented within the agreement. They are unable to demand specific treatments to be instigated if clinicians feel it is not in the patient's best interests.

Clinicians should consult the patient's family, friends or carers to aid identification of the patient's wishes if none of these are present. If patients lack an advocate, an IMCA should be consulted when making decisions regarding major medical treatments. Information and opinions gained from these discussions should then aid clinicians to make treatment decisions according to the best interests of the patient.

Reference

Mental Capacity Act 2005 Code of Practice. The Stationery Office on behalf of the Department for Constitutional Affairs, 2007. https://www.justice.gov.uk/downloads/protecting-the-vulnerable/mca/mca-code-practice-0509.pdf (accessed October 2014).

E3. A patient with a history of cirrhotic liver disease is referred to critical care for organ support.

Which of the following features suggests the worst prognosis in terms of 1-year mortality?

A. Child-Pugh score 6
B. MELD score 8
C. Spontaneous bacterial peritonitis (SBP)
D. Presence of oesophageal varices
E. Hepatic encephalopathy

Answer: C

Short explanation

Decompensated cirrhosis has a worse prognosis than compensated cirrhosis. Ascites, portal hypertensive bleeding, encephalopathy or jaundice are features of decompensation. Complications of one of these features (e.g. SBP, hepatorenal syndrome) are associated with an even worse prognosis. A Child-Pugh score of 6 and a MELD score of 8 are low and associated with good 1-year survival.

Long explanation

Hepatic cirrhosis is the end-stage of chronic liver disease causing fibrosis and formation of nodules. Once established, it is effectively irreversible and treatable only by transplantation. The commonest causes of cirrhosis in the West are viral hepatitis, alcohol abuse and non-alcoholic steatohepatitis (NASH). Patients with cirrhosis are susceptible to many complications and their life expectancy is dramatically reduced.

Initially following the development of cirrhosis, patients may remain asymptomatic for many years. This phase is known as compensated cirrhosis and has a median survival time of more than 12 years. During the phase of compensated cirrhosis, portal venous pressure may be normal or slightly raised but will increase as the disease progresses. This can lead to the formation of oesophagogastric varices which are initially asymptomatic.

The development of ascites, variceal bleeding, encephalopathy or jaundice are markers of decompensated disease. Patients progress from compensated to decompensated cirrhosis at a rate of around 5 to 7% per year. Median survival of patients with decompensated cirrhosis is approximately 2 years.

Complications of decompensated cirrhosis markedly worsen prognosis. Such complications include spontaneous bacterial peritonitis, hepatorenal syndrome and hepatopulmonary syndrome. Median survival in such patients is usually less than 6 months.

The Child-Pugh score has been shown to be the best predictor of mortality in hepatic cirrhosis. Scoring is based on five factors: bilirubin, albumin, international normalized ratio (INR) and the presence of ascites and encephalopathy. Patients with scores of 5–6 (class A) have a 100% 1-year survival according to the model. The MELD score is similar but based on bilirubin, creatinine and INR. A MELD score of ≤9 is associated with a predicted 3-month mortality of 1.9%

Reference

D'Amico G, Garcia-Tsao G, Pagliaro L. Natural history and prognostic indicators of survival in cirrhosis: a systematic review of 118 studies. *J Hepatol.* 2006;44(1): 217.

E4. A 48-year-old man has just been admitted to the intensive care unit (ICU) with bacterial pneumonia and multiorgan failure. He is significantly hypoxaemic and acidaemic, and so a decision is made to intubate him. A rapid sequence induction of anaesthesia (RSI) is performed, but intubation proves impossible over three attempts. It is not possible to ventilate him via face mask, nor via supraglottic airway. A surgical airway is required. You have a scalpel, a tracheal dilator and a size 6 tracheostomy tube.

How should tracheal access be achieved?

A. Send for percutaneous tracheostomy kit
B. Horizontal skin incision at laryngeal prominence; blunt dissection with dilator; vertical incision through cricothyroid membrane
C. Stab incision through skin at level of tracheal rings 1–2; enlarge incision with dilator
D. Vertical skin incision from below laryngeal prominence; blunt dissection with dilator; horizontal incision through cricothyroid membrane
E. Stab incision through skin and cricothyroid membrane; enlarge incision with dilator

Answer: E

Short explanation

This is a 'can't intubate, can't oxygenate' (CICO) situation in a critically ill patient. It is an extreme medical emergency, and a rapid surgical airway is required. The Difficult Airway Society (DAS) guidelines for this situation is to identify the cricothyroid membrane, use a stab incision through skin and cricothyroid membrane, enlarge the hole and intubate it.

Long explanation

The Difficult Airway Society (DAS) publish guidelines for management of the airway in a variety of scenarios. Most patients in a critical care setting require RSI due to critical illness, gastrointestinal pathology or the risk of aspiration. The DAS guidelines for RSI contain three strategic levels or plans. Plan A is the initial intubation plan, with a list of suggested manoeuvres to improve the chances of successful intubation, such as positioning and equipment changes. A maximum of three attempts is allowed. Plan B is maintenance of oxygenation using a supraglottic airway device.

Plan C is the maintenance of oxygenation with face-mask ventilation for the time it takes for induction drugs to wear off and the patient to wake up and regain control of his or her airway and breathing. This is rarely appropriate in ICU settings. Initial oxygenation attempts are with a face mask and airway adjuncts as required. If oxygenation is not possible, then a supraglottic airway device should be inserted and oxygenation attempted again. If oxygenation is not possible by any means, then the situation has become a CICO scenario. This is a medical emergency and further appropriate help should be sought.

Cannula cricothyroidotomy is no longer recommended by the DAS, which now recommends only surgical cricothyroidotomy in instances of CICO. The recommended technique for surgical cricothyroidotomy is to identify the cricothyroid membrane and make a stab incision through the skin and membrane. This hole should be enlarged with forceps, a tracheal dilator or the handle of the scalpel, and intubated, preferably whilst maintaining upwards traction of the caudal edge of the trachea with a tracheal hook, if available. In less emergent circumstances, a more measured approach to the cricothyroidotomy may be taken. This involves making a vertical skin incision from below the laryngeal prominence, blunt dissection and then a horizontal incision through the cricothyroid membrane.

References

Andersson ML, Møller AM, Pace NL. Emergency cricothyroidotomy for airway management. *Cochrane Database of Systematic Reviews* 2014: 1.

Difficult Airway Society intubation guidelines. http://www.das.uk.com/guidelines/das_intubation_guidelines (accessed February 2016).

Paix BR, Griggs WM. Emergency surgical cricothyroidotomy: 24 successful cases leading to a simple 'scalpel–finger–tube' method. *Emerg Med Australas*. 2012:24; 1:23–30.

E5. Which of the following intensive care unit (ICU) scoring systems has the highest sensitivity and specificity as measured by the best (i.e. closest to 1.0) area under the receiver-operating curve?

A. ICNARC 2007
B. SAPS II
C. APACHE III
D. APACHE IV
E. $MPMII_0$

Answer: C

Short explanation

The area under the receiver-operating curves are: APACHE III = 0.9, APACHE IV = 0.88, ICNARC 2007 = 0.87, SAPS II = 0.86 and $MPMII_0$ = 0.82

Long explanation

The area under the receiver-operating curve (AUROC) is used to compare a test's ability to distinguish true and false positives. A high (100%) AUROC indicates a test will detect all of the true positives (100% sensitivity) without any false positives (100% specificity).

Scoring systems have been developed in ICU to attempt to predict outcomes and model mortality rates from different physiological parameters. Those listed are some of the commonest used worldwide and have been developed using large patient databases.

ICNARC 2007 was developed using data from 216,626 UK ICU patients, the APACHE III score used 17,440 US patients in 1991, whereas APACHE IV used 110,558 US patients in 2006. SAPS II and $MPMII_0$ were developed across Europe and America in 1993.

These large databases are collected over several years either in one country or one part of the world. Because of the rapid improvements in ICU care over the past few decades and the differences in ICU care worldwide, care must be taken to appreciate the differences from the population the model was developed for and the patient in front of you.

There are always challenges when trying to fit a universal scoring system to a patient population and even the best scoring systems in use have only a 0.9 or 90% AUROC – that is, can only correctly identify a survivor from a non-survivor 90% of the time. In addition, these models have been developed primarily for use in research (e.g. comparing severity of illness in both arms of an RCT) or in benchmarking (e.g. to calculate standardized mortality ratios). Therefore none of the scoring systems should be used to base clinical decisions on their predicted scores.

Reference
Palazzo M. Severity of illness and likely outcome from critical illness. In Bersten AD, Soni N. *Oh's Intensive Care Manual*. 6th edition. Edinburgh: Butterworth- Heinemann, 2009, pp 17–30.

E6. A 24-year-old female patient with a background of mild asthma is admitted to the intensive care unit with Guillain-Barré syndrome. She is complaining of finding it difficult to breathe.

Which of the following measurements is it BEST guide to monitor this patient's respiratory function and identify her need for intubation and ventilation?

A. Peak flow measurement
B. Patient's self report of symptoms
C. Pulse oximetry (SpO_2)
D. Arterial blood gases (ABGs)
E. Measurement of vital capacity

Answer: E

Short explanation
All of these measurements monitor respiratory function. This patient may require ventilation due to Guillain-Barré syndrome rather than her mild asthma. Desaturation and hypercapnia are late signs of respiratory compromise. Self-reporting of symptoms is subjective, and peak flow measurements assess airflow obstruction in asthma. Vital capacity is thought to be one of the best tests to guide the need for intubation and ventilation in patients with neuromuscular disease.

Long explanation
Up to about one-third of patients with Guillain-Barré syndrome will need intubation and ventilation for respiratory support or airway protection from bulbar dysfunction. Respiratory function and monitoring for bulbar dysfunction should be monitored in all these patients.

Desaturation and hypercapnia are signs of established respiratory failure. Therefore both ABGs and SpO_2 fail to give an early indication of worsening respiratory function and to predict the need for increased respiratory support until patients are already in a compromised state.

Although patients can give an indication of their respiratory symptoms and any progressive deterioration, their reports are subjective and so cannot be relied on to guide the need for intubation and ventilation.

Objective measurements often quoted for their use in monitoring the degree of muscle weakness and to predict the need for intubation and ventilation include:

- Spirometry, specifically vital capacity. This is a measure of the maximal volume of air that patients exhale after full inspiration, that is total lung capacity minus the residual volume.
- Maximal inspiratory pressure (MIP) is the maximal negative pressure generated during inspiration from functional residual capacity (FRC) against an occluded airway. It can be used as a marker to reflect the strength of the diaphragm, the intercostals and accessory inspiratory muscles.
- Maximal expiratory pressure (MEP) is the maximal pressure generated during expiration following full inspiration. Although expiration at rest is passive, forced expiration uses accessory expiratory muscles such as those of the abdominal wall. It can be used as a marker of the strength of these expiratory muscles.

Values thought to be associated with predicting the need for intubation include:

- Forced vital capacity <15–20 mL/kg
- MIP $\geq$30 cmH$_2$O
- MEP <40 cmH$_2$O
- A reduction of >30% from baseline in any of the above

Serial test should be performed (initially 6 hourly) to review not only the actual value of these measurements but also their rate of decline in assessing the onset and speed of development of respiratory insufficiency.

Peak flow measurements assess the reduction in maximal airflow due to obstructive airways disease.

References

BMJ Best Practice. Guillain Barre syndrome. Treatment: step-by-step. http://best practice.bmj.com/best-practice/monograph/176/treatment/step-by-step.html (accessed October 14).

West JB. *Respiratory Physiology: The Essentials*. 8th edition. Philadelphia: Lippincott Williams & Wilkins, 2008, pp 13–24.

E7. A 64-year-old patient has been confirmed as brain-stem dead after a massive intracerebral bleed, and the decision has been made to proceed to organ retrieval. The transplant team are en route. He is not on any vasoactive drugs and is euvolaemic. His heart rate is 74 bpm, blood pressure is 74/48 mmHg, cardiac index is 3.1 l/min/m^2 and systemic vascular resistance index is 1014 dynes.sec/cm^5/m^2.

Which of the following is the BEST treatment for his hypotension?

A. Noradrenaline
B. Vasopressin
C. Adrenaline
D. Dobutamine
E. Tri-iodothyronine

Answer: B

Short explanation

This patient has an adequate cardiac index but a reduced systemic vascular resistance, suggesting the hypotension is as a result of vasodilatation. After ensuring adequate fluid status, a vasoconstrictor would be the drug of choice, rather than an inotrope such as dobutamine or tri-iodothyronine (T3). Although adrenaline vasoconstricts, it also has beta activity. Vasopressin is preferable to noradrenaline as first line for these patients.

Long explanation

There are numerous cardiovascular changes that occur in association with brain-stem death. Cushing's response, presenting with hypertension and bradycardia, can occur around the moment of brain-stem death. There is often a subsequent period of sympathetic stimulation, with vasoconstriction, tachycardia and hypertension, which can vary in severity and duration. After this, a loss of sympathetic tone occurs, causing marked vasodilatation and relative hypovolaemia along with myocardial depression. Organ function rapidly deteriorates if these abnormalities are not adequately controlled.

In the presence of cardiovascular abnormalities, cardiac output monitoring should be used to guide treatment. The period of intense sympathetic stimulation can be treated with a reduction in the level of inotropic and vasoactive support that may be pre-existing. It may be of short duration so if further treatment is required for hypertension or tachycardia, short-acting agents are preferred.

The aim of treatment for the subsequent phase where hypotension predominates is to maintain the cardiac output whilst minimizing any increase in cardiac work and myocardial oxygen demand. This is to avoid depleting cardiac adenosine triphosphate (ATP) stores, which occurs with the use of inotropes and vasopressors. Administration of drugs with increased beta activity have been associated worse graft outcomes in recipients.

Haemodynamic monitoring provides useful information to guide decisions regarding fluid, inotrope and vasopressor requirements. Initial treatment involves ensuring an adequate circulating volume is achieved. Vasodilation, which is a common cause of hypotension, should be treated with vasopressors such as vasopressin or noradrenaline. Guidelines regarding the management of brain-stem-dead organ donors from the Intensive Care Society suggest vasopressin to be superior to noradrenaline because it is less likely to cause pulmonary hypertension and metabolic acidosis. Canadian guidelines also recommend it as the first-line agent for haemodynamic support in these patients. Adrenaline has vasoconstrictor activity but also stimulates beta-adrenoceptors and should be avoided if possible. Low cardiac output states may be treated with inotropes including dobutamine, adrenaline, dopamine and tri-iodothyronine.

References

Intensive Care Society. Intensive Care Society Guidelines for Adult Organ and Tissue Donation. 2005 http://www.ics.ac.uk/ics-homepage/guidelines-and-standards (accessed December 2014).

Intensive Care Society of Ireland. Diagnosis of Brain Death and Medical Mangement of the Organ Donor. Guidelines for Adult Patients. 2010. https://www.anaesthesia.ie/archive/ICSI/ICSI%20Guidelines%20MAY10.pdf (accessed August 15).

Shemie SD, Ross H, Pagliarello J, et al. Organ donor management in Canada: recommendations of the forum on Medical Management to Optimize Donor Organ Potential. *CMAJ*. 2006;174(6):S13–S30.

E8. A 40-year-old woman presents to the emergency department with a 3-week history of lethargy, generalized weakness and pain on movement with polyuria and polydipsia. She has upper abdominal pain and nausea. She appears mildly confused. Past medical history includes a breast lump removed 5 years ago, but she is otherwise well. Her electrocardiogram (ECG) demonstrates a shortened QT interval.

What is the most important treatment to initiate first for this condition?

A. Intravenous (IV) calcium gluconate
B. IV saline with furosemide
C. IV magnesium replacement
D. IV potassium replacement
E. Insulin

Answer: B

Short explanation

The case describes hypercalcaemia, which typically presents with a short history of renal, neurological and abdominal symptoms with bone pain. ECG typically demonstrates a shortened QT interval in severe cases. A common cause of hypercalcaemia would include breast cancer in this age group. First-line therapy is fluid replacement and loop diuretics followed by bisphosphonates and treatment of the underlying cause.

Long explanation

The commonest causes of hypercalcaemia are primary hyperparathyroidism and malignancy – classically breast, lung or myeloma. It can also occur secondary to drugs such as diuretics, chronic kidney disease or endocrine disorders. Mild hypercalcaemia is relatively common depending on age and the normal local laboratory reference values. Severe hypercalcaemia presents less frequently but often includes neurological, cardiac and renal symptoms.

A serum calcium and albumin level should be checked to allow calculation of an adjusted serum calcium. Other investigations are necessary to identify a cause and include parathyroid hormone levels, imaging (e.g. chest radiography) and the checking of other electrolyte levels, particularly magnesium.

The initial management should include IV saline to ensure adequate hydration and to dilute the calcium. Patients are often sodium deplete as a result of excess sodium loss. Therefore treatment can be enhanced by introducing a loop diuretic such as furosemide. Once the patient is well hydrated, an IV bisphosphonate such as pamidronate should be administered to reduce osteoclast activity and reduce calcium release from bone. Oral bisphosphates reduce the gut absorption of calcium. Haemofiltration may be required to remove calcium from the plasma on ICU.

Treatment of the underlying cause may include surgical resection of parathyroid glands or malignancies secreting parathyroid hormone related peptide (PTHrP).

References

National Institute for Health and Care Excellence (NICE). *Clinical knowledge summary: hypercalcaemia.* = 2014. http://cks.nice.org.uk/hypercalcaemia (accessed April 2015).

Parikh M, Webb S. Cations: potassium, calcium, and magnesium. *Contin Educ Anaesth Crit Care Pain.* 2012; doi: 10.1093/bjaceaccp/mks020.

E9. A patient suffered a subarachnoid haemorrhage (SAH) 6 days earlier, and the aneurysm has been successfully secured by interventional radiology. Her Glasgow Coma Score (GCS) has dropped acutely from 13 to 10 and does not improve over the next hour. Transcranial Doppler (TCD) ultrasonography is performed and the ratio of flow velocity in the right middle cerebral artery compared with the right internal carotid artery is 3.7.

What is the MOST likely cause of her reduction in GCS?

A. Hydrocephalus
B. Seizure
C. Cerebral oedema
D. Vasospasm
E. Rebleeding

Answer: D

Short explanation

All of the preceding are causes of neurological deterioration post SAH. Rebleeding is unlikely because the aneurysm has been secured. There were no signs of seizure, and it is hoped that post-ictal symptoms would improve over an hour. A computed tomography (CT) scan would be required to exclude hydrocephalus or cerebral oedema, but the ratios of flow velocities identified on transcranial Doppler in the patient are consistent with a diagnosis of vasospasm.

Long explanation

Complications of SAH include the following:

- Neurological
  - Focal neurological signs due to the effect of haemorrhage itself.
  - Rebleeding rates are highest immediately after rupture and decrease with time, so securing of the ruptured aneurysm should occur as soon as possible to minimize this risk.
  - Delayed ischaemic neurological deficit (DIND), any neurological deterioration that lasts >1 hour that is due to ischaemia and has no other cause. This occurs in about 60% of patients.
    - Vasospasm is thought to be the cause of most cases of DIND and occurs between day 4 and 14 (although maximally between day 7 and 10) and lasts for several days. It occurs in up to 70% of patients, but only about half of all episodes are associated with neurological deficit. Transcranial Doppler ultrasonography can be used to detect vasospasm by measuring flow velocities in the cerebral arteries. A flow velocity >120 cm/s or a ratio of flows >3 between the middle cerebral artery compared with the ipsilateral internal carotid artery indicates vasospasm.
    - Microvascular thrombosis.
  - Seizures occur in <10% of patients
  - Cerebral oedema
  - Raised intracranial pressure
  - Hydrocephalus occurs in approximately 20 to 30% of patients, most commonly in those with more severe grades of SAH and those patients who have a high 'blood load' in the subarachnoid space and ventricles. It is most likely to occur early (within the first 3 days after SAH), but its presentation can be delayed.

- ○ Neurological decline can also occur secondary to systemic complications:
  - Hypoxia
  - Hypotension
  - Pyrexia and infection
  - Electrolyte disturbance
- Systemic complications
  - ○ Cardiovascular complications occurring secondarily to a surge in sympathetic stimulation and catecholamine release
    - Stunned myocardium
    - LV dysfunction (systolic and diastolic)
    - Regional wall motion abnormalities
    - Cardiogenic shock
    - Arrhythmias
    - Release of troponin
  - ○ Respiratory
    - Neurogenic/cardiogenic pulmonary oedema
    - Pneumonia
    - ARDS
    - Atelectasis
  - ○ Electrolyte derangements
    - Hyponatraemia
      - Syndrome of inappropriate antidiuretic hormone secretion
      - Haemodilution
      - Cerebral salt wasting syndrome
    - Hypernatraemia
      - Diabetes insipidus
    - Hypokalaemia
    - Hypomagnesaemia
  - ○ Fever
  - ○ Anaemia
  - ○ Hyperglycaemia
  - ○ Thromboembolic disease
    - Deep vein thrombosis/pulmonary embolism
  - ○ Heparin induced thrombocytopenia
  - ○ Infection
  - ○ Acute kidney injury
  - ○ Gastrointestinal/hepatic dysfunction

References

Caricato A, Maviglia R, Antonelli M. Systemic complications after subarachnoid haemorrhage. In Vincent J-L. *Yearbook of Intensive Care and Emergency Medicine: Annual Update.* Germany: Springer, 2010, pp 419–427.

Luoma A, Reddy U. Acute management of aneurysmal subarachnoid haemorrhage. *Contin Educ Anaesth Crit Care Pain.* 2013;13(2):52–58.

E10. A 35-year-old is admitted with a cough productive of green sputum. Initial observations demonstrate he is pyrexial at 39.5°C and has a heart rate of 120, a blood pressure of 76/36, a respiratory rate of 28 and oxygen saturations of 88% in air. Investigations reveal a lactate of 4, a C-reactive protein of 289, a white cell count (WCC) of $18 \times 10^9/l$, a platelet count of $98 \times 10^9/l$ and chest X-ray demonstrated left lower lobe consolidation.

Which is the BEST term to describe his condition?

A. Hypovolaemic shock
B. Sepsis
C. Systemic inflammatory response syndrome
D. Septic shock
E. Severe sepsis

Answer: E

Short explanation

This patient has severe sepsis secondary to likely pneumonia. Sepsis is defined as a systemic inflammatory response to the presence of confirmed or suspected infection. This patient has evidence of the presence of a systemic inflammatory response syndrome but also of organ dysfunction and tissue hypoperfusion, thus making this severe sepsis. Although he has evidence of sepsis induced hypotension he has not been adequately fluid resuscitated to confirm whether he has septic shock yet.

Long explanation

The Surviving Sepsis Committee (SSC) have produced guidelines regarding the management of severe sepsis and septic shock.

- Sepsis is the presence of systemic upset developing from a suspected or confirmed source of infection. The SSC define systemic manifestations of infection as:
  - Observations:
    - Temperature >38.3°C or <36°C
    - Heart rate >90 bpm
    - Tachypnoea
    - Reduced consciousness or confusion
    - Fluid filling (>20 ml/kg over 24 hours) or oedema
  - Investigations:
    - WCC $>12 \times 10^9/l$ or $<4 \times 10^9/l$
    - >10% immature leucocytes on blood film
    - CRP/procalcitonin >2 SD above normal
    - Plasma glucose >7.7 mmol/l (unless diabetic)
- Severe sepsis occurs when sepsis-induced organ dysfunction or tissue hypoperfusion results from the underlying sepsis.
  - Sepsis-induced tissue hypoperfusion is evidenced by any of the following:
    - Sepsis-induced hypotension
      - Systemic blood pressure <90 mmHg or decrease >40 mmHg (adults) or less than 2 SD below normal for age
      - Mean arterial pressure <70 mmHg
    - Urine output <0.5 ml/kg/hr lasting for >2 hours despite appropriate fluid filling
    - Lactate >1

- o Variables demonstrating organ dysfunction:
  - Arterial hypoxemia (PaO_2/FiO_2 <300)
  - Creatinine >176.8 μmol/l or increased by >44.2 μmol/l
  - Oliguria (as above)
  - Platelet count <100 × 10^9/l
  - Coagulopathy (international normalized ratio >1.5 or activated partial thromboplastin time >60 s)
  - Bilirubin >34.2 μmol/l
  - Ileus
- Septic shock is the presence of sepsis-induced hypotension existing following adequate fluid resuscitation.
- Systemic inflammatory response syndrome (SIRS) is defined as ≥2 of the following variables:
  - o Temperature >38°C or <36°C
  - o Heart rate >90 bpm
  - o Respiratory rate >20/min
  - o WCC >12 × 10^9/l or <4 × 10^9/l or >10% immature form
  - o SIRS can be a physiological syndrome and can be caused by many triggers such as trauma, inflammation, ischaemia, trauma, surgery and not just infection.

References
Bone RC, Balk RA, Cerra FB, et al. Definitions for sepsis and organ failure and guidelines for the use of innovative therapies in sepsis. The ACCP/SCCM Consensus Conference Committee. American College of Chest Physicians/Society of Critical Care Medicine. *Chest*. 1992;101(6)1644–1655.

Dellinger RP, Levy MM, Rhodes A, et al. Surviving Sepsis Campaign: international guidelines for management of severe sepsis and septic shock, 2012. *Intensive Care Med*. 2013;39(2):165–228.

E11. A 43-year-old is admitted to the intensive care unit at midnight requiring intubation and ventilation for acute respiratory distress syndrome (ARDS) secondary to acute severe pancreatitis.

Following stabilization, what feeding regime should be commenced initially for this patient?

A. Oral nutrition
B. Nil by mouth
C. Nasogastric feed
D. Nasojejunal feed
E. Total parenteral nutrition

Answer: C

Short explanation
This patient has acute severe pancreatitis and should receive nutritional support. Enteral nutrition should be given first line. This patient is intubated, so oral nutrition is not an option. Nasogastric (NG) and nasojejunal (NJ) feeding have been shown to have comparable outcomes; however, NG feeding is technically easier to achieve.

Long explanation
The goals surrounding nutrition in patients with pancreatitis has changed significantly over the past couple of decades. The traditional belief that the pancreas needed to be rested has now been superseded by the recognition of the importance of enteral

nutrition in improving patient outcomes. Patients with mild acute pancreatitis may not require additional nutritional support, but nutritional support is recommended for those with severe disease.

The American College of Gastroenterology has produced guidelines regarding the management of patients with acute pancreatitis. Along with the International Consensus Guidelines for Nutrition Therapy in Pancreatitis they recommend enteral feeding as first line in all patients with severe acute pancreatitis, and total parenteral nutrition (TPN) should be reserved for those patients who are intolerant to enteral feeding. This is due to the improved morbidity (decreased organ failure and infection rates) and mortality seen with enteral feeding compared with TPN.

Nasojejunal (NJ) feeding was previously thought to be superior to alternative forms of enteral feeding; however, more recent evidence suggests that NG feeding is comparable to NJ feeding in terms of patient outcomes and complication rates. NG feeding is also technically easier to achieve because it does not require specialist skills to insert the feeding tube. Therefore a reasonable strategy would be to trial NG feeding, proceed to NJ feeding if that is unsuccessful, and TPN should be reserved for those patients who continue to fail to meet their nutritional requirements by enteral means or if enteral nutrition is contraindicated for another reason.

References

Chang YS, Fu HQ, Xiao YM, Liu JC. Nasogastric or nasojejunal feeding in predicted severe acute pancreatitis: a meta-analysis. *Crit Care*. 2013,17(3):R118.

Mirtallo JM1, Forbes A, McClave SA, et al.; International Consensus Guideline Committee Pancreatitis Task Force. International consensus guidelines for nutrition therapy in pancreatitis. *JPEN J Parenter Enteral Nutr*. 2012;36(3):284–291.

Tenner S, Baillie J, DeWitt J, Swaroop S. American College of Gastroenterology Guideline: Management of Acute Pancreatitis. *Am J Gastroenterol*. 2013;108(9):1400–1415.

E12. Research has been carried out into a new drug, measuring the effect on heart rate before and after administration of the drug in a population of healthy volunteers. Which of the following statistical tests would be most appropriate?

A. Analysis of variance (ANOVA)
B. Wilcoxon signed rank
C. Friedman
D. Kruskal-Wallis
E. Fisher's exact test

Answer: A

Short explanation

This is paired, parametric, quantitative data and application of either the paired *t* test or ANOVA would be most appropriate. It would be possible, but less appropriate, to apply non-parametric paired tests to parametric paired data (Wilcoxon signed rank or Friedman). Tests for non-paired but multiple groups could also be applied but would again not be most appropriate method – for example, Kruskal-Wallis. The Fisher's exact test is used for small samples of qualitative data.

Long explanation

Heart rate is a continuous, quantitative variable. The test is described in the same population, before and after an intervention. The values are therefore paired. It is also reasonable to assume that a population of healthy volunteers will have a normal distribution of heart rate. Tests applied to a normal distribution are known as parametric and those that apply to a skewed population are non-parametric. In general, it is possible to apply non-parametric tests to all populations, that is, skewed and normally distributed data. However, it is preferable to carry out parametric tests if possible because they are more accurate, precise and therefore more powerful if their assumptions are met.

In any experiment like this, we create two populations – the baseline and the intervention. In this example, they happen to be the same people, but the question is – ~~that the populations with some difference between them that we in there a difference~~ between the populations before and after the drug? Our null hypothesis is that there is no difference. All the statistical tests that can be applied ask the same specific statistical question – what is the probability that the null hypothesis is correct? The tests will generate a number between 0 and 1, and we generally regard values less than 0.05 as significant – that is, only a 5% chance that the null hypothesis is correct, and there was no difference between the groups.

The following table outlines common tests applied to different types of data.

Qualitative data	Less than 10 in the sample	Fisher's exact test
	More than 10 in the sample	Chi-squared test
Quantitative data	Parametric	Student's t test
	Non-parametric	Mann-Whitney U test
Quantitative multiple groups	Parametric	ANOVA
	Non-parametric	Kruskal-Wallis
Quantitative, paired data	Parametric	Paired t test/ANOVA
	Non-parametric	Wilcoxon signed rank/ Friedman

Reference
Kirkwood BR, Sterne JAC. *Medical Statistics*. 2nd edition. Oxford: Blackwell Science.

E13. Which of the following methods of oxygen analysis is most appropriate for measuring partial pressure of oxygen in an arterial blood gas (ABG) sample?

A. A Severinghaus electrode
B. A mass spectrometer
C. A paramagnetic analyzer
D. A Clark electrode
E. A system based on the ratio of light absorption in oxygenated and deoxygenated haemoglobin

Answer: D

Short explanation
The Severinghaus electrode is used to measure CO_2 in an ABG machine. The Clark electrode measures the partial pressure of oxygen in a liquid sample. The paramagnetic analyzer measures the partial pressure of oxygen in a gas mixture. A mass spectrometer is an expensive option for analyzing oxygen and other molecules, often only used in research. Answer E describes the basis of the saturation probe, measuring the ratio of reduced and oxidized haemoglobin.

Long explanation
The ABG machine takes a liquid sample (blood) and measures the partial pressures of oxygen and carbon dioxide and the pH of the sample. Other variables on a standard ABG, including bicarbonate and base excess, are then calculated from these values. The machine contains a separate electrode for each measurement.

The pH analyzer consists of a silver/silver chloride electrode within pH-sensitive glass containing a buffer solution referenced against a Calomel electrode. The CO_2 electrode (also known as a Severinghaus electrode) is a modified version of the pH electrode. It consists of a silver/silver chloride electrode within pH sensitive glass, referenced against an outer electrode separated from the sample by a bicarbonate solution and a CO_2 permeable membrane. Essentially, this acts in a similar manner to the blood-brain barrier with CO_2 diffusing across the membrane instead of H^+ ions, exerting a pH change in the solution, which is detected by pH sensors.

The oxygen electrode is a Clark electrode and consists of a platinum cathode within a glass body separated by a salt solution from a silver/silver chloride anode. Oxygen passes through a semi-permeable membrane from the blood sample into the salt solution, resulting in a detectable change in potential difference.

The paramagnetic analyzer measures the partial pressure of oxygen in a gaseous state and is found within modern ventilator circuits. They utilize the paramagnetic properties of oxygen molecules within an induced electromagnetic field to either deflect a beam of light or to create a pressure difference across a membrane. This system is rapid to respond to changes in partial pressure of oxygen and can be used in breath-by-breath analysis.

The mass spectrometer is a large and expensive piece of equipment that will accurately provide information on the molecular composition of a sample. It is relatively slow to use and predominantly restricted to research settings. The molecules in the sample are ionized by free electrons and fired toward photo-voltaic receptors using a cathode plate. Molecules of different weights stimulate different peaks on the readout, from which the molecular composition and relative abundance can be calculated.

References
Al-Shaikh B, Stacey S. *Essentials of Anaesthetic Equipment*. 3rd edition. London: Churchill Livingstone, 2007.
Langton JA, Hutton A. Respiratory Gas Analysis. *Contin Educ Anaesth Crit Care Pain*. 2009;9(1):19–23.

E14. A 66-year-old woman has been brought in to hospital after several days of vomiting and diarrhoea. Her vital signs are as follows: pulse rate 95; blood pressure 110/50; respiratory rate 24; capillary refill time 4 seconds; cool skin; core temperature 36.3°C. An intravenous cannula is sited and routine blood tests sent.

Which IV fluid prescription is appropriate in this scenario?

A. 1 l 0.18% NaCl with 4% glucose and 27 mmol/l potassium over 12 hours
B. 500 ml 0.9% NaCl over 15 minutes
C. 3 l Hartmann's solution (compound sodium lactate) over 12 hours
D. 500 ml 6% tetrastarch (hydroxyethyl starch 130/0.4) over 30 minutes
E. 500 ml 4.5% human albumin solution (HAS)

Answer: B

Short explanation
This woman is dehydrated with complex electrolyte losses from vomiting and diarrhoea. However, her vital signs indicate that she requires resuscitation, not maintenance fluids. Initial resuscitation fluids should be in the form of 250- to 500-ml boluses of crystalloid (containing Na^+ 130–154 mmol/l) followed by reassessment. Tetrastarch should not be used for resuscitation; 4.5% human albumin solution may be used for resuscitation in severe sepsis. Glucose solutions are for maintenance fluids only.

Long explanation

Fluid and electrolyte management is crucial for good patient care but is often left to the most inexperienced medical staff with deleterious consequences. Recent guidelines have been published by National Institute for Health and Care Excellence (NICE) to help to address this shortcoming. Patients' fluid and electrolyte needs should be assessed and managed on every ward round by skilled, trained professionals.

There are four steps involved: assessment, resuscitation, routine maintenance and replacement. The first step is assessment for hypovolaemia and the requirement for resuscitation. Clinical signs suggesting a need for resuscitation include heart rate >90, systolic blood pressure <100 mmHg, respiratory rate >20 or oliguria. Resuscitation should take the form of 500-ml boluses of crystalloid with a sodium content of 130 to 154 mmol/l, such as Hartmann's solution or 0.9% NaCl. These boluses should be given quickly, for example, over 15 minutes, and then the patient reassessed for ongoing signs of hypovolaemia. Smaller boluses may be required in the elderly or patients with known heart failure. Repeated boluses may be required. If patients do not improve after two to three boluses, then critical care review should be considered.

Once fluid resuscitation has achieved haemodynamic stability, then patients should be assessed for their fluid and electrolyte needs. There are three components to this: replacement of fluid and electrolyte deficits from presentation; replacement of ongoing losses and redistribution; and routine maintenance fluids. Assessment of requirements will need to take into account examination findings, laboratory results and the composition of body fluids being lost, such as vomit, drain outputs, diarrhoea or sweat.

In this clinical scenario, the patient has a fluid and electrolyte requirement due to vomiting and diarrhoea in the time before arrival in hospital and will have ongoing losses of a similar nature until her symptoms resolve. Vomit contains high concentrations of potassium, chloride and hydrogen ions, with relatively little sodium. Diarrhoea has highly variable composition but commonly contains potassium, sodium and bicarbonate ions. Laboratory test results should be used to guide complex fluid management. In addition to resuscitation and replacement, the patient will also require routine fluid maintenance until she can fulfil her requirements enterally. This will be an additional 25 to 30 ml/kg/day of fluid with 1 mmol/kg/day each of sodium, potassium and chloride ions and 50 to 100 g/day of glucose to prevent ketosis.

Reference

National Institute for Health and Care Excellence (NICE). *Intravenous fluid therapy in adults in hospital* (CG174). 2013. http://www.nice.org.uk/nicemedia/live/14330/66015/66015.pdf (accessed May 2015).

E15. Which of the following components of the APACHE IV scoring system has the greatest impact on the final score?

A. Acute physiological parameters
B. Acute diagnosis
C. Chronic health conditions
D. Age
E. Admission source and previous length of stay

Answer: A

Short explanation

Acute physiology makes up 65.5% of the final score with the remainder as follows: diagnosis – 16.5%; chronic health conditions – 5%; age – 9.4%; admission source and length of stay – 2.9%; and requirement for mechanical ventilation – 0.6%.

Long explanation

The APACHE (Acute Physiology, Age and Chronic Health Evaluation) scoring system is one of the most commonly used intensive care unit (ICU) scoring systems in the world. It was first developed in 1981 by Knaus. The score is derived based on the components described and provides an estimate of ICU mortality. It is calculated at 24 hours after admission to ICU.

The APACHE I system was developed using 805 patients in US ICUs. It was replaced in 1985 by APACHE II, based on 5815 patients in 13 US ICUs. APACHE III was developed in 1991 using 17,440 patients in 40 ICUs and APACHE IV in 2006 using 110,558 in 104 US ICUs.

The APACHE II score used an acute physiology score based on 12 parameters, a chronic health score and the patient's age as well as an underlying diagnosis picked from a list of 50 broad diagnostic categories. APACHE III expanded the physiological variables from 12 to 17 and the diagnostic categories from 50 to 78 as well as including the admission source and previous length of stay.

References

Palazzo M. Severity of illness and likely outcome from critical illness. In Bersten AD, Soni N. *Oh's Intensive Care Manual*, 6th edition. Edinburgh: Butterworth-Heinemann, 2009, pp 17–30.

Zimmerman JE, Kramer AA, McNair DS, Malila FM. Acute Physiology and Chronic Health Evaluation (APACHE) IV: Hospital mortality assessment for today's critically ill patients. 2006. *Critical care medicine*, 34(5), 1297–1310.

E16. A 73-year-old patient who is anticoagulated for a mechanical aortic valve has just been admitted to the intensive care unit with severe bilateral pneumonia and has required intubation.

Which is the MOST appropriate stress ulcer regime for this patient to be commenced on admission?

A. Omeprazole (enterally)
B. Ranitidine (intravenously)
C. Enteral feed
D. Sucralfate (enterally)
E. None required

Answer: B

Short explanation

This patient should receive stress ulcer prophylaxis as he has major risk factors for their development. Proton pump inhibitors (PPIs) or histamine-2 receptor blockers (H2RBs) are first-line treatments for high-risk patients. This patient has not been commenced on enteral feed yet so intravenous medication would be the preferred initial route of administration.

Long explanation

This patient has two major risk factors for the development of stress ulcers (the presence of coagulopathy and mechanical ventilation likely to be required for >48 hours), so treatment for stress ulcer prophylaxis would be recommended.

Options for stress ulcer prophylaxis:

- Histamine-2 receptor blockers (H2RBs; e.g. ranitidine) competitively inhibit histamine binding to gastric parietal cells thus decreasing the volume and acidity of gastric secretions. When compared with placebo, they significantly reduce the risk of clinically important bleeding; however, there are concerns regarding tachyphylaxis.
- Proton pump Inhibitors (PPIs) e.g. lansoprazole, pantoprazole or omeprazole) irreversibly bind and inhibit the hydrogen/potassium ATPase enzyme on parietal cells, which are responsible for producing gastric acid. The result is a more sustained increase in the pH of gastric secretions compared with that from H2RBs.
- Sucralfate acts by forming a physical barrier between gastric secretions and mucosal epithelial cells. It has been demonstrated to be superior to placebo but inferior to H2RBs, and there are also concerns regarding the absorption of feed and enteral medications, so it is not routinely recommended as stress ulcer prophylaxis.
- Antacids are now less commonly used because of the frequency of administration, gastrointestinal side effects and derangement in electrolyte levels that may result from treatment.
- Enteral feed helps neutralize the acidity of gastric secretions; it also helps promote mucosal blood flow and the release of prostaglandins and mucus, which may all be beneficial in the reduction of stress ulcers and gastrointestinal bleeding.

There are concerns regarding the side effects and increased incidence of *Clostridium difficile* infections and infection-related ventilator-associated complications that may occur as a result of increasing gastric pH.

Current practice favours PPIs (recommended by the Surviving Sepsis Campaign) or H2RBs as first-line agents in at risk patients. Further research is still required, however, to establish the best stress ulcer prophylaxis regime for patients admitted to the intensive care unit and in whom it is required. This patient has not yet been established on nasogastric feed, so enteral absorption of medications cannot be guaranteed; therefore intravenous drugs, either PPIs or H2RBs, would be preferable in this patient.

Reference
Plummer MP, Reintam Blaser A, Deane AM. Stress ulceration: prevalence, pathology and association with adverse outcomes. *Crit Care.* 2014;18(2):213.

E17. You have just intubated a patient in the intensive care unit for airway protection due to a low Glasgow Coma Score. The patient is haemodynamically stable.

In conjunction with clinical assessment, which of the following methods is the best way to confirm placement of the endotracheal tube in the airway?

A. Colometric end tidal carbon dioxide detection
B. Chest X-ray
C. Visualization of tube passing through the cords
D. Oesophageal detection device
E. Continuous end tidal capnography

Answer: E

Short explanation

Capnography is recommended for the confirmation of endotracheal tube placement by the Intensive Care Society. Visualization of the tube passing through the cords is reassuring although is not possible in all patients. Oesophageal detection devices and colorimetric end tidal carbon dioxide ($ETCO_2$) are other methods that have been used. Chest X-ray can be used to confirm the position of the tip of the endotracheal tube once presence in the airway has been confirmed.

Long explanation

Methods used for assessing endotracheal tube placement include:

- Clinical evaluation
  - Visualization of the tube passing through the cords during intubation
  - Visualization of the chest rising and falling and absence of the abdomen distending with ventilation
  - Auscultation of bilateral air entry and no air entry heard over the stomach
  - Misting of the endotracheal tube on expiration
- Oesophageal detection devices
- Detection of end tidal carbon dioxide ($ETCO_2$)
  - Colorimetric $ETCO_2$ measurement
  - Digital $ETCO_2$ measurement
  - Continuous end tidal capnography
  - Oesophageal intubation may produce a small amount of $ETCO_2$ to be released, especially in the presence of recent carbonated beverage ingestion, which may be detected by any $ETCO_2$ detector. If continuous capnography is used in this situation, abnormal CO_2 curves will be seen, diminishing in size and disappearing within six breaths; therefore a one-off demonstration of the presence of CO_2, such as from colometric $ETCO_2$ detection, is less reliable at confirming correct endotracheal tube placement than capnography.
  - The presence of $ETCO_2$ requires the production of CO_2 in the tissues, an adequate circulation to transport it to the lungs, a patent airway and the presence of ventilation. Alteration in the continuous capnography trace can indicate problems with any of these.
- Bronchoscopy and visualization of the endotracheal tube in the trachea
- Chest X-ray will confirm position of ETT in trachea/bronchial tree once placement in the airway has been confirmed
- Computed tomography will demonstrate the exact position of the endotracheal tube but is not practical to use it for this purpose.

Confirmation of endotracheal tube placement by a combination of clinical evaluation and capnography has become the gold standard. Capnography guidelines from the Intensive Care Society recommend its use during the process of intubation or tracheostomy insertion for all patients in the intensive care unit as well as during all intra- and inter-hospital transfers in ventilated patients. They also recommend that continuous capnography is used for the duration of ventilation for all patients who have an endotracheal tube or tracheostomy in situ.

Reference

Thomas AN, Harvey DJR, Hurst T. Intensive Care Society. Capnography guidelines 2014. http://www.ics.ac.uk/ics-homepage/guidelines-and-standards (accessed 10/11/14).

E18. In which condition is treatment with magnesium LEAST useful?

A. Pre-eclampsia
B. Epilepsy
C. Refeeding syndrome
D. Asthma
E. Ventricular tachyarrhythmias

Answer: B

Short explanation

Treatment with magnesium sulphate is useful for patients with severe pre-eclampsia, eclampsia, hypomagnesaemia (common in refeeding syndrome), asthma and ventricular tachyarrhythmias (especially torsades de pointes). Although magnesium has been used to treat seizures secondary to eclampsia or hypomagnesaemia, it is not routinely used to manage patients with epilepsy.

Long explanation

Magnesium sulphate has a wide spectrum of physiological actions and hence clinical uses.

Respiratory

The British Thoracic Guidelines recommend the use of magnesium sulphate for patients with life-threatening asthma due to its bronchodilatory effect. They recommend 1.2 to 2 g to be given intravenously over 20 minutes.

Cardiovascular

Magnesium sulphate is useful in the treatment of a range of cardiac arrhythmias, especially ventricular tachyarrhythmias, those associated with long Q-T intervals (such as torsade de pointes) or digoxin-induced arrhythmias. It has also been used in the perioperative management of phaeochromocytoma surgery to improve haemodynamic instability.

Obstetrics

The MAGPIE trial demonstrated a reduced rate of eclampsia with the use of magnesium sulphate for patients with severe pre-eclampsia or eclampsia. Magnesium is the first-line treatment for the treatment of eclamptic seizures. The National Institute for Health and Clinical Excellence recommends a 4-g bolus dose over 5 minutes followed by administration of an infusion of 1 g/hr for 24 hours. They also recommend further bolus doses of 2 to 4 g for recurrent seizures.

Magnesium can also be used as a tocolytic as it causes uterine relaxation.

Nervous system

Magnesium sulphate infusion is used to treat spasms and the autonomic instability seen in patients with tetanus. It has also been used to prevent cerebral vasospasm after subarachnoid haemorrhage. There is evidence to suggest it reduces the incidence of vasospasm, although there is no evidence of improved patient outcome. Although magnesium has been used to treat seizures secondary to eclampsia or hypomagnesaemia, it is not routinely used to manage patients with epilepsy.

Pain

Magnesium is a physiological N-methyl-D-aspartate receptor antagonist and has been used perioperatively as an adjunct analgesic.

Gastrointestinal

Magnesium is a laxative and antacid and so can be used to treat constipation and dyspepsia, respectively.

Hypomagnesaemia can present secondary to decreased intake due to dietary deficiencies or malabsorption, increased loss via the kidneys or gastrointestinal tract or from its redistribution, which occurs with massive transfusion of citrated blood, hyperparathyroidism or the administration of insulin. During refeeding syndrome, there may be an underlying deficiency and then redistribution of the magnesium occurs with an increase in the body's natural insulin level. Magnesium replacement is therefore important in the treatment.

References

British Thoracic Society and Scottish Intercollegiate Guidelines Network. British guideline on the management of asthma. A national clinical guideline [SIGN 141]. 2014. https://www.brit-thoracic.org.uk/document-library/clinical-information/asthma/btssign-asthma-guideline-2014 (accessed 12/10/14).

Herroeder S, Schönherr ME, De Hert SG, Hollmann MW. Magnesium – essentials for anaesthesiologists. *Anaesthesiology*. 2011:114(4);971–993.

National Institute for Health and Care Excellence (NICE). *Hypertension in pregnancy. The management of hypertensive disorders during pregnancy* (CG107). London: NICE, 2010. http://www.nice.org.uk/guidance/cg107 (accessed October 14).

E19. Which of the following is LEAST likely to potentiate (prolong) the effect of a bolus of cis-atracurium?

A. Acidosis
B. Low plasma cholinesterase levels due to liver disease
C. Cyclosporin
D. Gentamicin
E. Verapamil

Answer: B

Short explanation

Cis-atracurium is mainly degraded by Hofmann degradation and therefore not reliant on plasma cholinesterase levels. Acidosis slows all enzyme breakdown systems, and drugs C, D and E all have the potential to prolong the effect of cis-atracurium on the neuromuscular junction.

Long explanation

Neuromuscular blocking drugs (NMBDs) may be classified as depolarizing or non-depolarizing. Depolarizing drugs (suxamethonium) initiate activation of the neuromuscular junction on binding, before then blocking further action. Non-depolarizing drugs simply block the post-synaptic acetylcholine receptor without activating it. All NMBDs compete with acetylcholine for the receptor and therefore duration of action depends on the relative concentrations of acetylcholine and of the drug within the neuromuscular junction. Metabolism and elimination of the drugs occur predominantly in the blood, and as levels in the blood fall, the drug moves from the active site back to the bloodstream. Many factors influence both the metabolism and excretion of the NMBDs and the concentration of acetylcholine at the receptor site, and therefore the duration of action.

Suxamethonium and mivacurium are metabolized by plasma cholinesterases, which may be increased or decreased in a number of disease states, such as pregnancy and chronic liver disease.

Atracurium is degraded by Hofmann degradation (spontaneous degeneration) into laudanosine and also hydrolyzed via nonspecific esterases. Cis-atracurium is an enantiopure preparation of 1R-cis 1'R-cis atracurium, which is predominantly metabolized by Hofmann degradation. Because this is a spontaneous chemical reaction and the rate limiting step, the duration of action is affected by physicochemical characteristics such as temperature and pH.

Electrolyte disturbances, drugs and critical illness can all interact with the production, release and re-uptake of acetylcholine by the neuromuscular junction and therefore all increase the duration of action of all NMBDs. Many drugs increase their effects, including antibiotics such as aminoglycosides, metronidazole and tetracyclines; calcium channel blockers, including verapamil; immunosuppressants; and chemotherapy agents.

It is reasonable to use NMBDs in the intensive care unit (ICU) with caution, and significant dose reductions compared with anaesthetic practice may be appropriate, particularly when these drugs are used in infusions or repeated dosing. The use of NMBDs in ICU is associated with many complications, including retention of respiratory secretions and subsequent infection. Patients who receive NMBDs for a significant period of time are at increased risk of pressure sores, venous thromboembolism and critical care neuromyopathy. It is also vital to ensure that any patient who may still be experiencing prolonged neuromuscular blockade is appropriately sedated.

References
Khirwadkar R, Hunter JM. Neuromuscular physiology and pharmacology: an update. *Contin Educ Anaesthesia, Crit Care Pain*. 2012;12:237–244.
Tripathi SS, Hunter JM. Neuromuscular blocking drugs in the critically ill. *Contin Educ Anaesthesia, Crit Care Pain*. 2006;6(3):119–123.

E20. A 64-year-old patient is in the intensive care unit undergoing haemofiltration for acute renal failure. She is receiving a blood transfusion because her haemoglobin is 67 g/l.

Which of the following complications is the MOST likely to occur from the blood transfusion?

A. Transfusion-related acute lung injury (TRALI)
B. Transfusion-associated circulatory overload (TACO)
C. Acute transfusion reaction (ATR)
D. Transfusion-transmitted infection (TTI)
E. Haemolytic transfusion reaction (HTR)

Answer: C

Short explanation
The 2013 report by SHOT (Serious Hazards of Transfusion) documents the frequency of adverse events and reactions to blood products as well as near misses. It reported 1571 events in 2013. Acute transfusion reaction (n = 320) was the commonest physiological reaction to occur. Ten patients experienced TRALI, 96 patients had TACO and 49 experienced HTR, and there were no cases of transfusion transmitted infections.

Long explanation
A number of complications can develop after blood transfusion, including immunological and non-immunological conditions:

Immunological complications
- Haemolytic reactions:
  - Acute haemolytic reactions, usually due to ABO incompatible transfusion. This generally presents as an early acute severe reaction.

- o Delayed haemolytic reactions. These are usually due to antibodies to non ABO, RhD antigens.
- Non-haemolytic reactions. These occur due to sensitization of the recipient to donor white cells, platelets or plasma proteins. Types of reaction include the following:
  - o ATR (occur within 24 hours of transfusion):
    - febrile, urticarial, anaphylaxis/allergy type reactions, hypotensive reactions
  - o TRALI
  - o Transfusion associated graft-versus-host disease (TA-GvHD)
  - o Alloimmunization
  - o Transfusion-related immunomodulation
  - o Post-transfusion purpura

Non-immunological complications

- Infectious complications – TTIs:
  - o Hepatitis
  - o HIV/AIDS
  - o Other viral agents (cytomegalovirus, Epstein-Barr, human T-lymphotropic virus)
  - o Bacterial/parasitic contamination
  - o Creutzfeldt-Jakob disease
- Dilutional coagulopathy
- TACO
- Hypothermia
- Left shift of oxyhaemoglobin dissociation curve due to depleted stores of 2,3-DPG (resolves within 24 hours)
- Metabolic
  - o Hypocalcemia resulting from citrate toxicity
  - o Hyperkalaemia from potassium release from stored red blood cells
  - o Alkalosis secondary to conversion of citrate to bicarbonate
  - o Acidosis
    - From raised partial pressure of carbon dioxide
    - Lactic acidosis secondary to tissue hypoxia from inadequate resuscitation may also be present
  - o Decreased adenosine triphosphate
  - o Haemosiderosis (iron overload) from chronic repeated blood transfusion

Serious Hazards of Transfusion (SHOT) is a voluntary independent haemovigilance scheme that collects information about adverse events and reactions to blood transfusions and near misses in the United Kingdom. It publishes a yearly report that includes information about events that have occurred that year along with recommendations to improve the safety of blood transfusions. The latest report demonstrates acute transfusion reactions to be the commonest (both reported in 2013 and based on cumulative data since 1996–1997) pathological reactions that occurs in response to transfusion.

Since reporting started in 1996–1997, the commonest adverse event surrounding transfusion administration is that the incorrect blood component is transfused. Other adverse events that occur by error include avoidable or delayed transfusion or undertransfusion, errors in the administration of anti-D immunoglobulin and handling and storage errors. Human factors have been identified as a major cause for these errors occurring, and these adverse events are avoidable.

Reference

Bolton-Maggs PHB(Ed), Poles D, Watt A and Thomas D, on behalf of the Serious Hazards of Transfusion (SHOT) Steering Group. The 2013 Annual SHOT Report. 2014. http://www.shotuk.org/wp-content/uploads/74280-SHOT-2014-Annual-Report-V12-WEB.pdf (accessed July 2015).

E21. A 56-year-old man presents after an out-of-hospital cardiac arrest. He suffered an unwitnessed cardiac arrest. His initial rhythm was asystole, and return of spontaneous circulation (ROSC) was achieved after a further 34 minutes. His Glasgow Coma Score (GCS) is 3 in the emergency department (ED). He is admitted to the intensive care unit for targeted temperature management.

Which of the following is MOST useful in prognosticating poor neurological outcome in this patient?

A. Electroencephalogram (EEG) with an unreactive baseline at 72 hours post-rewarming

B. Absent bilateral N20 somatosensory evoked potentials (SSEPs) 24 hours post-rewarming

C. Absent corneal and pupillary reflexes, bilateral N20 SSEPs and GCS 3 at 72 hours post-arrest

D. Clinical history, fixed dilated pupils and a GCS motor score of 1 on admission to ED

E. Loss of grey-white matter differentiation on admission computed tomography (CT)

Answer: C

Short explanation

The European Resuscitation Council (ERC) and the European Society of Intensive Care Medicine (ESICM) recommend a multimodal approach to assessment of predicted neurological recovery post cardiac arrest. All of the foregoing are poor prognostic signs, but the absence of pupillary and corneal reflexes and/or the absence of bilateral N20 SSEPs in patients who remain unconscious with a GCS motor score of 1 or 2 ≥72 hours post-ROSC have been recommended as predicting poor neurological outcome with the best possible accuracy.

Long explanation

Factors associated with poor neurological outcome:

- Clinical history
  - Initial non-shockable rhythm
  - Factors that increase time to ROSC (unwitnessed arrest, no bystander or prolonged cardiopulmonary resuscitation)
- Clinical examination
  - Absent pupillary and corneal reflexes are not specific enough to use immediately but are useful ≥72 hours post-arrest
  - Absent or extensor motor response to pain (M 1–2 on GCS scoring) especially ≥72 hours post–cardiac arrest
  - Confirmed myoclonic status (≥30 minutes not just the presence of myoclonic jerks) that develops ≤48 hours post-ROSC
- Neurophysiological investigations
  - EEG findings of an unreactive baseline in response to external stimuli, presence of burst suppression or refractory status epilepticus ≥72 hours post-arrest

- o The bilateral absence of N20 SSEPs ≥72 and ≥24 hours post-arrest from those patients treated with and without TTM, respectively
- Imaging:
  - o Either CT ≤24 hours demonstrating evidence of cerebral oedema: sulcal effacement and loss of grey-white matter differentiation
  - o Magnetic resonance imaging (MRI) 2 to 5 days post-ROSC, compared with CT, provides more detailed information regarding the hypoxic-ischaemic brain injury
- Biomarkers:
  - o Elevated biomarker (neurone specific enolase [NSE] and S-100ß) levels are associated with poor neurological outcomes but are not sensitive or specific enough to be used in isolation

The ERC/ESICM advice recommends a multimodal approach to prognostication. Initial clinical assessment should occur on day 3 post-arrest, following rewarming after TTM and after the cessation of residual effects of sedation and neuromuscular blocking drugs and exclusion of other reversible causes. If the patient remains unconscious, with a GCS motor score of 1 to 2 ≥72 hours post-arrest and with absent pupillary and corneal reflexes and/or bilaterally absent N20 SSEPs they are likely to experience a poor neurological outcome. Repeat assessment should occur >24 hours later if the preceding criteria are not met. The patient's likely neurological outcome is poor if his or her clinical assessment remains unchanged and he or she has two or more of the following:

- Myoclonic status ≤48 hours post-ROSC
- High NSE levels 48–72 hours post-ROSC
- Unreactive malignant EEG pattern
- Diffuse anoxic injury on CT ≤24 hours or MRI at 2 to 5 days

Continued care and ongoing observation is recommend for patients not meeting these criteria.

References

Sandroni C, Cariou A, Carvallaro F, et al. Prognostication in comatose survivors of cardiac arrest: an advisory statement from the European Resuscitation Council and the European Society of Intensive Care Medicine. *Intensive Care Med.* 2014;40:1816–1831.

Temple A, Porter R. Predicting neurological outcome and survival after cardiac arrest. *Contin Educ Anaesth Crit Care Pain.* 2012;12(6):283–287.

E22. You are asked by the nurses in the intensive care unit to review a 50-year-old man who was admitted earlier in the day. He has septic shock secondary to a community-acquired pneumonia and is intubated and ventilated. He has just developed a tachyarrhythmia, with a heart rate of 175 and a blood pressure of 70/30. The electrocardiogram (ECG) shows fast atrial fibrillation (AF).

Which of the following is the most appropriate initial management?

A. Amiodarone IV
B. Digoxin IV
C. Beta-blockade IV
D. Synchronized DC shock
E. Non-synchronized DC shock

Answer: D

Short explanation

The Resuscitation Council (UK) recommends that for patients who have a tachycardia with adverse features present (BP <90 mmHg, syncope, heart failure or myocardial ischaemia), the initial treatment should be three attempts at synchronized DC cardioversion.

Long explanation

AF is commonly seen in the Intensive care unit and can have a profound effect on already critically unwell patients because of the physiological changes associated with arrhythmia. Atrial activity becomes chaotic and leads to irregularly irregular ventricular contractions. The ventricular rate, if untreated, can be very fast.

The chaotic atrial activity of AF results in a loss of the effect of atrial systole on ventricular filling, which leads to a reduction in stroke volume. This alone might have a minimal effect if ventricular filling time was sufficient to allow passive filling. The rapid ventricular rate seen in untreated AF further compounds this and leads to a further reduction in cardiac output. The loss of atrial contraction can have marked effects in patients with pre-existing cardiac disease with impaired ventricular diastolic function.

Synchronized cardioversion should be used in those with adverse signs or those who are deteriorating. Initial energy should be 50 to 100 J, although AF is frequently resistant to lower energies and may require 150 J biphasic (200 J monophasic). Important points to remember include the requirement for synchronization of the defibrillation (to avoid the risk of precipitating VF) and the use of sedation. If defibrillation fails to restore sinus rhythm, then amiodarone 300 mg IV should be given over 15 minutes, and then synchronized cardioversion re-attempted.

In patients with fast AF who are stable, treatment can be rate control or rhythm control. Beta-blockade, diltiazem, digoxin and amiodarone can all be used, although amiodarone should not be used for long-term rate control. Goals of treatment include anticoagulation where appropriate, to reduce the risk of systemic embolization, and either conversion to sinus rhythm or ventricular rate control.

References

Holt A. Chapter 18: Management of cardiac arrhythmias. In *Oh's Intensive Care Manual*. Sixth Edition. Bersten A, Soni N. Edinburgh: Butterworth Heinemann Elsevier, pp 204–205.

Nolan J, Soar J, Lockey A, et al., for the Resuscitation Council (UK). Chapter 11: Peri-arrest arrhythmias. *Advanced Life Support*. 6th Edition. 2011, pp 105–110. https://www.resus.org.uk/resuscitation-guidelines (accessed April 2015).

E23. With regard to oxygen delivery to the tissues, assuming cardiac output remains the same, which of the following patients will have the greatest oxygen delivery?

A. SaO_2 100%, Hb 100 g/l, PaO_2 12 kPa
B. SaO_2 100%, Hb 150 g/l, PaO_2 12 kPa
C. SaO_2 100%, Hb 100 g/l, PaO_2 85 kPa
D. SaO_2 80%, Hb 150 g/l, PaO_2 12 kPa
E. SaO_2 80%, Hb 150 g/l, PaO_2 85 kPa

Answer: B

Short explanation

Oxygen delivery is calculated as cardiac output multiplied by the oxygen content of blood. The calculation for oxygen content is as follows:

$$CaO_2 = (1.39 \times Hb \times SaO_2) + (0.0225 \times PaO_2) \text{ [all units as above]}$$

Option B comes out highest, at roughly 212 ml/l, with the next closest option E at roughly 187 ml/l.

Long explanation

Oxygen delivery (DO_2) is the amount of oxygen delivered to the tissues per unit time. It is calculated from the cardiac output (CO) and the oxygen content of arterial blood (CaO_2):

$$DO_2 = CO \times CaO_2$$

The oxygen content of arterial blood (CaO_2) is described using the preceding equation and is the sum of the two forms in which oxygen is carried – bound to haemoglobin and dissolved in the plasma. The vast majority of oxygen, in health, is carried bound to haemoglobin. The affinity of haemoglobin for oxygen is represented by a constant, roughly 1.39 ml/g (some books use 1.34 or 1.31, to reflect the fact that haemoglobin does not carry its maximal potential).

A small amount of oxygen is carried dissolved in the plasma and is calculated by the solubility coefficient of oxygen, roughly 0.0225 ml/l/kPa. Even at high FiO_2 (and therefore high PaO_2), this oxygen is insignificant at normal atmospheric pressure. However, in some circumstances, the oxygen dissolved in plasma is highly significant. For example, a patient with severe carbon monoxide poisoning being treated in a hyperbaric chamber will rely on dissolved oxygen for the majority of their oxygen delivery.

The DO_2 must be sufficient to meet the oxygen demand of the body, or an oxygen deficit will result. When attempting to optimize oxygen delivery, the individual components of DO_2 can be assessed and manipulated. With regard to CaO_2, the Hb and SpO_2 have the most significant effect on this value and must be optimized. Although theoretically a higher Hb will be beneficial, this must be weighed against the risks of blood transfusion, and the flow dynamic effects of a higher Hb resulting in more sluggish flow. With regard to cardiac output, the heart rate has an optimal range, above which the stroke volume will be impaired, and its three constituents – preload, contractility and afterload – although affecting each other, can also be optimized individually.

Reference

McLellan S, Walsh T. Oxygen delivery and haemoglobin. *Contin Educ Anaesth Crit Care Pain.* 2004;(4):123–126.

E24. You are called urgently to help the emergency department with a 20-year-old gentleman who has taken a mixture of recreational drugs including MDMA (ecstasy), cocaine, LSD and alcohol. He was markedly agitated in the community and has already been intubated. He has a temperature of 39.5°C, is tachycardic and flushed, and had a seizure before intubation. You diagnose likely serotonin syndrome (SS).

Which of the following is the best way to manage him?

A. Supportive treatment only
B. Dantrolene 2.5 mg/kg
C. Benzodiazepine infusion
D. Sedation with propofol and fentanyl
E. Bicarbonate infusion

Answer: C

Short explanation

Treatment of SS includes stopping trigger agents, supportive care, sedation with benzodiazepines and treatment with serotonin antagonists if available. There is no evidence to support the use of dantrolene in SS. Fentanyl can trigger SS and should not be used in its management. You are not given enough information to suggest a bicarbonate infusion is indicated.

Long explanation

SS is a potentially fatal condition that occurs when synaptic serotonin levels are raised and can be caused by any drug that increases serotonin levels. Selective serotonin reuptake inhibitors and monoamine oxidase inhibitors both cause a rise in serotonin levels, and their use in combination is the commonest cause of SS. Other drugs that can trigger SS include pethidine, fentanyl and tramadol, as well as recreational drugs such as ecstasy, cocaine and LSD.

The clinical presentation of SS of often difficult to differentiate from neuroleptic malignant syndrome (NMS), which has a higher mortality, but the speed of onset is usually much quicker for SS (hours, rather than days to weeks for NMS). Dantrolene can feature in the management of NMS, but has no role in the management of SS. The Hunter Serotonin Toxicity Criteria and Sternback criteria both exist for the diagnosis of SS, and both feature clinical signs such as clonus, agitation, fever and hyper-reflexia.

Treatment of SS includes stopping all drugs with serotoninergic effects (including fentanyl, tramadol and ondansetron). Seizures and agitation should be treated with benzodiazepines. Autonomic instability is particularly challenging to treat because both hypertension and hypotension may occur. Short-acting agents are of value in these circumstances.

The treatment options for NMS include bromocriptine and dantrolene, but the treatment options for SS are more limited. Cyproheptadine is a serotonin antagonist, the administration of which may be of help in SS, although evidence is limited to case reports. An intravenous formulation does not exist, so cyproheptadine must be given orally or via a nasogastric tube.

Reference

Iqbal M. et al. Overview of serotonin syndrome. *Ann Clin Psychiatry*. 2012;24(4):310–318.

E25. All of the following patients have sustained a head injury. Which of them is LEAST likely to require an urgent (<1 hour) computed tomography (CT) head scan?

A. A 40-year-old with a Glasgow Coma Score (GCS) of 142 hours after falling down a flight of stairs
B. A 30-year-old with a GCS of 12 on arrival to the emergency department after falling down a flight of stairs
C. A 35-year-old with a GCS of 15 who has vomited twice after falling down a flight of stairs
D. A 30-year-old with a GCS of 15 who had a brief generalized seizure that self-terminated after falling down a flight of stairs
E. A 40-year-old with a GCS of 15 who takes warfarin for atrial fibrillation and fell down a flight of stairs

Answer: E

Short explanation

The NICE guidance for head injury has criteria that determine which adults should have a CT head scan within 1 hour. Options A through D all feature in those criteria. Patients who are taking warfarin but do not fulfil any other criteria for CT head scan within 1 hour should have a CT head scan within 8 hours of the injury.

Long explanation

The NICE guidance offers clear criteria for performing a CT head scan. In adults who have sustained a head injury, and have any of the following risk factors, a CT head scan should be performed within 1 hour:

– Two or more episodes of vomiting
– GCS 12 or below on first presentation to the emergency department
– GCS 14 or below 2 hours after the injury
– Likely depressed or open skull fracture
– Seizure after the injury
– Possible basal skull fracture (look for panda eyes, Battle sign, haemotympanum, or cerebrospinal fluid leak)
– Focal neurological signs

Once the scan has been performed, a provisional written report should be produced within 1 hour. If the preceding criteria are not met, then a CT head scan may still be indicated but on a less urgent basis. Patients with a head injury who have had either loss of consciousness or some amnesia may be appropriate for a CT head within 8 hours if they are aged over 64 years, have any coagulopathy, had a dangerous mechanism for their injury (e.g. ejected from a car) or have amnesia of events for at least half an hour before the injury.

The mechanism of injury features for a nonurgent CT head scan but alone is not an indication for a CT within 1 hour. This is in contrast to the guidance on cervical spine imaging in which the mechanism of injury does still feature in the list of indications for a CT of the cervical spine within 1 hour.

Reference

National Institute for Health and Care Excellence (NICE). *Head injury: triage, assessment, investigation and early management of head injury in children, young people and adults* (NICE Clinical Guideline 176). London: NICE, 2014.

E26. You are called to the emergency department to assist with the management of a 25-year-old man who is fitting. He is a known epileptic and takes regular levetiracetam. He has been fitting for 25 minutes and has not responded to diazepam 10 mg rectally given pre-hospital, nor to two 4-mg intravenous (IV) doses of lorazepam. His airway is not currently at risk, and his SpO2 is 97% on 15 l oxygen via a non-rebreathe mask.

Which of the following is the most appropriate next treatment?

A. Lorazepam 0.1 mg/kg IV bolus
B. Phenytoin 18 mg/kg IV infusion
C. Thiamine 250 mg IV bolus
D. Thiopentone 4 mg/kg IV bolus and intubation
E. Glucose 50% 50 ml IV bolus

Answer: B

Short explanation

The National Institute for Clinical Excellence (NICE) has published guidelines for the treatment of convulsive status epilepticus (CSE) in adults. In early CSE, lorazepam can be given twice, 10 to 20 minutes apart, and thiamine or glucose can be given if there is suggestion of alcohol abuse or impaired nutrition. In established CSE, phenytoin should be given, and in refractory CSE, general anaesthesia is required, with propofol, thiopentone or midazolam.

Long explanation

The NICE guidance gives a protocol for the treatment of CSE. This splits the management of CSE into stages with theoretical timings from the onset of seizure activity.

Stage 1: This relates to the first 10 minutes of seizure activity, and priorities should be maintenance of the airway, delivering oxygen, and securing IV access.

Stage 2: This is the first 30 minutes of seizure activity, and full monitoring should be established, urgent bloods sent and emergency treatment commenced (described subsequently). Specific treatment with glucose or thiamine should be instituted in patients with alcohol abuse or malnutrition.

Stage 3: This is the first 60 minutes of seizure activity, during which time a likely cause should be established, and the anaesthetic/intensive care team should be involved.

Stage 4: This equates to 30 to 90 minutes after the seizure commenced, and priorities are transfer to the intensive care unit, appropriate investigations (e.g. CT, lumbar puncture) and commencement of electroencephalogram (EEG) monitoring.

The treatment of CSE should be as follows:

Stage 0 (pre-hospital): midazolam 10 mg buccally or diazepam 10 to 20 mg rectally

Stage 1: lorazepam 0.1 mg/kg IV (max 4 mg) can be repeated once after 15 minutes. Give usual anti-epileptics if any doses missed of usual treatment.

Stage 3: Options are as follows:
a. phenytoin 15–18 mg/kg infused at 50 mg/min
b. phenobarbital 10–15 mg/kg infused at 100 mg/min
c. fosphenytoin equivalent dose and rate as phenytoin

Stage 4: Options for inducing general anaesthesia are propofol, midazolam and thiopentone. After the last seizure (clinical or EEG monitored), the anaesthetic should be continued for another day before the dose is gradually reduced. Propofol in particular has a short half-life, and abrupt cessation of the infusion will lead to a rapid fall in plasma concentration, and therefore rapid removal of its anti-epileptic activity. This can lead to rebound seizures.

Reference

National Institute for Health and Clinical Excellence (NICE). *Appendix F: Protocols for treating convulsive status epilepticus in adults and children; adults* (CG137). London: National Institute for Health and Clinical Excellence, 2012.

E27. You have been asked to transfer an intubated 64-year-old with a sub-arachnoid haemorrhage to the regional neuro intensive care unit for decompressive surgery. His Glasgow Coma Score was 4/15 before intubation (M2, V1, E1). He is sedated and ventilated.

Which of the following is the LEAST important to monitor on the transfer?

A. Central venous pressure (CVP)
B. Invasive blood pressure (BP)
C. End-tidal CO_2 (ET CO_2)
D. Temperature
E. Airway pressure

Answer: A

Short explanation

The Intensive Care Society has produced guidelines for the transport for the critically ill adult. The minimum monitoring must include continuous electrocardiogram ECG, non-invasive BP, SpO_2, $ETCO_2$, and temperature. Non-invasive BP is usually inaccurate during transfer, and so invasive BP monitoring should be used. Ventilated patients require airway pressure to be monitored. CVP monitoring may be of assistance pre-transfer but is unlikely to be of use during the transfer itself.

Long explanation

The Intensive Care Society's guidelines for the transport of the critically ill adult (2011) cover all sections of the transfer of patients, from organization within networks, standards for ambulances, training and competencies, and equipment.

Patients should have their clinical condition optimized before transfer, and a checklist should be used to ensure adequate preparation. For the transfer itself, the monitoring and care should of the same standard as in the hospital. As an absolute minimum, all patients should have ECG, SpO_2, non-invasive saturation and temperature monitoring, and ventilated patients also need monitoring of $ETCO_2$, FiO_2, ventilator settings and airway pressure.

Invasive BP measurement should normally be used in preference to non-invasive measurement because it will allow more accurate measurement with less drain on the monitor battery. Central venous pressure will be inaccurate during patient movement, but the central venous access itself may be useful for pre-transfer optimization or inotrope administration.

Reference

Intensive Care Society (ICS). *Guidelines for the transport of the critically ill adult*. 3rd edition. London: ICS, 2011. http://www.ics.ac.uk/ics-homepage/guidelines-and-standards (accessed May 2015).

E28. You are administering prophylactic heparin to an intensive care patient as thromboprophylaxis.

Which of the following complications of heparin is associated with the highest mortality?

A. Haemorrhage
B. Type I thrombocytopenia (non-immune mediated)
C. Type II thrombocytopenia (immune mediated)
D. Arrhythmias
E. Osteoporosis

Answer: C

Short explanation

Although haemorrhage due to a relative overdose is the most common complication of heparin use, immune-mediated thrombocytopenia (type II) has the highest mortality associated with it. This is due to the thrombotic complications, either venous or arterial. Osteoporosis is associated with long-term use, and it is hypotension, rather than arrhythmias, that can complicate rapid injection.

Long explanation

Heparins are widely used in hospitals as thromboprophylaxis, but they are not without significant side effects. Heparin-induced thrombocytopenia (HIT) is the most serious of these. It can clinically be differentiated into two types. Type I is a thrombocytopenia without immune involvement, which is common and occurs within a few days, resolving without stopping heparin. Type II HIT is initiated by an immune response to heparin. This causes a thrombocytopenia but also a significant thrombosis risk, particularly pulmonary thromboembolism. Type II HIT typically occurs between 4 days and 2 weeks, although this can be sooner in patients who have previously been exposed.

Diagnosing HIT poses a challenge in the intensive care environment where a number of factors can cause a thrombocytopenia. The "Four T's" scoring system is based on four questions: Thrombocytopenia; Time to thrombocytopenia; Thrombosis or other complication; alternative explanation. It allows evaluation of the pre-test probability of HIT. Those with pre-test scores <4 are unlikely to test positive for HIT antibodies, and so performing the test is of little value.

	0 points	1 point	2 points
Platelet count decrease	<30%	30–50%	>50%
Time to thrombocytopenia	≤3 days without recent exposure	≥5 days, but uncertain	5–10 days
Sequelae	None	Suspected, progressive or recurrent thrombosis; other skin lesions	New thrombosis; skin necrosis
Alternative explanation for thrombocytopenia	Definite	Possible	None

If HIT is suspected, no further heparin should be given to the patient. Once heparin has been stopped, the risk of thrombosis slowly decreases, but it remains high for the first few days, and so alternative anticoagulation is required. Options include direct factor Xa inhibitors (e.g. danaparoid), direct thrombin inhibitors (e.g. lepirudin and argatroban) and fondaparinux, which, despite producing heparin antibodies, do not seem to induce HIT.

References

Peck T, Hill S, Williams M. Chapter 23: Drugs affecting coagulation, in Section 4: Other important drugs. In *Pharmacology for Anaesthesia and Intensive Care*. 3rd edition. Cambridge: Cambridge University Press, pp 340–341.

Sakr Y. Heparin-induced thrombocytopenia in the ICU: an overview. *Crit Care Med*. 2011;15:211.

E29. You are called to the emergency department as part of the major trauma team. A 33-year-old woman has been run over by a bus and has an unstable pelvic fracture. She has the following findings on primary survey:

Airway:	Intact
Breathing:	Respiratory rate 30/min; SpO$_2$ 97% on 15 l oxygen; bilateral air entry, no added sounds
Circulation:	Heart rate: 135; blood pressure 75/45; cold peripheries; no external haemorrhage; deformed pelvis
Disability:	GCS 15/15; pupils equal and reactive; in pain

Whole body computed tomography (CT) scan shows a isolated, "open-book" pelvic injury. She has had intravenous access established and oxygen given but no other treatment.

What is the most appropriate next step in her management?

A. Interventional radiology for angio-embolization
B. Immediate transfer to theatre for pelvic fixation
C. Attempted stabilization of pelvic fracture using an external binder
D. Rapid transfusion of 1000 ml crystalloid before blood products
E. Further plain radiographs of her pelvis including Judet views

Answer: C

Short explanation

This patient is haemodynamically compromised with an unstable pelvic fracture. Pelvic binding reduces the volume of a disrupted pelvic ring and aids tamponade of bleeding. If she remains unstable, rapid further intervention is necessary, which may include external pelvic fixation or angio-embolization. Judet views involve abduction of the lower limb and are clearly inappropriate here. Transfusion of blood products should commence immediately.

Long explanation

Pelvic fractures causing haemodynamic instability usually occur as the result of blunt trauma and often associated with other significant injuries. Mortality estimates for patients presenting with shock and isolated pelvic fractures vary in different case series but are in the region of 10% to 40%. The primary mechanisms of death are exsanguination (within 24 hours of injury) and multiorgan failure (after 24 hours).

'Open-book pelvic fracture' is a general term describing an injury leading to disruption of the pelvic ring, usually involving widening of the symphysis pubis. Mechanically unstable pelvic fractures typically incorporate multiple bony or ligamentous injuries to the pelvis, owing to its structure as a bony ring. The greater the degree of disruption, the greater the risk of major haemorrhage from damaged veins and arteries. In the majority of cases, venous disruption predominates.

Mechanically unstable pelvic injuries lead to an increase in the volume of the pelvis, which leads to a reduction of the tamponade effect of the pelvic viscera and other tissues. Reduction of this volume by pelvic binding, clamps or external fixation devices is thought to reduce haemorrhage. The exact type of device used is unimportant, but the time taken for application and pelvic reduction is key. Hence, simple measures such as binding with a bedsheet secured with towel clips is likely to be as effective as another pelvic orthotic device.

If haemodynamic stability is achieved with binding and fluid resuscitation, then further imaging may be appropriate. If the patient remains unstable, then he or she should undergo surgery to include external pelvic fixation/pelvic packing or emergent angio-embolization, depending on availability and clinical findings.

References

Cullinane DC, Schiller HJ, Zielinski MD, et al. Eastern Association for the Surgery of Trauma practice management guidelines for hemorrhage in pelvic fracture – update and systematic review. *J Trauma Inj Infect Crit Care*. 2011;71(6):1850–1868.

Rajab TK, Weaver MJ, Havens JM. Technique for temporary pelvic stabilization after trauma. *N Engl J Med*. 2013;369(17):e22.

Smith W, Williams A, Agudelo J, et al. Early predictors of mortality in hemodynamically unstable pelvis fractures. *J Orthop Trauma*. 2007;21(1):31–37.

E30. You have just inserted a radial arterial line into a patient, but when you connect it to the monitor, the display shows a hyper-resonant trace:

Which of the following is the most likely reason for this?

A. Using a long arterial cannula
B. Using a wide arterial cannula
C. Using a stiff cannula
D. Clot formation in the cannula
E. Air bubble in the catheter tubing

Answer: A

Short explanation

Clot formation or air in the catheter tubing will lead to damping of the system, while using a wide, stiff, and short cannula may raise the resonant frequency of the system above that seen in arterial pressure waveform. Therefore, using a long arterial cannula is most likely to cause resonance.

Long explanation

The arterial pressure monitoring system is susceptible to a number of factors that can influence its accuracy, and appearance of the waveform. If the system has been set up and connected correctly, with the transducer at the correct height, then two of the other potential sources of error are from damping or resonance.

Damping is the process in which there is restriction or resistance between the arterial line in the artery and the transducer diaphragm. This is a common occurrence, and frequent causes are kinks in the cannula or tubing, clot formation or air bubbles. Air bubbles are much more compressible than fluid, and so these act as shock absorbers, compressing during systole, and relaxing during diastole. The result of damping is reduced movement of the transducer diaphragm, which leads to reduced spread of the arterial waveform, although the mean arterial pressure remains largely accurate.

Resonance is a more challenging concept. All systems have a resonant frequency, which is a frequency at which oscillations seem to become amplified. One way of visualizing this is to imagine a pendulum or swing. If the pendulum weight, or person on the swing, is pushed intermittently at around the resonant frequency, the oscillations will propagate and get larger. A similar process can be applied to the invasive pressure monitoring system. It is being subjected to pressure changes from the arterial pressure waveform, which is usually up to 40 Hz, and if the resonant frequency of the pressure monitoring system falls within this range, then resonance will be evident on the displayed waveform. If the resonant frequency can be elevated above this level, then resonance should be avoided. This is best achieved by using a short, wide cannula with non-distensible walls.

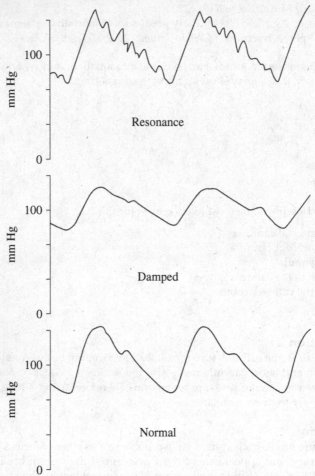

Reference

Davis P, Kenny G. *Basic Physics and Measurement in Anaesthesia*. 5th edition London: Elsevier, 2003.

Exam F: Questions

F1. You are asked to review a previously fit and well 60-year-old woman on the medical admissions ward who has been admitted complaining of shortness of breath and lethargy, she is tachycardic (110 bpm) and blood tests reveal anaemia. There is no evidence of external or internal bleeding.

Hb 75 g/l	(110–140 g/l)
MCV 122 fl	(77–95 fl)
Reticulocytes – 0.8%	(<2%)
Platelets – 189 × 10⁹/l	(150–400 × 10⁹/l)
WCC – 8.4 × 10⁹/l	(4–11 × 10⁹/l)
Ferritin – 100 ng/ml	(12–150 ng/ml)
LDH – 400 U/l	(240–480 U/l)

Given these results, what is the most appropriate first line treatment for the underlying condition?

A. IV Iron
B. Folate
C. Vitamin B$_{12}$
D. IV methylprednisolone
E. Packed red blood cells

F2. According to international consensus, which of the following guidelines is most appropriate for reviewing a meta-analysis of papers looking at ventilator associated pneumonia interventions on ICU?

A. QUORUM
B. SQUIRE
C. CONSORT
D. MOOSE
E. STROBE

F3. You have been asked to review an inpatient who has developed signs of infection. Which of the following is the most common health care–associated infection (HCAI) in adult patients?

A. Respiratory tract infection
B. Surgical site wound infection
C. Urinary tract infection
D. Catheter-related bloodstream infection (CRBSI)
E. *Clostridium difficile* diarrhoea

F4. A 68-year-old diabetic woman presents to hospital with nonspecific symptoms of being generally unwell. Her blood glucose is 42 mmol/l, ketones are 1.5 mmol/l and pH 7.33. Her osmolality is 354 mOsm/l and Na$^+$ is 150 mmol/l. She has evidence of an acute kidney injury, and her creatinine is 254 μmol/l.

Regarding her management which of the following is LEAST appropriate?

A. A fixed rate insulin infusion at 0.05 units/kg/hr once fluids have been commenced should be started
B. This patient needs consideration for ongoing care on the high dependency unit
C. Intravenous 0.9% saline should be administered initially for fluid replacement
D. Prophylactic anticoagulation should be administered unless contraindicated
E. Fluid replacement with 0.45% saline should be started once the osmolality stops decreasing with 0.9% saline

F5. Excessive use of which of the following agents is LEAST likely to cause diarrhoea?

A. Senna
B. Ispaghula husk
C. Sodium docusate
D. Magnesium hydroxide
E. Lactulose

F6. In which of the following groups is the Modification of Diet in Renal Disease (MDRD) calculation for estimating glomerular filtration rate most likely to give an accurate value?

A. The morbidly obese
B. Pregnant women
C. Afro-Caribbean populations
D. Bodybuilders
E. Those with chronic neuromuscular disease

F7. Which of the following is the most important mechanism in preventing health care–associated infections (HCAI)?

A. Ensuring adequate staffing levels
B. Antibiotic stewardship
C. Routine surveillance cultures
D. Good hand hygiene
E. Use of personal protective equipment

F8. A 67-year-old patient has been treated for pneumonia in your intensive care unit (ICU) and is now recovering. He has been demonstrating signs of agitation, and so you decide to initiate the CAM-ICU screening tool for delirium. You explain the test to him, asking him to squeeze your hand when he hears the letter 'A'. You then read out the letter string 'S A V E A H A A R T', as detailed in your unit policy for delirium screening. He squeezes your hand four times, three times in response to 'A' and once to 'E'.

What is the most appropriate step to take now?

A. Repeat the test, giving him a further opportunity to improve his responses
B. Stop the test, the patient does not have delirium
C. Stop the test, the patient has delirium
D. Ask the four 'disordered thinking' questions
E. Assess the patient's Richmond Agitation Sedation Score (RASS)

F9. You are performing fibre-optic bronchoscopy on an intubated, sedated patient in your intensive care unit. After insertion of the scope into the endotracheal tube and advancement, you are looking to identify structures to orientate yourself. Advancing the scope, you can see, immediately ahead, a trifurcation of three equal-sized openings arranged like a Mercedes Benz logo.

Where is the scope tip most likely to be?

A. Left lower lobe
B. Left main bronchus
C. Right lower lobe
D. Right bronchus intermedius
E. Right upper lobe

F10. A 63-year-old previously completely fit and well female patient is admitted with left-sided weakness. Neurological assessment is consistent with a right-sided cerebrovascular accident. At 4 hours after onset of symptoms, her blood pressure is 178/100, and a computed tomography (CT) scan demonstrates evidence of an acute stroke, but not of raised intracranial pressure or intracerebral haemorrhage.

Which of the following is the BEST initial treatment for this patient?

A. Commence treatment with statins
B. Perform a decompressive hemicraniectomy
C. Commence treatment with aspirin
D. Commence antihypertensives
E. Thrombolyze with alteplase

F11. A 52-year-old patient is on the high dependency unit, 5 days after an emergency abdominal aortic aneurysm (AAA) repair. He has a persistent blood pressure of more than 185/120 mmHg with no infusions running. He appears calm and denies any pain.

Which of the following treatment strategies would be most appropriate?

A. Intravenous (IV) sodium nitroprusside infusion 0.25 to 0.5 µg/kg/min titrated to effect
B. IV labetalol 20 mg bolus followed by 1 to 2 mg/min titrated to effect
C. PO ramipril 5 mg OD
D. IV alfentanil infusion, titrated to effect
E. IV clonidine 150-µg bolus followed by 1.2 to 7.2 µg/min infusion titrated to effect

F12. An intensive care unit (ICU) with 12 beds currently has 10 patients. Six of the patients are ventilated, and two more are requiring renal replacement therapy. One other patient has no organ support requirements but has an arterial line and central line for monitoring, and the final patient has just returned from major elective surgery but is not requiring any organ support or invasive monitoring.

Which of the following MINIMUM staffing arrangements is most appropriate?

A. One consultant, one registrar, a total of 10 nurses, including one band 6/7
B. Two consultants, two registrars, a total of 10 nurses, including one band 6/7
C. One consultant, two registrars, a total of 12 nurses, including one band 6/7
D. One consultant, two registrars, a total of 11 nurses, including one band 6/7
E. One consultant, one registrar, a total of 11 nurses, including one band 6/7

F13. A 65-year-old is admitted with an anterior ST-elevation myocardial infarction (STEMI). He had no lesion amenable to percutaneous intervention at angiogram, and his echo revealed left ventricular (LV) dysfunction with an ejection fraction of 35%. His renal function is normal.

Which of the following would NOT be appropriate to start as first-line secondary prevention treatments for this patient?

A. Eplerenone
B. Atorvastatin
C. Losartan
D. Bisoprolol
E. Dual antiplatelet therapy

F14. A 35-year-old male patient with known oesophageal varices presents with a profuse ongoing massive gastrointestinal bleed. He has a heart rate of 120 and BP is 88/65 on presentation. He has no known allergies.

Fluid resuscitation is commenced. Which of the following should be instigated NEXT?

A. Administer terlipressin
B. Insert a Sengstaken-Blakemore tube
C. Commence beta-blockers
D. Perform a shunt procedure (transjugular intrahepatic portosystemic shunt; TIPSS)
E. Perform an endoscopy (oesophago-gastro-duodenoscopy)

F15. You are called to review a patient in the resuscitation room of ED who has a working diagnosis of viral pneumonia secondary to suspected infection with influenza virus H1N1. He is in significant respiratory distress and has saturations of 82% on 15 l oxygen via a non-rebreathe mask. You decide that he requires immediate intubation and ventilation.

Of the following, which is the LEAST important to implement in this case to protect staff and stop the spread of his suspected H1N1 influenza?

A. Wearing a fluid repellent gown
B. Wearing goggles or a visor
C. Ensuring good hand hygiene
D. Wearing nonsterile gloves
E. Intubating the patient in a side room

F16. A patient in the intensive care unit is recovering from abdominal sepsis and has an albumin of 18 mmol/l. He has developed a unilateral pleural effusion, detected on chest X-ray. What is the most appropriate course of action?

A. Insert a large-bore ('surgical') chest drain to drain the effusion
B. Perform a diagnostic pleural tap and await results before deciding whether to drain
C. Perform a diagnostic pleural tap and insert a small-bore ('Seldinger') chest drain
D. Obtain a chest ultrasound
E. Obtain chest computed tomography (CT)

F17. You have just intubated a 25-year-old woman with a life-threatening acute exacerbation of asthma.

Which of the following initial ventilator settings, with reference to the tidal volume (V_T), set respiratory rate (RR), and inspiratory to expiratory ratio (I:E), would be best?

A. A V_T of 600 ml, RR of 10, I:E of 1:4
B. A V_T of 400 ml, RR of 10, I:E of 1:4
C. A V_T of 600 ml, RR of 16, I:E of 1:4
D. A V_T of 400 ml, RR of 16, I:E of 1:6
E. A V_T of 600 ml, RR of 16, I:E of 1:6

F18. You are called to urgently review a 60-year-old man on the coronary care unit who underwent coronary angiography 5 hours ago, with stents to both his left anterior descending (LAD) and right marginal arteries. He has a heart rate of 120, a blood pressure of 75/40, and a central venous pressure (CVP) of 25 mmHg. His cardiac monitor shows electrical alternans.

Which is the following is the best investigation to determine the underlying cause of his current state?

A. 12-lead electrocardiogram (ECG)
B. Portable chest X-ray (CXR)
C. Central venous SpO_2
D. Transthoracic echocardiogram
E. Repeat angiography

F19. You are asked by the nurses to review a patient who has been receiving drugs via a 20-gauge cannula placed yesterday into a vein in the dorsum of the right hand. The cannula has 'tissued', and there has been extravasation of the drugs.

Which of the following statements is the most accurate?

A. Atracurium usually causes no effect, and no further action is needed
B. Pain on injection of propofol is likely to suggest extravasation
C. Cytotoxic agents should be managed using the 'spread and dilute' technique with a saline washout
D. Ketamine causes ischaemia and should be managed using the 'spread and dilute' technique with a saline washout
E. Phenytoin causes necrosis and should be managed with a stellate ganglion block

F20. You are about to discharge a patient from the intensive care unit (ICU) back to a general ward at the weekend.

Which of the following statements about ICU discharge is the LEAST accurate?

A. Discharges between 22:00 and 07:00 should be documented as an adverse incident
B. Patients discharged from ICU should have access to an ICU follow-up clinic
C. Discharge to the ward should occur within 6 hours of the decision being made
D. Handover must include a plan for ongoing treatment
E. Handover must include psychological and emotional needs

F21. A nurse has just inserted a nasogastric (NG) tube into an intubated patient and has asked you to confirm its position. You are able to aspirate some fluid.

Which of the following is the recommended way to confirm the position?

A. pH ≤5.5
B. pH <6 in a patient on omeprazole
C. Chest X-ray showing NG tube in stomach
D. Aspiration of 20 ml of brown fluid
E. Positive 'whoosh' test

F22. A patient is likely to need total parenteral nutrition (TPN) for 7 to 10 days. The patient does not otherwise need central venous access.

Which of the following is the LEAST acceptable route of administering this?

A. A dedicated peripherally inserted central catheter (PICC)
B. A dedicated internal jugular vein central catheter
C. A dedicated lumen of a 4-lumen internal jugular vein central catheter
D. A dedicated peripheral venous catheter
E. A dedicated tunnelled subclavian central catheter

F23. You are called urgently to the emergency department to assist with the management of a 53-year-old who suffered an out of hospital cardiac arrest. Before his collapse he reported crushing central chest pain with radiation to his left arm. His initial rhythm was pulseless electrical activity, but after 1-mg intravenous adrenaline, he showed signs of life, and cardiopulmonary resuscitation was stopped. He is now unresponsive to painful stimuli, has a heart rate of 40 bpm and complete heart block on the electrocardiogram (ECG). His blood pressure is 60/30 mmHg, and his oxygen saturations are 95% on 15 l O_2 via a non-rebreathe mask.

Which coronary artery is most likely to be affected?

A. Left anterior descending artery (LAD)
B. Circumflex artery
C. Right marginal artery
D. Posterior interventricular artery
E. Right coronary artery

F24. You are called urgently to the emergency department to help with the initial assessment and management of a 6-year-old boy who has been involved in a house fire. He has 50% total body surface area burns including his face, and he is unconscious. You need to intubate the child.

Which of the following is the most appropriate?

A. Rapid sequence induction (RSI) with propofol and atracurium inserting a cut 5.5 mm endotracheal tube to 15 cm at the lips

B. RSI with propofol and atracurium inserting an uncut 5.0 mm endotracheal tube to 17 cm at the lips

C. RSI with propofol and suxamethonium inserting a cut 5.5 mm endotracheal tube to 17 cm at the lips

D. RSI with propofol and suxamethonium inserting an uncut 5.5 mm endotracheal tube to 15 cm at the lips

E. RSI with propofol and atracurium inserting a cut 6.0 mm endotracheal tube to 17 cm at the lips.

F25. You have just intubated a patient in the intensive care unit. Although you have seen the endotracheal tube pass through the cords, you want to confirm the presence of carbon dioxide on expiration.

Which of the following is the most appropriate device to do this?

A. Severinghaus electrode
B. Gas chromatograph
C. Paramagnetic analyzer
D. Infrared analyzer
E. Mass spectrometer

F26. You are ventilating a 40-year-old male patient who has a persistent bronchopleural fistula and collapse of the left lung despite two chest drains correctly placed on that side. You decide to insert a double-lumen endotracheal tube and commence one-lung ventilation.

Which of the following cases is the WEAKEST indication for one-lung ventilation?

A. Bronchopleural fistula compromising ventilation
B. Traumatic left main bronchus rupture
C. Large left sided endobronchial haemorrhage
D. Large unilateral apical bullae at risk of rupture
E. Emergency thoracic aortic aneurysm repair

F27. You are commencing total parenteral nutrition in a malnourished patient.
Which of the following patients is MOST at risk of refeeding syndrome?

A. A 35-year-old with a body mass index (BMI) of 15.5 kg/m^2
B. A 35-year-old with unintended weight loss of 17% in the past 6 months
C. A 35-year-old alcoholic with no nutritional intake for 12 days
D. A 35-year-old with low potassium, phosphate and magnesium before commencement of feeding
E. A 35-year-old with a BMI of 17 kg/m^2 with unintended weight loss of 8% in the past 3 months

F28. You are asked by the nursing staff to examine a patient. He is a 33-year-old man who was admitted 3 days ago with bilateral pneumonia. He is intubated and ventilated according to lung-protective ventilation strategies and is receiving appropriate antimicrobial therapy. His ventilator settings are as follows: volume control ventilation; FiO_2 0.8%; V_T 450 ml; respiration rate 30/min; positive end expiratory pressure 14 cmH$_2$O. His latest arterial blood gas analysis shows the following: pH 7.26; pO_2 6.85 kPa; pCO_2 7.51 kPa.

Which pharmacological therapy has the best evidence of benefit in such cases?

A. Methylprednisolone
B. Cis-atracurium
C. Prostacyclin (inhaled)
D. Nitric oxide (inhaled)
E. Surfactant

F29. A 16-year-old male patient is admitted to hospital with pyrexia, headache, neck pain and photophobia.

What is the most appropriate treatment regimen?

A. IV cefotaxime and amoxicillin
B. IV ceftriaxone and amoxicillin
C. IV cefotaxime only
D. IV ceftriaxone, amoxicillin and dexamethasone
E. IV ceftriaxone and dexamethasone

F30. A 38-year-old man has been involved in a road traffic collision and has sustained abdominal injuries. He has linear bruising horizontally and diagonally across his abdomen in keeping with seatbelt positioning. Intravenous (IV) access has been secured, and he has already received 2 units of packed red cells and 500 ml of crystalloid. His primary survey findings are as follows:

Airway:	Intact, talking
Breathing:	Respiratory rate: 28/min, SpO$_2$ 100% on 15 L oxygen; no signs of thoracic injury
Circulation:	Heart rate: 135 bpm, blood pressure: 70/56 mmHg, cool peripheries; seatbelt sign; slight abdominal tenderness
Disability:	Glasgow Coma Score 15/15; pupils size 4 mm, reactive; in pain
Exposure:	No signs of bony or spinal injuries

An ultrasound (focused assessment with sonography in trauma [FAST] scan) of his abdomen is performed, which shows significant intraperitoneal fluid.

What is the next step of his management?

A. Immediate transfer to theatre for laparotomy
B. Further boluses of IV crystalloid
C. Computed tomography (CT) scan of chest, abdomen and pelvis
D. Diagnostic peritoneal lavage (DPL)
E. Plain abdominal radiograph

Exam F: Answers

F1. You are asked to review a previously fit and well 60-year-old woman on the medical admissions ward who has been admitted complaining of shortness of breath and lethargy, she is tachycardic (110 bpm) and blood tests reveal anaemia. There is no evidence of external or internal bleeding.

Hb 75 g/l	(110–140 g/l)
MCV 122 fl	(77–95 fl)
Reticulocytes – 0.8%	(<2%)
Platelets – 189 × 10⁹/l	(150–400 × 10⁹/l)
WCC – 8.4 × 10⁹/l	(4–11 × 10⁹/l)
Ferritin – 100 ng/ml	(12–150 ng/ml)
Lactate dehydrogenase – 400 U/l	(240–480 U/l)

Given these results, what is the most appropriate first line treatment for the underlying condition?

A. IV iron
B. Folate
C. Vitamin B₁₂
D. IV methylprednisolone
E. Packed red blood cells

Answer: C

Short explanation
The diagnosis here is megaloblastic anaemia, for which the initial management should include replacement of folate and B₁₂ stores. Replacement of B₁₂ should occur first to avoid a sudden drop in serum B₁₂ and the development of neurological sequelae.

Long explanation
In an otherwise healthy person, who is not acutely bleeding and haemodynamically stable, a transfusion trigger of below 70 g/l is appropriate. If acute coronary syndrome or severe angina were present, transfusion may be appropriate but should be done cautiously so as not to lead to fluid overload.

Anaemia is common and can be due to increased losses or decreased red cell production. Losses may be acute, occult or intravascular (haemolysis). Sources of acute blood loss should be sought through history, examination and investigations. Occult blood loss should be considered, with the most common cause in this age group being a bowel malignancy. This would typically lead to anaemia due to iron deficiency, with microcytosis, reticulocytosis and a low ferritin level.

Causes of decreased red cell production can be divided between those which affect all cell lines such as leukaemia and bone marrow failure and those affecting only red cell production. Diseases affecting all cell lines would demonstrate reduced platelet and leucocyte counts in addition to anaemia.

The mean corpuscular volume (MCV) denotes the average size of the red blood cell in circulation. As cells mature and divide in the marrow, they shrink. Therefore a high MCV is indicative of circulating premature red cells, or those which have failed to fully divide. A common cause for this includes a lack of the elements needed for cell division – including B_{12} and folate. Small red cells with a low MCV are typically those which are lacking in their main constituent, haemoglobin, most commonly because of a deficiency in iron.

Iron deficiency anaemia may be treated with iron replacement. Typically this is enteral, although IV iron replacement is increasingly common. Both B_{12} and folate deficiencies may be effectively treated by replacement of the deficient nutrient. If both are low, B_{12} should be replaced first because replacing folate alone may precipitate a further fall in B_{12}, leading to neurological sequelae. A lack of B_{12} may cause subacute degeneration of the spinal cord or damage to peripheral or optic nerves. Onset is often with paraesthesia or numbness in the peripheries and may degenerate to include weakness and ataxia.

Reference
Craig JIO, McClelland DBL, Ludlam CA. Blood disorders. In *Davidson's Principles and Practice of Medicine.* 20th edition. Edinburgh: Churchill Livingstone Elsevier, 2006, p 1029.

F2. According to international consensus, which of the following guidelines is most appropriate for reviewing a meta-analysis of papers looking at ventilator-associated pneumonia interventions in the intensive care unit?

A. QUORUM
B. SQUIRE
C. CONSORT
D. MOOSE
E. STROBE

Answer: A

Short explanation
CONSORT (Consolidated Standards of Reporting Trials) guidelines are for randomized trials, QUORUM (Quality of Reporting of Meta-analyses) supports meta-analyses of trials such as those given in the example and MOOSE (Meta-analysis Of Observational Studies in Epidemiology) is for meta-analyses specific to epidemiology, STROBE (Strengthening the Reporting of Observational Studies in Epidemiology) is for observational epidemiological studies and SQUIRE (Standards for Quality Improvement Reporting Excellence) for quality improvement work.

Long explanation
In an attempt to improve the standards of publication, the International Committee of Medical Journal Editors (ICMJE) has published guidelines for authors and editors

entitled 'Uniform Requirements for Manuscripts Submitted to Biomedical Journals: Writing and Editing for Biomedical Publication'. The purpose of these guidelines is to standardize and structure the reporting of scientific research.

Subsequently, more specific guidelines have been published to outline the structure and minimum information required to be published for different types of research. These are listed above. Probably the best known of these is the CONSORT guidelines, which apply to randomized control trials. The CONSORT guideline explains every step that researchers are required to make when planning a randomized controlled trial (RCT). It outlines the data that should be published, and many journals now require written confirmation that these guidelines have been followed. Many RCTs will publish the CONSORT diagram, which shows the steps of the research project in a flow-chart format, accounting for every patient recruited, followed up, lost and included.

Subsequent reporting guidelines have been developed for other models of trial design including quality improvement work. Many of these guidelines are under constant review, and updated versions appear. It is often helpful to review the guidelines before starting a research project. The guidelines are designed to minimize bias and variations in reporting, making it easier to search through published work and perform subsequent meta-analyses. Some publications will also pre-approve publication of larger pieces of research in an attempt to reduce reporting bias and encourage the reporting of results regardless of a positive or negative outcome.

The biggest challenge reported has been encouraging the uptake of the guidelines and ensuring researchers are both aware of the guidelines and have the resources to follow them appropriately.

Reference
Brand RA. Editorial: Standards of Reporting: The CONSORT, QUORUM, and STROBE Guidelines. *Clin Orthop Relat Res.* 2009;467(6):1393–1394.

F3. You have been asked to review an inpatient who has developed signs of infection. Which of the following is the most common health care–associated infection (HCAI) in adult patients?

A. Respiratory tract infection
B. Surgical site wound infection
C. Urinary tract infection
D. Catheter-related bloodstream infection
E. *Clostridium difficile* diarrhoea

Answer: A

Short explanation
Respiratory tract infection is the commonest HCAI identified on the most recent statistics from the Health Protection Agency.

Long explanation
Definition of HCAI is an infection occurring directly from contact with health care facilities or from medical interventions. Not only do they result in increased length of stay and morbidity and mortality for individual patients, they are also a financial burden due to the additional cost involved in diagnosing and treating them.

The latest statistics from the Health Protection Agency (2011) state that 6.4% of all inpatients in hospital are suffering from a HCAI at any one time. This rate increases to 23.4% in patients admitted on ICU. It is estimated that about 25 to 40% of these are preventable. Intervention programmes such as Matching Michigan have been implemented to try and prevent the development of HCAI. The Matching Michigan campaign reported a 60% reduction in the rate of BSI related to central venous catheter (central-line–associated and CRBSIs) in adult intensive care units in England.

The Health Protection Agency identified that respiratory tract infections were the commonest HCAIs and reported rates of different infections as follows:

- Respiratory tract infections 22.8%
- Urinary tract infections 17.2%
- Surgical site infections 15.7%
- Clinical sepsis 10.5%
- Gastrointestinal infections 8.8%
- Bloodstream infections 7.3%
- Others 17.7%

References

Bion J, Richardson A, Hibbert P, et al. 'Matching Michigan': a 2-year stepped interventional programme to minimise central venous catheter-blood stream infections in intensive care units in England. *BMJ Qual Saf.* 2013;22(2):110–123.

Health Protection Agency. English National Point Prevalence Survey on Healthcare-associated Infections and Antimicrobial Use. 2011. https://www.gov.uk/government/uploads/system/uploads/attachment_data/file/331871/English_National_Point_Prevalence_Survey_on_Healthcare_associated_Infections_and_Antimicrobial_Use_2011.pdf (accessed March 2015).

F4. A 68-year-old diabetic woman presents to hospital with nonspecific symptoms of being generally unwell. Her blood glucose is 42 mmol/l, ketones are 1.5 mmol/l and pH 7.33. Her osmolality is 354 mOsm/l and Na^+ is 150 mmol/l. She has evidence of an acute kidney injury, and her creatinine is 254 μmol/l.

Regarding her management which of the following is LEAST appropriate?

A. A fixed rate insulin infusion at 0.05 units/kg/hr once fluids have been commenced
B. This patient needs consideration for ongoing care on the high dependency unit
C. Intravenous 0.9% saline should be administered initially for fluid replacement
D. Prophylactic anticoagulation should be administered unless contraindicated
E. Fluid replacement with 0.45% saline should be started once the osmolality stops decreasing with 0.9% saline

Answer: A

Short explanation

This patient is in a hyperosmolar hyperglycaemic state. The mainstay of treatment is with fluids: 0.9% saline until the osmolality stops decreasing, at which point continual fluid replacement should be with 0.45% saline. Fixed rate intravenous insulin infusion should not be routinely started after fluids have been commenced, but only started if the glucose level is not decreasing with fluids alone. Unless contraindicated, patients should receive prophylactic anticoagulation.

Long explanation

Hyperosmolar hyperglycaemic state (HHS) is a condition that presents with hyperglycaemia, often with blood glucose levels >30 mmol/l. These patients have large fluid deficits (10–20 l water deficit) and an associated hyperosmotic state with a typical serum osmolality of >320 mOsm/kg. However, unlike diabetic ketoacidosis (DKA), significant hyperketonaemia or acidosis are not a feature of this condition, with usual ketone levels of <3 mmol/l, pH >7.3 and bicarbonate >15 mmol/l. Compared with patients with DKA, these patients have greater fluid deficits but less electrolyte derangement, the onset is more insidious, and they experience higher morbidity and mortality rates.

The mainstay of treatment for these patients is restoring the fluid deficit. Intravenous 0.9% saline should be used to restore the circulating volume and reverse dehydration. It can be swapped to 0.45% saline once the osmolality stops changing in response to fluids. The sodium concentration should decline by more than 10 mmol/l in 24 hrs. There may be an initial rise in the sodium concentration, but this should not trigger a swap to hypotonic fluids. By 12 hours, the patient should have a 3 to 6 l positive fluid balance with the remaining fluid deficit replaced over the next 12 hours. Full biochemical recovery may take up to 72 hours.

The glucose concentration should not decline by >5 mmol/l/hr. Insulin at 0.05 unit/kg/hr should be added if the glucose value is not decreasing with fluids alone or there is the presence of significant ketonaemia (>1 mmol/l).

Patients are at high risk of venous thromboembolism so prophylactic anticoagulation is required unless the patient has other contraindications.

Indications for consideration of management on critical care include the presence of the following:

- Biochemical abnormalities:
  o Osmolality >350 mOsm/kg
  o Na^+ >160 mmol/l
  o pH <7.1
  o K^+ derangement (<3.5 mmol/l or >6 mmol/l)
  o Creatinine >200 μmol/l
- Clinical condition:
  o Glasgow Coma Score <12
  o Oxygen saturations <92%
  o Systolic blood pressure <90 mmHg
  o Heart rate <60 or >100 beats/min
  o Urine output <0.5 ml/kg/hr
  o Hypothermia
  o Macrovascular event such as myocardial infarction or stroke
  o Other serious co-morbidity

Reference

Joint British Diabetes Societies Inpatient Care Group. The management of the hyperosmolar hyperglycaemic state (HHS) in adults with diabetes. 2012. http://www.diabetes.org.uk/Documents/Position%20statements/JBDS-IP-HHS-Adults.pdf (accessed October 14).

F5. Excessive use of which of the following agents is LEAST likely to cause diarrhoea?

A. Senna
B. Ispaghula husk
C. Sodium docusate
D. Magnesium hydroxide
E. Lactulose

Answer: B

Short explanation

Ispaghula husk is a bulk-forming laxative that increases stool volume and stimulates peristalsis. It can also be used occasionally in diarrhoea. Senna and docusate are stimulant laxatives that induce peristalsis and can cause diarrhoea. Lactulose and magnesium are osmotic agents that draw water into the gut lumen and can also cause diarrhoea.

Long explanation

Laxatives can be divided into bulk-forming, stimulants, faecal softeners and osmotic agents. There are separate bowel cleansing agents that include oral magnesium salts and contrast agents such as gastrografin.

Bulk-forming agents such as ispaghula husk work in a similar way to fibre and draw water in to the gut whilst increasing stool volume. Adequate hydration is required, and they can cause faecal impaction if patients are dehydrated.

Stimulants include senna, sodium docusate and sodium picosulphate. These agents increase motility and cause cramping. Symptoms can be worse if the stool is hard, and a softening agent may need to be co-administered. They can be dangerous in intestinal obstruction.

Sodium docusate also acts as a softening agent, along with drugs given as enemas, such as glycerol and arachis oil.

Osmotic laxatives include lactulose, magnesium hydroxide, phosphate enemas and macrogols. These large polymers draw water in to the bowel. Lactulose produces a low faecal pH and reduces the growth of ammonia-forming bacteria, hence its use in hepatic encephalopathy.

Reference

Joint Formulary Committee. *British National Formulary*. 69th ed. London: BMJ Group and Pharmaceutical Press; 2015.

F6. In which of the following groups is the Modification of Diet in Renal Disease (MDRD) calculation for estimating glomerular filtration rate most likely to give an accurate value?

A. The morbidly obese
B. Pregnant women
C. Afro-Caribbean populations
D. Bodybuilders
E. Patients with chronic neuromuscular disease

Answer: C

Short explanation

The MDRD calculation is commonly used by hospital labs to estimate glomerular filtration rate. The variables it considers are sex, creatinine and ethnicity, because Afro-Caribbeans have a higher muscle mass per body surface area. The calculation is not validated in pregnancy and will give incorrect values in patients with a high or low fat-to-muscle ratio, such as the morbidly obese, bodybuilders or those with muscle wasting.

Long explanation

Since 2006, estimated GFR (eGFR) has been calculated in the United Kingdom using the MDRD (Modification of Diet in Renal Disease) calculation, which takes in to account age, sex, creatinine and ethnicity (Afro-Caribbean or not). The Schwartz formula is used in children and takes in to account height, age and creatinine. Recently, the Chronic Kidney Disease Epidemiology Collaboration (CKD-EPI) formula has been developed, which is more accurate in patients with an eGFRs >60 ml/min/ 1.73 m^2.

GFR is the volume of fluid filtered through Bowman's capsule per unit time. Creatinine clearance is defined as the amount of creatinine removed from a volume of blood per unit time. Because creatinine is freely filtered, this is equivalent to the GFR. The Cockcroft-Gault formula was previously used to calculate creatinine clearance and takes into account age, sex, weight and creatinine.

When eGFR is estimated using creatinine, it is important to consider other factors that affect creatinine levels in the blood such as muscle mass. Creatinine is predominantly released from muscle, and blood creatinine levels will be much higher in patients with a large muscle mass, irrespective of renal function. To avoid these complications, other measures have been developed to estimate GFR. Cystatin C is a protein secreted by most cells in the body and freely filtered at the glomerulus. Calculations have been validated using adjusted Cystatin C, age, sex and race.

GFR is measured in millilitres/minute but is often corrected for body surface area and as such given in units millilitres/min/1.73 m^2.

References

Abdallah E, Waked E, Nabil M, El-Bendary O. Cystatin C as a marker of GFR in comparison with serum creatinine and formulas depending on serum creatinine in adult Egyptian patients with chronic kidney disease. *J Am Sci*. 2014;10(6):162– 169.

The Renal Association. About eGFR. http://www.renal.org/information-resources/ the-uk-eckd-guide/about-egfr#sthash.bdeBhUZI.dpbs (accessed April 2015).

F7. Which of the following is the most important mechanism in preventing health care–associated infections (HCAI)?

A. Ensuring adequate staffing levels
B. Antibiotic stewardship
C. Routine surveillance cultures
D. Good hand hygiene
E. Use of personal protective equipment

Answer: D

Short explanation

Antibiotic stewardship helps prevent the development of antibiotic resistance. Routine surveillance cultures help direct antibiotic choice once patients develop active infection. Adequate staffing levels, correct use of personal protective equipment and ensuring good hand hygiene are all important methods to prevent HCAI. It has been suggested that the latter is the single most important intervention to reduce the rate of HCAIs.

Long explanation

The epic3 guidelines have a number of recommendations to minimise the development of HCAIs. These recommendations include the following:

- Ensuring the hospital environment is clean and appropriately decontaminated according to the specific requirements of each different area as well as the presence of known or suspected contamination with specific pathogens.

- Medical equipment needs to be appropriately decontaminated between uses.
- The transmission of micro-organisms by hand transfer has been identified as being a major factor in the spread of HCAIs. It has been suggested that good hand hygiene is the single most important intervention to reduce this.
  o Hand decontamination is recommended
    - immediately before patient contact or performing clean or aseptic procedures;
    - immediately after patient contact, activities that may have resulted in exposure to body fluids and after contact with the patients environment; and
    - Immediately after removal of gloves.
  o Hand decontamination requires effective hand-washing technique. Alcohol-based hand rub is an appropriate alternative to soap and water unless hands are visibly soiled or patients have an illness associated with diarrhoea and vomiting.
  o Jewellery and false nails should be avoided to ensure hands can be effectively cleaned and decontaminated.
- Staff should use appropriate personal protective equipment.
- Sharps should be used and disposed of safely.
- Asepsis should be ensured during any invasive procedure.
- These guidelines also cover specific guidance regarding catheters and intravascular devices to minimize the risk of infection associated with their use.
  o Guidelines include ensuring a definite need for the device before insertion, choosing the correct site and device to be inserted, ensuring strict asepsis during insertion and whilst the device remains in situ, minimizing the risk of complications during insertion (such as use of the ultrasound for central lines), careful ongoing care and monitoring for complications as well as daily consideration regarding their continued need so timely removal is achieved.
  o Ventilator care bundle for intubated patients

Other institutional factors:

- Ensuring adequate staffing levels
- Appropriate isolation of infected patients
- Antibiotic stewardship helps prevent the development of antibiotic resistance
- Surveillance cultures and antimicrobial skin decontamination – chlorhexidine bathing patients colonized with methicillin-resistant *Staphylococcus aureus*

References

Dhillon MS, Shah KB, Rimawi RH. ICU infection control and preventive measures. http://esciencecentral.org/ebooks/bedside-critical-care-guide/icu-infection-control-and-preventive-measures.php (accessed March 15).

Loveday HP, Wilson JA, Prat RJ, et al. epic3: National evidence-based guidelines for preventing healthcare-associated infections in NHS hospitals in England. *J Hosp Infect*. 2014;86(Suppl 1):S1–S70.

F8. A 67-year-old patient has been treated for pneumonia in your intensive care unit (ICU) and is now recovering. He has been demonstrating signs of agitation, and so you decide to initiate the CAM-ICU screening tool for delirium. You explain the test to him, asking him to squeeze your hand when he hears the letter 'A'. You then read out the letter string 'S A V E A H A A R T', as detailed in your unit policy for delirium screening. He squeezes your hand four times, three times in response to 'A' and once to 'E'.

What is the most appropriate step to take now?

A. Repeat the test, giving him a further opportunity to improve his responses.
B. Stop the test, the patient does not have delirium
C. Stop the test, the patient has delirium
D. Ask the four 'disordered thinking' questions
E. Assess the patient's Richmond Agitation Sedation Score (RASS)

Answer: B

Short explanation
The patient has made two mistakes and scores 8/10. This is a pass, and the test should stop because the patient does not have delirium. If he had made more than two errors, the next step would be to assess his RASS before moving on to the questions to identify disordered thinking.

Long explanation
Delirium is an acute and fluctuating disorder of consciousness and cognition, which is reversible. It can be subdivided into hypoactive, hyperactive or mixed and is thought to be present in the majority of ICU patients. It is now recommended to routinely assess ICU patients for delirium alongside pain and agitation. Recent guidelines for assessment and management of these three conditions were published by the American College of Critical Care Medicine in 2013. The Intensive Care Society is due to update its 2006 guidelines in the near future.

Delirium is a source of increased morbidity and mortality in the ICU. Patients who develop ICU delirium have increased ICU and hospital length of stay and mortality and are more likely to develop a post-ICU cognitive impairment. There has been a recent focus on predicting, identifying and treating delirium. The four baseline risk factors are a history of dementia, alcoholism or hypertension and severity of illness at admission. Duration and type of sedation is linked to the development of delirium, with benzodiazepines particularly associated with increased risk of developing delirium.

The CAM-ICU (Confusion Assessment Method for ICU) questionnaire is a validated screening tool for delirium in ICU. It should be carried out daily on appropriate patients and has four parts. First, consider whether the patient has shown an acute fluctuation in mental status in the past 24 hours. If so, continue on to test for inattention (as detailed earlier). If the patient makes more than two errors (out of 10), proceed to test for an altered level of consciousness with the RASS score. If this is anything other than zero, the patient has delirium. If it is zero, move on to test for disordered thinking with four questions such as 'Will a stone float on water?' and 'Are there fish in the sea?' If the patient makes more than one error, he or she has delirium.

References
Barr J, Fraser GL, Puntillo K, et al. Clinical practice guidelines for the management of pain, agitation, and delirium in adult patients in the intensive care unit. *Crit Care Med*. 2013;41(1):263–306.

Ely EW, Margolin R, Francis J, et al. Evaluation of delirium in critically ill patients: validation of the Confusion Assessment Method for the Intensive Care Unit (CAM-ICU). *Crit Care Med*. 2001;29(7):1370–1379.

F9. You are performing fibre-optic bronchoscopy on an intubated, sedated patient in your intensive care unit. After insertion of the scope into the endotracheal tube and advancement, you are looking to identify structures to orientate yourself. Advancing the scope, you can see, immediately ahead, a trifurcation of three equal-sized openings arranged like a Mercedes Benz logo.

Where is the scope tip most likely to be?

A. Left lower lobe
B. Left main bronchus
C. Right lower lobe
D. Right bronchus intermedius
E. Right upper lobe

Answer: E

Short explanation
The opening of the right upper lobe is off the right main bronchus and immediately divides into three equal segments – apical, posterior and anterior segments. This is the only place in the upper bronchial tree where an equal trifurcation is commonly found. The right lower and middle lobes divide from the right bronchus intermedius and the left upper and lower lobes divide from the left main bronchus.

Long explanation
Understanding the bronchial anatomy is key to bronchoscopy. The trachea is formed of C-shaped cartilaginous rings with their opening on the posterior aspect of the trachea. It is over these openings that the trachealis muscle runs longitudinally down the trachea and into the bronchi. The trachealis muscle can thus be used for orientation while the bronchoscope is within the trachea.

The trachea divides at the carina into the left and right main bronchi (LMB and RMB). At this level, the oesophagus is immediately posterior to the trachea and the posterior wall of the trachea appears flattened compared with the semi-circular anterior wall. The LMB is much longer (5 cm) than the RMB, which gives off a branch to the right upper lobe (RUL) almost immediately. I a small percentage of patients their RUL comes directly off the trachea. From this point the RMB becomes the bronchus intermedius. The RUL bronchus opens into three equal segments at a trifurcation. These are the apical, anterior and posterior segments. The bronchus intermedius further divides into the right lower and middle lobe (RLL and RML) bronchi. The opening of the RLL is usually slightly larger than the RML. The RML consists of two segments – medial and lateral. The RLL consists of five segments: one superior and four basal (anterior, posterior, lateral and medial).

On the left, the LMB divides equally into the left lower and upper lobe (LLL and LUL) bronchi. The LUL divides into the upper and lower divisions. The upper division is analogous to the RUL and consists of the fused anterior/posterior segments and an apical segment. The inferior division is the lingula and consists of superior and inferior segments. The LLL bronchus gives off the superior segment first and then divides into either three or four basal segments – with some population variance. These are analogous to the four RLL basal segments, but in some cases the anterior and medial lobes are fused into one.

There are therefore 10 segments in the right lung and some variation between 9 or 10 segments in the left lung.

Reference

KL Moore, AMR Agur, AF Dalley. *Essential Clinical Anatomy*. 5th edition. Baltimore: Lippincott Williams & Wilkins.

http://www.thoracic-anesthesia.com (accessed April 2015).

F10. A 63-year-old previously completely fit and well female patient is admitted with left-sided weakness. Neurological assessment is consistent with a right-sided cerebrovascular accident. At 4 hours after onset of symptoms, her blood pressure is 178/100, and a computed tomography (CT) scan demonstrates evidence of an acute stroke, but not of raised intracranial pressure or intracerebral haemorrhage.

Which of the following is the BEST initial treatment for this patient?

A. Commence treatment with statins
B. Perform a decompressive hemicraniectomy
C. Commence treatment with aspirin
D. Commence antihypertensives
E. Thrombolyze with alteplase

Answer: E

Short explanation

Treatment with alteplase should be commenced because the patient is presenting within the recommended 4.5-hour time frame and has no contraindications to thrombolysis. Aspirin should be withheld for 24 hours if treatment with thrombolysis is initiated. Hypertension >185/110 mmHg requires treatment if thrombolysis is to be given. There are no indications for decompressive hemicraniectomy. Statins are not recommended to be started within 48 hours of an acute stroke.

Long explanation

The latest NICE guidance along with the American Heart Association/American Stroke Association (AHA/ASA) guidelines recommend thrombolysis for patients ≥18 years, presenting with acute ischaemic stroke within 4.5 hours from symptom onset. This is providing the absence of any secondary haemorrhage has been confirmed and the patient has no contraindications. Although the time frame for indication of thrombolysis from stroke onset has been increased in recent years, it should still be administered as soon as possible to decrease mortality and the rate of severe disability.

Contraindications to thrombolysis include the following:

- Patients at increased risk of bleeding that would result in mortality or significant morbidity, including active bleeding, intracranial neoplasm/arteriovenous malformation, stroke/significant head injury <3 months, recent intracranial/spinal surgery, abnormal clotting/thrombocytopenia, anticoagulant use, uncontrolled hypertension or massive infarct (hypodensity >1/3 cerebral hemisphere)
- Patients whose symptoms may be potentially reversible, such as hypoglycaemia
- Relative contraindications include pregnancy, recent surgery or myocardial infarction, post-ictal state or patients experiencing only minor or rapidly improving neurological symptoms

Guidelines for contraindications are stricter for patients thrombolyzed between 3 and 4.5 hours due to the risk/benefit ratio.

Thrombolysis should not be given if the BP is >185/110 mmHg. The BP should be controlled <185/110 mmHg (NICE guidance) and <180/110 mmHg (AHA/ASA

241

guidance) for 24 hours after the administration of thrombolysis. If thrombolysis is not indicated antihypertensives have not been recommended unless the BP is >200–220/120 mmHg. Although aspirin is recommended to start within 24 to 48 hours of ischaemic stroke, it should be withheld for 24 hours in patients who receive thrombolysis.

Younger patients with larger middle cerebral infarcts benefit from early decompressive surgery. Early decompressive hemicraniectomy to be performed <48 hours of symptom onset has been recommended by NICE for patients ≤60 years, who have a large middle cerebral artery infarcts with evidence of infarction of >50% of its territory and with neurological findings consisting of a reduced level of consciousness and a NIHSS score of >15.

Guidance also recommends good supportive treatment:

- Supplemental oxygen to maintain sats ≥95%
- Glucose control, maintaining glucose level between 4 and 11 mmol/l
- Nutrition:
  o Perform a swallow assessment before any oral intake. If any concerns exist, a formal SALT review should be sought within 72 hours (preferably 24 hours).
  o Feeding should be commenced within 24 hours (nasogastric if oral not acceptable).
  o Patients should be screened for malnutrition.
- Physiotherapy and early mobilization as soon as patients are able
- Treat fevers >38°C (treat the source and give antipyretics).

References

Jauch EC, Saver JL, Adams HP, et al. Guidelines for the early management of patients with acute ischemic stroke. A guideline for healthcare professionals from the American Heart Association/American Stroke Association. *Stroke* 2013;44(3):870–947.

National Institute for Health and Care Excellence (NICE). *Alteplase for treating acute ischaemic stroke* (TA264). London: NICE, 2012. http://nice.org.uk/guidance/ta264 (accessed March 2015).

National Institute for Health and Care Excellence (NICE). *Stroke: diagnosis and initial management of acute stroke and transient ischaemic attack (TIA)* (CG68). London: NICE, 2012. http://www.nice.org.uk/guidance/cg68 (accessed March 2015).

F11. A 52-year-old patient is on the high dependency unit, 5 days after an emergency abdominal aortic aneurysm (AAA) repair. He has a persistent blood pressure of more than 185/120 mmHg with no infusions running. He appears calm and denies any pain.

Which of the following treatment strategies would be most appropriate?

A. Intravenous (IV) sodium nitroprusside infusion 0.25 to 0.5 µg/kg/min titrated to effect
B. IV labetalol 20 mg bolus followed by 1 to 2 mg/min titrated to effect
C. PO ramipril 5 mg OD
D. IV alfentanil infusion, titrated to effect
E. IV clonidine 150-µg bolus followed by 1.2 to 7.2 µg/min infusion titrated to effect

Answer: B

Short explanation

This is post-operative hypertension at a level high enough to cause concern and therefore warranting treatment. Oral agents may be too slow to take effect and we have no information here about gastrointestinal function, which may be limited after AAA surgery. There is no pain and therefore no indication for alfentanil or opiates. IV SNP is not appropriate for first-line management. Both IV labetalol and clonidine could be used, but it is not recommended to give clonidine as a bolus in this setting because it can cause sudden hypotension.

Long explanation

Blood pressure control after vascular surgery can be difficult. The emergency AAA patient population often has underlying hypertension and may or may not have previously been on treatment. Complicating factors such as pain, anxiety and vasopressor infusions may also mask underlying blood pressure problems.

Hypertension is classified as stage 1 (SBP ≥140 mmHg or DBP ≥90 mmHg), stage 2 (SBP >160 mmHg or DBP >100 mmHg) or stage 3 (SBP >180 or DBP >110 mmHg). A hypertensive crisis or emergency is one featuring acute end-organ dysfunction and requires immediate treatment. Equally, other causes of acute hypertension (such as pain and anxiety) should be sought and treated first.

This case features stage 3 hypertension that probably requires treatment, but not as an emergency. Treatment should be given in such a way that facilitates a slow and titrated reduction in blood pressure, to avoid a sudden rise or fall.

Treatment options for intravenous infusion include alpha-2 agonists such as clonidine, which act on central adrenoceptors to reduce endogenous noradrenaline release. However, a bolus dose of 150 µg IV is large and likely to exert a rapid fall in blood pressure, possibly associated with rebound hypertension. Other options include nitrates such as GTN or sodium nitroprusside, which are dilators of both the arterial and venous systems, causing a drop systemic vascular resistance. They require careful monitoring and adjustment because the complications include exacerbation of pulmonary shunt, increase in intracranial pressure, coronary steal syndrome, tachyphylaxis and the production of active metabolites including cyanide. It is therefore often only used as a last-line treatment option.

Strategies to reduce heart rate or contractility may need to be considered after assessment of the patient's cardiovascular risk factors and heart rate. Labetalol is a mixed alpha- and beta-adrenoceptor antagonist that is useful for managing hypertension in the intensive care unit population but that should be used cautiously in the presence of heart failure.

Another approach may be to reduce pre-load with a diuretic. Angiotensin-converting enzyme inhibitors exert their action on the renin-angiotensin aldosterone system and a reduction in glomerular filtration rate is seen with an associated rise in creatinine. They should be used with caution in patients with an acute kidney injury, as is commonly seen following emergency AAA repair.

Calcium channel blockers all act on L-type calcium channels causing a relaxation of smooth muscle. They fall into two groups – the dihydropyridines form the majority and include nifedipine and act predominantly to reduce smooth muscle tone in peripheral vessels, thus reducing systemic vascular resistance. The non-dihydropyridines include diltiazem and verapamil and predominantly act on heart rate by slowing the influx of calcium and are therefore also class 4 anti-arrhythmic agents.

References

Peck TE, Hill SA, Williams M. *Pharmacology for Anaesthesia and Intensive Care*. 3rd edition. Cambridge: Cambridge University Press, 2008.

Salgado DR, Silva E, Vincent J-L. Control of hypertension in the critically ill: a pathophysiological approach. *Ann Inten Care*. 2013, 3:17.

F12. An intensive care unit (ICU) with 12 beds currently has 10 patients. Six of the patients are ventilated and two more are requiring renal replacement therapy. One other patient has no organ support requirements but has an arterial line and central line for monitoring, and the final patient has just returned from major elective surgery but is not requiring any organ support or invasive monitoring.

Which of the following MINIMUM staffing arrangements is most appropriate?

A. One consultant, one registrar, a total of 10 nurses, including one band 6/7
B. Two consultants, two registrars, a total of 10 nurses, including one band 6/7
C. One consultant, two registrars, a total of 12 nurses, including one band 6/7
D. One consultant, two registrars, a total of 11 nurses, including one band 6/7
E. One consultant, one registrar, a total of 11 nurses, including one band 6/7

Answer: C

Short explanation

The description is of a 12-bed unit with eight level 3 patients requiring 1:1 nursing and two level 2 patients, requiring 1:2 nursing. The unit is larger than 10 beds and therefore requires an extra supernumerary nurse to be available plus a supernumerary nurse-in-charge of at least band 6/7. There must be one consultant for units of between 8 and 15 patients and a minimum of one trainee per eight patients.

Long explanation

This question has two parts. The first is to recognize the number of patients at each level of care with level 3 patients requiring 1:1 nursing (or higher) and level 2 patients requiring at least 1:2 care.

- Level 3 – Patients requiring multiorgan support or with increased nursing needs. This includes those requiring advanced respiratory support or basic respiratory support plus support of two other organ systems.
- Level 2 – Patients requiring detailed observation or intervention or single organ support. It also includes those recently stepped down from level 3 care.
- Level 1 – Patients who can receive ward based care but either have recently stepped down from level 2 or 3 or are at risk of deteriorating. It often involves the critical care outreach team.
- Level 0 – Normal ward-based care.

This example identified a 12-bed unit with eight level 3 and two level 2 patients. There is therefore the space to admit up to two more patients, which should be taken into consideration in the shift staffing requirements.

The Intensive Care Society recently published guidelines for the provision of critical care. In the guidelines, there is the recognition that any ICU will require the capacity to accept emergency admissions and require supernumerary staff to facilitate leadership and management roles on the unit. Once the patient and management requirements are calculated, the provision of staff can then be allocated based on the guidelines.

- One consultant for 8 to 15 patients
- One junior doctor for up to 8 patients
- Level 3 patients – 1:1 nursing (minimum)
- Level 2 patients – 1:2 nursing (minimum)
- One supernumerary coordinating nurse (band 6/7)
- One additional supernumerary nurse for every 10 patients

Nursing and medical staffing ratios will often vary, and these guidelines outline a minimum requirement. The level of training and experience for nursing staff and recommendations of qualifications are also laid out in the guidance. There are other requirements not covered here, such as the requirement for an education lead nurse, associated health care practitioners, availability of staff for training etc.

Reference
Joint FICM/ICS Standards Committee. Guidelines for Provision of Intensive Care Services 2015. http://www.ics.ac.uk/ics-homepage/latest-news/guidelines-for-the-provision-of-intensive-care-services (accessed April 2015).

F13. A 65-year-old is admitted with an anterior ST-elevation myocardial infarction (STEMI). He had no lesion amenable to percutaneous intervention at angiogram and his echo revealed left ventricular (LV) dysfunction with an ejection fraction of 35%. His renal function is normal.

Which of the following would NOT be appropriate to start as first-line secondary prevention treatments for this patient?

A. Eplerenone
B. Atorvastatin
C. Losartan
D. Bisoprolol
E. Dual antiplatelet therapy

Answer: C

Short explanation

All of these treatments are recommended as secondary prevention for STEMI patients. Losartan is an angiotensin receptor blocker, and these should be used in those patients who are intolerant to angiotensin-converting enzyme inhibitors (ACEIs; recommended as first-line treatment).

Long explanation

Patients who have presented with a STEMI not only require treatment of the acute event and revascularization but also ongoing secondary prevention. The European Society of Cardiology (ESC) and the National Institute for Clinical Excellence (NICE) recommend a number of treatments as secondary prevention.

Non-pharmacological interventions:

- Smoking cessation
- Advice regarding a healthy diet, alcohol consumption, weight-loss and exercise
- Exercise-based cardiac rehabilitation programme

Pharmacological treatments:

- Antiplatelets
  - Low-dose aspirin should be continued indefinitely (or clopidogrel for those with aspirin intolerance)
  - Dual antiplatelet therapy (DAPT) for up to 1 year post-STEMI. Treatment with aspirin and ticagrelor or prasugrel is preferred to aspirin and clopidogrel in patients treated with PCI if there are no contraindications
  - Additional gastric protection with proton pump inhibitors of those on DAPT and with a high bleeding risk
- Anticoagulation is recommended for patients with an LV thrombus, mechanical valves or for patients with atrial fibrillation who have been identified as being at high risk for developing stroke
- Beta-blockers are recommended for all patients post-STEMI to reduce mortality unless there are specific contraindications

- ACEIs are recommended for all patients post-STEMI, especially those with evidence of heart failure, LV systolic impairment, who have had an anterior infarct or are diabetic
- Angiotensin receptor blockers (ARBs) have not been demonstrated to be superior in terms of outcome but are better tolerated than ACEIs. They are recommended for use for those patients who are intolerant to ACEIs.
- Aldosterone antagonists:
  - Eplerenone is recommended in patients who have evidence of heart failure, have LV impairment with an ejection fraction of <40% or are diabetic
  - Aldosterone antagonists act as a potassium-sparing diuretic so are not recommended for patients with renal failure or hyperkalaemia
- Statins: all patients should receive high-dose statins unless contraindicated
- Calcium antagonists (verapamil/diltiazem)
  - These should not be given acutely but may be used long term as secondary prevention in patients in whom beta-blockers are contraindicated
  - They can only be given to those who have no evidence of heart failure.

References

National Institute for Health and Care Excellence (NICE). *MI – secondary prevention: secondary prevention in primary and secondary care for patients following a myocardial infarction* (CG172; last revised in September 2014) (accessed March 2015).

The Task Force on the Management of ST-Segment Elevation Acute Myocardial Infarction of the European Society of Cardiology (ESC). ESC guidelines for the management of acute myocardial infarction in patients presenting with ST-segment elevation. *Eur Heart J.* 2012;33:2569–2619. http://eurheartj.oxford journals.org/content/ehj/33/20/2569.full.pdf (accessed January 2016).

F14. A 35-year-old male patient with known oesophageal varices presents with a profuse ongoing massive gastrointestinal bleed. He has a heart rate of 120 and BP is 88/65 on presentation. He has no known allergies.

Fluid resuscitation is commenced. Which of the following should be instigated NEXT?

A. Administer terlipressin
B. Insert a Sengstaken-Blakemore tube
C. Commence beta-blockers
D. Perform a shunt procedure (transjugular intrahepatic portosystemic shunt; TIPS)
E. Perform an endoscopy (oesophago-gastro-duodenoscopy)

Answer: A

Short explanation

Terlipressin should be initiated immediately to reduce portal pressure. Once fluid resuscitated, the patient should undergo endoscopic interventions to stop the bleeding. A Sengstaken-Blakemore tube and shunt procedures are rescue treatments. Beta-blockers reduce rebleed rates in patients with varices but should not be used in patients acutely bleeding or with haemodynamic instability.

Long explanation

Treatment of acute variceal bleed:

- Pharmacological:
  - Fluid resuscitation
  - Transfusion

- Aiming for haemoglobin >70 g/l for non-exsanguinating haemorrhage
  - Blood products to correct coagulopathy or with massive transfusion
  ○ Drugs to decrease portal pressure:
    - Terlipressin, a relatively selective splanchnic vasoconstrictor, reduces variceal bleeding, rebleed rates as well as mortality rates.
    - Somatostatin or its analogue octreotide have been used. Somatostatin may be superior to octreotide.
  ○ Prophylactic broad spectrum antibiotics (such as aminoglycosides, piperacillin with tazobactam or cephalosporins) reduce bacterial infections, rebleed rates and mortality so should be commenced in all patients with acute variceal bleeds.
  ○ Lactulose to reduce the risk of developing encephalopathy
- Endoscopic:
  ○ Endoscopic therapy should be performed after initial resuscitation has taken place. This should ideally be once the patient has been stabilized, although this may not be possible with massive ongoing haemorrhage. Band ligation therapy has been shown to be preferable to sclerotherapy in controlling bleeding and improving mortality. However, this may be more technically difficult in patients who are actively bleeding, so sclerotherapy is an accepted alternative in these patients.
- Rescue treatments:
  ○ Sengstaken-Blakemore tubes should only be used in patients with ongoing massive haemorrhage and failure or unavailability of the preceding treatments. The oesophageal balloon should only be used if the gastric balloon fails to adequately control the bleeding. Because of side effects and the high percentage of patients who rebleed after the cessation of the procedure, it can only be used for <24 hours as a temporary bridge to definitive treatment (repeat endoscopy, TIPS or transplant).
  ○ Stent procedures provide definitive treatment to reduce portal pressure in patients who continue to bleed despite the above treatment. The insertion of a stent (between the portal and hepatic veins) is usually performed percutaneously (TIPS) or less commonly surgically, and is associated with increased development of encephalopathy. There is emerging evidence that patients at high risk of conventional treatment failure may have a better outcome if this procedure is performed earlier rather than as a rescue strategy.

References

LaBrecque D, Khan AG, Sarin SK, Le Mair AW. World Gastroenterology Organisation global guidelines. Esophageal varices. 2014. http://www.world gastroenterology.org/assets/export/userfiles/2014_FINAL_ESOPHAGEAL-VARICES.pdf (accessed February 2015).

Reverter E, Garcia-Pagan JC. Management of an acute variceal bleeding episode. *Clin Liver Dis*. 2012:1(5);151–154.

Villanueva C, Colomo A, Bosch A, et al. Transfusion strategies for acute upper gastrointestinal bleeding. *N Eng J Med*. 2013;368(1):11–21.

F15. You are called to review a patient in the resuscitation room of ED who has a working diagnosis of viral pneumonia secondary to suspected infection with influenza virus H1N1. He is in significant respiratory distress and has oxygen saturations of 82% on 15 l oxygen via a non-rebreathe mask. You decide that he requires immediate intubation and ventilation.

Of the following, which is the LEAST important to implement in this case to protect staff and stop the spread of his suspected H1N1 influenza?

A. Wearing a fluid repellent gown
B. Wearing goggles or a visor
C. Ensuring good hand hygiene
D. Wearing nonsterile gloves
E. Intubating the patient in a side room

Answer: E

Short explanation
Intubation may result in the generation of aerosolized particles so all of the preceding personal protective equipment should be worn before performing this procedure. Good hand hygiene is imperative to stop the spread of infection. Although it is ideal to intubate this patient in a side room, it is noted that this may not always be possible because of the urgency of the intervention, as is the case here.

Long explanation
The epic3 guidelines have produced recommendations on the use of personal protective equipment (PPE) to prevent healthcare–associated infections. Recommendations include wearing gloves for all contact with mucous membranes, non-intact skin and sterile sites, where there is risk of exposure to blood or body fluids or when handling sharps or performing invasive procedures. When there is risk of contaminating clothing with bodily fluids or pathogenic microorganisms, the use of disposable plastic aprons is recommended. However, if extensive splashing/exposure of skin and clothes is likely, then full body fluid repellent gowns are preferable. If there is risk of splashing onto the face, eye protection is required, and an appropriate face mask depending on the pathogenic organism. This may include a fluid-repellent surgical face masks or appropriate masks incorporating filtration devices.

The level of PPE that is required to protect staff as well as prevent the spread of H1N1 varies on the level of patient contact. Patients with suspected H1N1 influenza should be isolated as soon as possible once it has been considered as a differential diagnosis.

For those coming into close contact (<1 m) with patients, the following PPE is required:

- Gloves (sterile gloves are only required for sterile procedures)
- Fluid-repellent gown is required if contact with skin or bodily fluids is anticipated; otherwise a plastic apron is adequate
- A fluid-repellent surgical facemask is adequate as long as the patient is not generating aerosolized particles, such as coughing, physiotherapy etc.

If aerosolized particles are likely to be produced during contact with the patient, for example, during intubation, manual ventilation, suctioning (not closed in-circuit suction), bronchoscopy or physiotherapy, the following PPE is required:

- Gloves
- Fluid-repellent gown
- Well-fitting FFP3 respirator/mask
- Visor/goggles for eye protection

If these are non-disposable items it needs to be ensured that they are appropriately decontaminated.

References

Department of Health and the Health Protection Agency. Pandemic (H1N1) 2009 influenza – a summary of guidance for infection control in healthcare settings. https://www.gov.uk/government/uploads/system/uploads/attachment_data/file/061997/Pandemic_influenza_guidance_for_infection_control_in_critical_care.pdf (accessed August 2015).

Loveday HP, Wilson JA, Prat RJ, et al. epic3: National Evidence-Based Guidelines for Preventing Healthcare-Associated Infections in NHS Hospitals in England. *Journal of Hospital Infection* 2014;86(Suppl 1):S1–S70.

F16. A patient in the intensive care unit is recovering from abdominal sepsis and has an albumin of 18 mmol/l. He has developed a unilateral pleural effusion, detected on chest X-ray.

What is the most appropriate course of action?

A. Insert a large-bore chest drain to drain the effusion
B. Perform a diagnostic pleural tap and await results before deciding whether to drain
C. Perform a diagnostic pleural tap and insert a small-bore ('Seldinger') chest drain
D. Obtain a chest ultrasound
E. Obtain chest computed tomography (CT)

Answer: D

Short explanation

The BTS 2010 guidelines suggest that in patients where the pleural effusions are likely to be transudates, then the underlying cause should be identified and treated. In this case, hypoalbuminaemia is highly likely to cause a transudate. It is appropriate to obtain a chest ultrasound as the best method of quantifying the volume of the effusion and identifying any septations or other pathology that might help isolate a cause.

Long explanation

Pleural effusions occur due to an accumulation of fluid within the pleural space and may be an exudate or transudate. The distinction is made by applying Light's criteria. An effusion is an exudate if it meets one or more of the following criteria:

- Pleural fluid protein/serum protein ratio of >0.5
- Pleural fluid lactase dehydrogenase (LDH)/serum LDH is >0.6
- Pleural fluid LDH greater than two-thirds of the upper limit of normal range for serum LDH

The commonest causes of transudates include left ventricular failure, cirrhosis, hypoalbuminaemia, hypothyroidism, peritoneal dialysis, nephrotic syndrome and mitral stenosis. The commonest causes of exudates include: cancer, pneumonia, tuberculosis, pulmonary embolism, autoimmune diseases, pancreatitis and ischaemic heart disease.

Initial management should start with a review of the clinical history and examination, including a drug history. Drugs associated with effusions include methotrexate, phenytoin, amiodarone, beta-blockers and nitrofurantoin.

If the effusion is likely to be a transudate, the underlying cause should be treated and the clinical course reviewed. If it does not resolve or an exudate is thought more likely, then a respiratory opinion should be sought and a diagnostic tap performed.

It is often helpful to obtain a chest ultrasound early in the assessment of an effusion, especially in ventilated or critically ill patients, because it may help to quantify the size and identify septations. Diagnostic pleural taps should be performed under ultrasound guidance with a 21-gauge needle and a 50-ml syringe. Samples should always be sent for microscopy, Gram stain, biochemistry and cytology.

If the patient develops sepsis or a possible exudative cause was present, it may become appropriate to diagnostically tap the fluid as part of a 'septic screen' to exclude a source of infection. If the patient developed respiratory embarrassment, it may be appropriate to drain the fluid simply to improve the respiratory function.

Reference
Hooper C, Lee YCG, Maskell N. Investigation of a unilateral pleural effusion in adults: British Thoracic Society Pleural Disease Guideline 2010. *Thorax*. 2010;65(Suppl 2):ii4–i17.

F17. You have just intubated a 25-year-old woman with a life-threatening acute exacerbation of asthma.

Which of the following initial ventilator settings, with reference to the tidal volume (V_T), set respiratory rare (RR), and inspiratory to expiratory ratio (I:E), would be best?

A. A V_T of 600 ml, RR of 10, I:E of 1:4
B. A V_T of 400 ml, RR of 10, I:E of 1:4
C. A V_T of 600 ml, RR of 16, I:E of 1:4
D. A V_T of 400 ml, RR of 16, I:E of 1:6
E. A V_T of 600 ml, RR of 16, I:E of 1:6

Answer: B

Short explanation
The principles of initial mechanical ventilation in asthma are to avoid dynamic hyperinflation and barotrauma. Therefore, the best initial ventilator settings would be a TV of 5 to 7 ml/kg, a respiratory rate of 10 to 14 breaths/min, and a short inspiratory time to ensure an expiratory time of at least 4 seconds.

Long explanation
Once the decision has been made to intubate an asthmatic, a large endotracheal tube should be used. This is to improve gas flow, as flow is proportional to the fourth power of the radius. It also reduces the risk of partial tube occlusion by the thick secretions that often occur with asthma.

Mechanical ventilation in acute asthma is associated with a significant mortality. Trying to achieve a "normal" minute ventilation, or worse, a supra-normal minute ventilation to try and avoid hypercapnia, will result in pulmonary hyperinflation, barotrauma and increased risk of significant harm. Deliberate hypoventilation, using the settings preceding, will attempt to avoid worsening the existent state. Hypercapnia will be permitted, and an associated acidosis should be expected.

Deep sedation and paralysis may be needed to assist ventilation, but any paralysis should be for the shortest duration possible because these patients are already at high risk of myopathy, from the steroid administration to treat their asthma.

The prolonged expiratory time required in severe asthma will necessitate a shorter inspiratory time. Although this will result in a high inspiratory flow rate (to achieve the same volume in a shorter time) and a rise in peak airway pressure, the overall effect is a reduction in plateau pressure, and reduced barotrauma.

PEEP use is fairly contentious, but it seems logical that it should be avoided during the initial stages of mechanical ventilation because it would worsen hyperinflation. Some authors advocate setting extrinsic positive end expiratory pressure at no more than 80% of measured intrinsic PEEP.

References
Oddo M, Feihl F, Schaller MD, Perret C. Management of mechanical ventilation in acute severe asthma: practical aspects. *Intensive Care Med.* 2006;32(4):501–510.

Tuxen D, Naughton M. Chapter 31: Acute severe asthma. In Bersten A, Soni N. *Oh's Intensive Care Manual*. Sixth Edition. Edinburgh: Butterworth Heinemann Elsevier, 2006.

F18. You are called to urgently review a 60-year-old man on the coronary care unit (CCU). He underwent coronary angiography 5 hours ago, with stents to both his left anterior descending (LAD) and right marginal arteries. He has a heart rate of 120 bpm, a blood pressure of 75/40 mmHg, and a central venous pressure (CVP) of 25 mmHg. His cardiac monitor shows electrical alternans.

Which is the following is the best investigation to determine the underlying cause of his current state?

A. 12-lead electrocardiogram (ECG)
B. Portable chest X-ray (CXR)
C. Central venous SpO_2
D. Transthoracic echocardiogram (TTE)
E. Repeat angiography

Answer: D

Short explanation
The likely diagnosis here is one of cardiac tamponade, which is a recognized complication of coronary intervention. The best investigation for this in an awake patient would be transthoracic echocardiography (TTE), which would show a fluid collection in the pericardial space. The fluid initially compresses the lowest pressure chamber – the right atrium – and this impairs atrial filling, causing hypotension but a raised CVP.

Long explanation
Cardiac tamponade after coronary intervention is an emergency. The classical signs are hypotension, distended neck veins (high CVP), and distant, muffled heart sounds – termed 'Beck's triad'. These signs can be predicted from the physiology of tamponade, with the fluid surrounding the heart and impairing right atrial filling. Other signs include tachycardia and occasionally electrical alternans. This is an ECG appearance where the QRS complex axis alternates between beats. This is thought to be due to movement of the heart within the fluid-filled pericardial sac.

A 12-lead ECG may demonstrate this unusual sign or may show low-voltage QRS complexes due to the fluid layer between heart and ECG electrodes. The most likely ECG finding, however, is that of sinus tachycardia. Transthoracic echocardiography has a high sensitivity for detecting pericardial effusions. The significant interface between the tissue and the fluid leads to a strong echocardiographic signal, meaning even small effusions can be identified.

A tension pneumothorax would cause hypotension and a raised CVP, but the absence of any respiratory symptoms or signs in the question make the CXR a less useful investigation in this scenario. Central venous SpO_2 would help estimate tissue perfusion, but would not be useful to determine the underlying cause. Repeat angiography may be useful if the concern is one of an occluded stent, which could produce the hypotension, tachycardia and elevated CVP, but in the absence of chest pain or appropriate ECG changes, this is less likely than cardiac tamponade. Electrical alternans is pathognomic of pericardial fluid.

References

Donovan K, Colreavy F. Echocardiography in intensive care. In *Oh's Intensive Care Manual*. 6th edition. Bersten A, Soni N. Edinburgh: Butterworth Heinemann Elsevier, Chapter 23.

F19. You are asked by the nurses to review a patient who has been receiving drugs via a 20-gauge cannula placed yesterday into a vein in the dorsum of the right hand. The cannula has 'tissued' and there has been extravasation of the drugs.

Which of the following statements is the most accurate?

A. Atracurium usually causes no effect, and no further action is needed.
B. Pain on injection of propofol is likely to suggest extravasation.
C. Cytotoxic agents should be managed using the 'spread and dilute' technique with a saline washout.
D. Ketamine causes ischaemia and should be managed using the 'spread and dilute' technique with a saline washout.
E. Phenytoin causes necrosis and should be managed with a stellate ganglion block.

Answer: D

Short explanation

Atracurium can cause ischaemia when extravasation occurs, and thus doing nothing would not be appropriate. Propofol often causes pain on injection and is not a reliable indicator of extravasation. Cytotoxic agents can cause extensive damage and should not be spread. Ketamine and phenytoin can both cause necrosis, but the 'spread and dilute' technique has much stronger evidence than a stellate ganglion block.

Long explanation

The accidental administration of substances into the tissues rather than into a blood vessel is termed 'extravasation'. This is reasonably frequent occurrence in hospitalized patients and is not without significant risk. Many drugs administered to patients are markedly toxic, and when unintentionally delivered into the extravascular tissues, they can cause tissue necrosis and even lead to amputation.

Drugs that particularly cause problems are largely grouped into four categories:

1. Vasoconstrictors: extravasation of these can lead to profound vasoconstriction of surrounding blood vessels leading to tissue hypoxia and necrosis.
2. Hyperosmolar/concentrated electrolytes: examples of these include high concentrations of glucose, calcium, potassium and total parenteral nutrition.
3. Extremes of pH: both acids and alkalis can cause profound damage. Examples include phenytoin and amiodarone.
4. Cytotoxics: all chemotherapy agents can cause significant damage.

In reference to other drugs frequently seen in the intensive care unit, thiopentone can cause ischaemia, due to its alkalinity, as can ketamine and atracurium.

If extravasation occurs, then any injection should be immediately stopped, and aspiration through the cannula attempted. Once advice has been sought (e.g. pharmacy, plastic surgery), if the cannula if not required for specialist treatment of the extravasation, then it should be removed. The limb should be elevated to reduce secondary injury from oedema, but it should be kept warm to encourage vascular dilation and drug absorption.

With the exception of cytotoxic agents, the spreading of which can exacerbate the damage, the majority of extravasation injuries can be managed with a 'spread and dilute' approach if risk of tissue necrosis is high. This utilizes a saline washout, under sterile conditions, in which a number of stab wounds are made around the affected area. Through a cannula inserted into one of these, saline can be flushed through the tissues, to exit through the other stab wounds. This reduces the concentration of drug left available and minimizes damage to tissues.

Reference
Lake C, Beecroft C. Extravasation injuries and accidental intra-arterial injection. *Contin Educ Anaesth Crit Care Pain*. 2010;10(4):109–113.

F20. You are about to discharge a patient from the intensive care unit (ICU) back to a general ward at the weekend.

Which of the following statements about ICU discharge is the LEAST accurate?

A. Discharges between 22:00 and 07:00 should be documented as an adverse incident
B. Patients discharged from ICU should have access to an ICU follow-up clinic
C. Discharge to the ward should occur within 6 hours of the decision being made
D. Handover must include a plan for ongoing treatment
E. Handover must include psychological and emotional needs

Answer: C

Short explanation
Both the Intensive Care Society (ICS) core standards document and NICE guidance agree on statements A, B, D and E. With regard to discharge timing, the ICS states discharge should be within 4 hours of the decision being made, whereas NICE states it should be as early as possible during the day.

Long explanation
Discharge from intensive care is a time of increased risk for patients, with a potential for information to get lost during handover to different clinical teams. In addition to this, there is an increased mortality and reduced patient satisfaction from ICU discharges between the hours of 22:00 and 07:00. To mitigate this risk, both NICE guidance and a subsequently produced ICS core standards document exist to help formalize the discharge of intensive care patients.

Handover of patients being transferred to ward must include a number of pieces of information, which should be in written format. First, there should be a summary of the patient's stay in ICU including treatment interventions, changes to admission medications and a diagnosis. There should be a plan for ongoing monitoring and treatment and a summary of the patient's ongoing needs. Finally, there should be a plan for follow-up of the patient.

Patients should be discharged between the hours of 07:00 and 21:59 because discharges outside these times are associated with an excess mortality, as well as patient disturbance and reduced satisfaction. Any discharges that occur outside these times should be documented as an adverse incident.

Patients discharged from ICU should have access to an ICU follow-up clinic, which can be delivered on a regional basis. This is to help address the physical and complex psychological issues that can persist after an ICU admission.

References
Danbury T, Gould T, Baudouin S, et al, on behalf of the Core Standards Working Party of the Intensive Care Society. Core standards for intensive care units. 2013. http://www.ics.ac.uk/ics-homepage/guidelines-and-standards (accessed April 2015).

National Institute for Health and Clinical Excellence (NICE). *Acutely ill patients in hospital: recognition of and response to acute illness in adults in hospital* (CG 50). London: NICE, 2007.

F21. A nurse has just inserted a nasogastric (NG) tube into an intubated patient and has asked you to confirm its position. You are able to aspirate some fluid. Which of the following is the recommended way to confirm the position?

A. pH ≤5.5
B. pH <6 in a patient on omeprazole
C. Chest X-ray showing NG tube in stomach
D. Aspiration of 20 ml of brown fluid
E. Positive 'whoosh' test

Answer: A

Short explanation
The National Patient Safety Agency decision tree for nasogastric (NG) tube placement checks in adults states that if an aspirate is obtained, a pH between 1 and 5.5 is sufficient to proceed to feed. Only if an aspirate cannot be obtained, or the pH is not between 1 and 5.5, a chest X-ray should be performed. A 'whoosh' test, in which air is injected down the NG tube, while auscultating over the stomach, is not recommended.

Long explanation
The National Patient Safety Agency (NPSA) and NHS England have issued a number of alerts with regard to NG tube insertion. Feeding through misplaced oro- or nasogastric tubes are on NHS England's list of 'never events'. The NPSA Alert 'Reducing the harm caused by misplaced nasogastric tubes in adults' (2011) states that a pH of 1 to 5.5 or an X-ray are the only acceptable methods for confirming initial placement of a nasogastric tube. A further alert in December 2013 confirmed that placement devices to assist NG tube insertion do NOT replace initial position checks.

The NPSA decision tree for NG tube placement checks in adults is as follows:

1. If an aspirate is obtained, test the aspirate on pH indicator paper. If an aspirate is not obtained, there are a number of techniques that can be implemented to try to attain an aspirate. These include advancing or withdrawing the tube by up to 20 cm, positioning the patient on his or her side or giving mouth care, which stimulates gastric secretion.
2. If the aspirate pH is <5.5, the NG tube can be used.
3. If no aspirate can be obtained, or the pH is >5.5, then perform an X-ray.
4. The X-ray will need to be reviewed to confirm the NG tube is correctly positioned.

Therefore, in this case, where aspirate is obtained, a pH of ≤5.5 would be sufficient to use the NG feed. To proceed to an X-ray would be unnecessary.

References
National Patient Safety Agency. CAS reference: NPSA/2011/PSA002. Nasogastric feeding tubes decision tree adults. March 2011. http://www.nrls.npsa.nhs.uk/EasySiteWeb/getresource.axd?AssetID=129697& (accessed May 2015).
NHS England. Patient safety alert (alert reference no. NHS/PSA/W/2013/001. December 2013. http://www.england.nhs.uk/wp-content/uploads/2013/12/psa-ng-tube.pdf (accessed May 2015).

F22. A patient is likely to need total parenteral nutrition (TPN) for 7 to 10 days. The patient does not otherwise need central venous access.

Which of the following is the LEAST acceptable route of administering this?

A. A dedicated peripherally inserted central catheter (PICC)
B. A dedicated internal jugular vein central catheter
C. A dedicated lumen of a 4-lumen internal jugular vein central catheter
D. A dedicated peripheral venous catheter
E. A dedicated tunnelled subclavian central catheter

Answer: E

Short explanation
NICE guidance for the route of access for TPN includes all of the preceding routes of access. A peripheral venous catheter can be used for short-term parenteral nutrition (<14 days) in patients who do not need a central line for other reason. Tunnelled lines are recommended for longer term use (>30 days).

Long explanation
The NICE guidance for nutritional support in adults (CG32) clarifies what routes are acceptable for TPN. These state that parenteral nutrition can be given via a PICC so long as the lumen is dedicated for the nutrition and not used for anything else. Other peripheral approaches for nutrition include a venous cannula, which again must be dedicated for the nutrition. This is a particular option in those who need TPN for less than 14 days and who do not otherwise need central venous access. If a central venous catheter is in situ, then a dedicated lumen of this can be used for TPN. Tunnelled lines are recommended for those likely to be used for more than 30 days but is not necessary for shorter durations of TPN.

Administration of TPN can be either continuous or intermittent. Continuous administration is the usually the preferred initial method, particularly in those who are significantly unwell, but intermittent is likely to be more physiologically appropriate, and so those requiring longer term TPN should be changed onto cyclical administration.

TPN is indicated if a patient is malnourished or at risk of malnutrition, but feeding via the enteral feeding route is not possible. This could be due to safety concerns (e.g. duodenal perforation) or due to inefficiency of enteral intake (e.g. non-functioning gastrointestinal tract).

Reference
National Institute for Health and Clinical Excellence (NICE). *Nutrition support in adults: oral nutrition support, enteral tube feeding and parenteral nutrition* (CG32). London: NICE, 2006.

F23. You are called urgently to the emergency department to assist with the management of a 53-year-old who suffered an out of hospital cardiac arrest. Before his collapse, he reported crushing central chest pain with radiation to his left arm. His initial rhythm was pulseless electrical activity, but after 1-mg intravenous adrenaline he showed signs of life, and cardiopulmonary resuscitation was stopped. He is now unresponsive to painful stimuli, has a heart rate of 40 bpm and complete heart block on the electrocardiogram (ECG). His blood pressure is 60/30 mmHg, and his oxygen saturations are 95% on 15 l O_2 via a non-rebreathe mask.

Which coronary artery is most likely to be affected?

A. Left anterior descending artery
B. Circumflex artery
C. Right marginal artery
D. Posterior interventricular artery
E. Right coronary artery

Answer: E

Short explanation

The right coronary artery (RCA) is the most likely territory affected here. The RCA provides a branch to the sinoatrial node in roughly 60% of the population and also supplies the atrioventricular node in approximately 80 to 90% of the population.

Long explanation

The blood supply to the myocardium is from the right and left coronary arteries (RCA and LCA). These originate from the ascending aorta, just after the aortic valve leaflets. The RCA gives off a nodal branch to the SA node before running in a groove between the right atrium and ventricle. As it reaches the right border of the heart, it gives off a large marginal branch, which runs down this heart border. The remainder of the RCA continues in the atrioventricular groove on the back of the heart. It gives off terminal branches to anastomose with the circumflex artery (from the LCA) and descends in the groove between the two ventricles to anastomose with the left anterior descending artery (also from the LCA).

The LCA initially runs in the groove between the left atrium and ventricle but splits into two large branches – the LAD (also sometimes called the anterior interventricular artery), and the circumflex artery. The LAD runs down the front of the heart, in a groove between the two ventricles. When it reaches the apex, it remains in this groove as it heads posteriorly, to anastomose with the terminal branch of the RCA (the posterior interventricular artery). The other branch of the LCA, the circumflex artery, continues in the groove between the atrium and ventricle, passing posteriorly to anastomose with the other terminal branch of the RCA.

Reference

Moore KL, Dalley AF, Agur AMR. Chapter 1: Thorax. In *Clinically Oriented Anatomy*. 7th edition. Baltimore: Lippincott, Williams and Wilkins, 2014.

F24. You are called urgently to the emergency department to help with the initial assessment and management of a 6-year-old boy who has been involved in a house fire. He has 50% total body surface area burns including his face, and he is unconscious. You need to intubate the child.

Which of the following is the most appropriate?

A. Rapid sequence induction (RSI) with propofol and atracurium inserting a cut 5.5 mm endotracheal tube to 15 cm at the lips

B. RSI with propofol and atracurium inserting an uncut 5.5 mm endotracheal tube to 17 cm at the lips

C. RSI with propofol and suxamethonium inserting a cut 5.5 mm endotracheal tube to 17 cm at the lips

D. RSI with propofol and suxamethonium inserting an uncut 5.5 mm endotracheal tube to 15 cm at the lips

E. RSI with propofol and atracurium inserting a cut 6.0 mm endotracheal tube to 17 cm at the lips

Answer: D

Short explanation

In a burn patient, suxamethonium may still be safely used in the first 24 to 36 hours; after this time, significant hyperkalaemia can occur, and suxamethonium should be avoided. Patients with facial burns can suffer significant swelling, so uncut endotracheal tubes should be used, often requiring re-tying several times as swelling progresses. Sizing of endotracheal tubes in paediatrics can be estimated as follows: size $=$ (age / 4) $+ 4$; length $=$ (age / 2) $+ 12$

Long explanation

Burns patients provide a number of challenges, including fluid management, temperature control, analgesia, a potentially difficult airway, and electrolyte disturbance. Children provide additional challenges for the adult intensivist in the form of consent, drug doses and different equipment.

The general principles of the immediate management of the burns patient should be along ATLS guidelines, including cervical-spine control where appropriate. Where the airway needs securing, standard RSI drugs may be used in the first 24 to 36 hours, but intubation may be difficult due to airway oedema. Swelling will get worse over the first few days, and so an uncut tube must be used and will need to be regularly re-tied to ensure it is positioned and secured correctly. After the first 24 to 36 hours, proliferation of extra-junctional acetylcholine receptors leads to a risk of significant hyperkalaemia if suxamethonium is used, which should be then avoided for up to 1 year.

Fluid resuscitation is usually based on the Parkland Formula. This calculates the first 24 hours of fluid replacement as follows:

$$\text{Fluid volume} = \text{Pt weight (kg)} \times 4 \times \text{\% body surface area burnt.}$$

Half of this volume is administered over the first 8 hours from the time of burn injury, and the rest delivered over the subsequent 16 hours. This is in addition to the normal daily fluid requirements of the child. This formula is a guide only and should be used in conjunction with regular clinical assessment, measurement of urine output and electrolytes.

Paediatric equipment may be less familiar to adult intensivists, and most emergency departments may have some basic calculations on the wall (e.g. weight, adrenaline dose) but may not have detailed anaesthetic calculations. The calculation for endotracheal tube size (internal diameter) is (age/4) $+ 4$ mm. Always have smaller and larger sizes available, anticipating the potential difficult intubation. Length is calculated as $(age/2) + 12$ cm for an oral endotracheal tube, but this is a guide, and position will need confirmation on a chest X-ray.

References

Bishop S, Maguire S. Anaesthesia and intensive care for major burns. *Contin Educ Anaesth Crit Care Pain*. February 2012. http://ceaccp.oxfordjournals.org/content/early/2012/02/23/bjaceaccp.mks001.full.pdf (accessed January 2016).
Fenlon S, Nene S. Burns in children. *Contin Educ Anaesth Crit Care*. 2007;7(3):76–80.

F25. You have just intubated a patient in the intensive care unit. Although you have seen the endotracheal tube pass through the cords, you want to confirm the presence of carbon dioxide on expiration.

Which of the following is the most appropriate device to do this?

A. Severinghaus electrode
B. Gas chromatograph
C. Paramagnetic analyzer
D. Infrared analyzer
E. Mass spectrometer

Answer: D

Short explanation

The infrared analyzer allows continuous recording of CO_2, displayed as a capnograph, for additional information. Severinghaus electrodes measure PCO_2 in a blood gas analyzer. Gas chromatography does not allow continuous measurement. A mass spectrometer, like the gas chromatograph, is bulky and less convenient outside the laboratory. A paramagnetic analyser could be used to measure a paramagnetic gas, such as oxygen. CO_2 is diamagnetic.

Long explanation

The Royal College of Anaesthetists' fourth National Audit Project (NAP4) was a year-long study of complications related to airway management. A quarter of all major airway events occurred in the intensive care unit (ICU) or emergency department, and failure to use capnography contributed to the majority of airway-related deaths in the ICU. No intubation should be undertaken without working capnography, allowing a continuous reading of CO_2 tension.

An infrared analyzer is the best tool for this purpose. This works on the basis that molecules with two or more dissimilar atoms (e.g. CO_2 or N_2O but not O_2 or N_2) will absorb infrared radiation, the amount of absorbance of which is related to the concentration of the substance. This is the same principle that is used in a pulse oximeter, as described by Beer's Law and Lambert's Law. A wavelength is chosen that will be absorbed solely by the substance of interest within the sample (4.28 μm for carbon dioxide).

There are two types of capnograph: side-stream and main-stream. In a side-stream capnograph, there is a time delay as the sample is aspirated from a connector near the patient's airway and passes down tubing to the analyzer, often housed with the monitoring module. In a main-stream analyzer, this delay is eliminated, but at the cost of increased dead space as a result of a special connector in the breathing system, which allows the infrared beam to pass directly through the gases in the system. It also adds a little bulk to the system, but modern main-stream analyzers are much smaller and lighter than previous incarnations.

References

Cook T, Woodall N, Frerk C. 4th National Audit Project (NAP4): Major complications of airway management in the United Kingdom. Report and findings. March 2011. http://www.rcoa.ac.uk/nap4 (accessed June 2015).

Davis P, Kenny G. *Basic Physics and Measurement in Anaesthesia*. 5th edition. London: Elsevier, 2003.

F26. You are ventilating a 41-year-old male patient who has a persistent bronchopleural fistula and collapse of the left lung despite two chest drains correctly placed on that side. You decide to insert a double-lumen endotracheal tube and commence one-lung ventilation.

Which of the following cases is the WEAKEST indication for one-lung ventilation?

A. Bronchopleural fistula compromising ventilation
B. Traumatic left main bronchus rupture
C. Large left sided endo-bronchial haemorrhage
D. Large unilateral apical bullae at risk of rupture
E. Emergency thoracic aortic aneurysm repair

Answer: E

Short explanation

Surgical access is a relative indication for one-lung ventilation, whereas preventing contamination of the contra-lateral lung (e.g. haemorrhage) or controlling distribution of ventilation (e.g. bronchopleural fistula, traumatic bronchus rupture or large bullae) would be absolute indications.

Long explanation

One-lung ventilation (OLV) can be achieved by placement of a double-lumen endo-bronchial tube (DLT), insertion of an endobronchial blocker, or, as a temporary measure, deliberate endobronchial intubation with a tracheal tube.

Indications for OLV can be either absolute or relative. Absolute indications include the following:

1. Protecting a lung, for example, from haemorrhage or bronchiectasis
2. Controlling the distribution of ventilation, for example, with a bronchopleural fistula, traumatic damage to a major bronchial airway or with a bullae which may rupture with IPPV.

Relative indications for OLV include the following:

1. Surgical exposure, for example, thoracic aortic aneurysm, pneumonectomy, oesophagectomy

Hypoxia is an inevitable effect of OLV in a healthy individual for elective surgery, but in many of the preceding scenarios, hypoxia may actually be improved by OLV. In the lateral position, on OLV (with the ventilated lung at the base), there is better ventilation-perfusion matching, and so hypoxia is not as profound as in the supine position, but this may not always be possible in the critically ill. There is still some perfusion to the non-ventilated lung, contributing to shunt, despite the effects of hypoxic pulmonary vasoconstriction (HPV). This is the process in which pulmonary arterioles constrict in response to low alveolar partial pressure of oxygen. If HPV is impaired, shunt will be worsened, and there are a number of drugs that will impair HPV, such as vasoconstrictors and vasodilators. If there is an increase in the pulmonary vascular resistance of the ventilated lung, this can also worsen shunt, and causes of this include the use of positive end expiratory pressure and atelectasis.

The mode of implementing OLV depends on the circumstances. Although a DLT is considered the gold standard, this may be difficult to site in the intensive care unit. Not only does it take longer to site, it is also significantly larger than a standard endotracheal tube (roughly 13 mm external diameter), but with smaller individual diameters for each lumen (roughly 5 mm). Advancing a standard endotracheal tube into an endobronchial position may allow control of the situation in the acute setting, although this should not usually be seen as more than a temporizing measure. Bronchial blockers allow the isolation of one lung without changing the endotracheal tube and so may be the most appropriate in intensive care unit.

Reference

Ng A, Swanevelder J. Hypoxaemia during one-lung anaesthesia. *Contin Educ Anaesthe Crit Care Pain*. 2010;10(4):117–122.

F27. You are commencing total parenteral nutrition in a malnourished patient. Which of the following patients is MOST at risk of refeeding syndrome?

A. A 35-year-old with a body mass index (BMI) of 15.5 kg/m^2
B. A 35-year-old with unintended weight loss of 17% in the past 6 months
C. A 35-year-old alcoholic with no nutritional intake for 12 days
D. A 35-year-old with low potassium, phosphate and magnesium before commencement of feeding
E. A 35-year-old with a BMI of 17 kg/m^2 with unintended weight loss of 8% in the past 3 months

Answer: C

Short explanation

NICE guidance identifying patients at risk of refeeding syndrome would include patients A through D, as the low BMI, unintended weight loss, lack of nutritional intake and pre-existing low levels of K^+, PO_4^{3-} and Mg^{2+}, all fulfil the criteria for patients at high risk of developing refeeding problems. The addition of the alcohol history for patient C adds an additional risk factor, making them even higher risk.

Long explanation

Refeeding syndrome can occur when feeding is commenced in malnourished patients. It is a metabolic state in which shifts in fluids and electrolytes result in a critical state with high associated mortality. During prolonged starvation, the body attempts to conserve muscle and protein by shifting metabolism to fatty acids, reducing insulin secretion and reducing gluconeogenesis. During this time, electrolyte levels can become severely depleted from the intracellular compartment, although serum levels may be normal. Once feeding is recommenced, insulin levels rise, stimulating synthesis of several substances but requiring electrolytes in the process. This can lead to a profound fall in levels of these electrolytes, particularly magnesium, potassium and phosphate.

Patients at high risk of refeeding syndrome need to be identified because they will require frequent monitoring of these electrolytes when feeding is commenced. The NICE guidance for nutritional support in adults (CG32) clarifies patients who are at high risk of refeeding syndrome. Major and minor risk factors are identified; patients with one major or two minor risk factors are more likely to develop refeeding syndrome on initiation of feeding:

	Major	Minor
BMI (kg/m^2)	<16	<18.5
Weight loss in 3–6 months	15%	10%
Days without food	10	5
Other	Low Mg^{2+}, K$^+$ or PO$_4$$^{3-}$	Alcohol abuse, diuretics, insulin, chemotherapy

Those patients who are at high risk of refeeding syndrome should have feeding started at much lower levels (starting at 5–10 kcal/kg/day), and this should be gradually increased to full requirements by day 7. Should electrolyte disturbance occur, a lower rate of feed may be required for a longer duration to ensure time for adequate replacement of electrolytes.

To coincide with the commencement of feeding, and continue during the first 10 days of feeding, a number of vitamin supplements should be given. These include thiamine, vitamin B complex and supplemental multivitamins.

Sodium and potassium must be measured daily in all patients in whom feeding is being commenced until levels are stable. Magnesium and phosphate need only be measured three times a week initially if patients are not at risk of refeeding syndrome. If the patients are at risk of refeeding syndrome, they must also be measured daily.

References

Mehanna H, Moledina J, Travis J. Refeeding syndrome: what is it, and how to prevent and treat it. *Br Med J.* 2008;336(7659):1495–1498.

National Institute for Health and Clinical Excellence (NICE). *Nutrition support in adults: Oral nutrition support, enteral tube feeding and parenteral nutrition* (CG32). London: NICE, 2006.

F28. You are asked by the nursing staff to examine a patient. He is a 33-year-old man who was admitted 3 days ago with bilateral pneumonia. He is intubated and ventilated according to lung-protective ventilation strategies and is receiving appropriate antimicrobial therapy. His ventilator settings are as follows: volume control ventilation; FiO$_2$ 0.8%; V$_T$ 450 ml; respiration rate 30/min; positive end expiratory pressure 14 cmH$_2$O. His latest arterial blood gas analysis shows the following: pH 7.26; pO$_2$ 6.85 kPa; pCO$_2$ 7.51 kPa.

Which pharmacological therapy has the best evidence of benefit in such cases?

A. Methylprednisolone
B. Cis-atracurium
C. Prostacyclin (inhaled)
D. Nitric oxide (inhaled)
E. Surfactant

Answer: B

Short explanation

This patient has severe acute respiratory distress syndrome (ARDS). Along with protective ventilation, the only therapies with evidence of reduced mortality are cis-atracurium infusions and prone positioning. Prostacyclin and nitric oxide have evidence of improved oxygenation but no evidence of mortality benefit. Surfactant has no evidence of benefit in adults and is rarely used, although common in neonatal and paediatric practice. Corticosteroids are controversial but have no strong evidence of benefit in early, severe ARDS.

Long explanation

This patient has severe ARDS. His PaO_2/FiO_2 ratio is 65.075 mmHg (or 8.56 kPa). In addition to ventilator and mechanical strategies, many pharmacological treatments have been studied in severe ARDS with little success. Inhaled vasodilators (prostacyclin and nitric oxide) have a sound basis in physiology in that they only act on blood vessels adjacent to aerated alveoli, thus improving V/Q mismatch. Although both therapies have evidence of improvement in oxygenation and other parameters with their use, evidence of a mortality benefit has proved elusive.

The use of corticosteroids in ARDS has an equally compelling physiological basis because much of the pathological process seen in ARDS is inflammatory. Many studies have been performed, although many of them are small or contain methodological flaws. Recent meta-analyses have concluded that there is no evidence of benefit for the use of corticosteroids to treat ARDS, although there may be a trend towards benefit in late ARDS. Two larger trials are ongoing at the time of writing.

Surfactant is widely used in neonates and premature infants, both prophylactically and for treatment of respiratory failure, with good results. The use of surfactant in adults has been studied and found to have short-term benefits in oxygenation but not to have any mortality or other outcome benefit.

Neuromuscular blocking agent infusions are thought to reduce patient-ventilator dyssynchrony as well as improving chest-wall compliance, allowing the use of lower tidal volumes. Reduction in tidal volumes to 6 ml/kg predicted body weight is well known to reduce the development of ventilator-induced lung injury and improve outcomes in ARDS. A large multicentre trial published in 2010 using infusions of cis-atracurium demonstrated improved mortality. It is not clear if this effect is limited to cis-atracurium or a wider class effect of neuromuscular blocking agents.

Many other pharmacological therapies have been suggested and/or studied in ARDS, although results have been negative or equivocal. These include beta-2 adrenergic agonists, statins, heparin, growth factors, neutrophil elastase inhibitors, aspirin and vitamin D. It is likely that many further pharmacological therapies will be suggested in the future.

References

Boyle A, MacSweeney R, McAuley D. Pharmacological treatments in ARDS; a state-of-the-art update. *BMC Med*. 2013;11(1):166.

Papazian L, Forel J-M, Gacouin A, et al. Neuromuscular blockers in early acute respiratory distress syndrome. *N Engl J Med*. 2010;363:1107–1116.

Peter JV, John P, Graham PL, et al. Corticosteroids in the prevention and treatment of acute respiratory distress syndrome (ARDS) in adults: meta-analysis. *BMJ*. 2008;336:1006–1009.

Slutsky AS, Ranieri VM. Ventilator-induced lung injury. *N Engl J Med*. 2013;369(22): 2126–2136.

F29. A 16-year-old male patient is admitted to hospital with pyrexia, headache, neck pain and photophobia.

What is the most appropriate treatment regimen?

A. IV cefotaxime and amoxicillin
B. IV ceftriaxone and amoxicillin
C. IV cefotaxime only
D. IV ceftriaxone, amoxicillin and dexamethasone
E. IV ceftriaxone and dexamethasone

Answer: E

Short explanation

This patient is likely to have bacterial meningitis. *Neisseria meningitidis* and *Streptococcus pneumoniae* are the commonest causes of bacterial meningitis in this age group, so empirical antibiotics should reflect this. NICE recommends IV ceftriaxone and dexamethasone for initial treatment of bacterial meningitis in patients older than 3 months.

Long explanation

Bacterial meningitis is a medical emergency. Worldwide, it is responsible for more than 100,000 deaths per year. The disease typically worsens rapidly, and so a high index of clinical suspicion is required, along with prompt diagnosis and treatment, to reduce morbidity and mortality.

Clinical features include the classical triad of fever, neck stiffness and altered mental status, although all three may not be present in all patients. Other common features include severe headache, photophobia, arthralgia, malaise, neurological deficits and seizures. Meningitis caused by *Neisseria meningitidis* also commonly causes a characteristic petechial rash. Patients with bacterial meningitis often present with severe sepsis.

In the developed world, the commonest pathogens responsible for bacterial meningitis vary according to the age of the patient. The age of the patient therefore dictates the choice of empiric antibiotics to treat patients with bacterial meningitis with. The following table includes the empirical antibiotics recommended by NICE.

Age	Common pathogens	First-line antibiotics
<3 months	Group B streptococci, *Escherichia coli*, *Listeria monocytogenes*, *Klebsiella*	Cefotaxime AND ampicillin/amoxicillin
3 month–2 years	*S. pneumonia*, *N. meningitidis*, group B streptococci, *Haemophilus influenzae*, *E. coli*	Ceftriaxone
2–50 years	*N. meningitidis*, *S. pneumonia*	Ceftriaxone
>50 years	*S. pneumonia*, *N meningitidis*, *L. monocytogenes*	Ceftriaxone AND ampicillin/amoxicillin

In addition to antibiotics, dexamethasone 0.15 mg/kg (max 10 mg) QDS should be given before or alongside the first dose of antibiotics in patients over 3 months of age with suspected bacterial meningitis. Dexamethasone has been shown to improve mortality and neurological status in patients with meningitis caused by *S. pneumonia*, although its use is not supported by evidence in meningococcal meningitis.

References

National Institute for Health and Clinical Excellence (NICE). *Bacterial meningitis and meningococcal septicaemia* (CG102). London: NICE, 2010. http://www.nice.org.uk/guidance/cg102/chapter/1-guidance (accessed February 23, 2016).

Tunkel A, Hartman B. Practice guidelines for the management of bacterial meningitis. *Clin Infect Dis*. 2004;39;1267–1284.

F30. A 38-year-old man has been involved in a road traffic collision and has sustained abdominal injuries. He has linear bruising horizontally and diagonally across his abdomen in keeping with seatbelt positioning. Intravenous (IV) access has been secured, and he has already received 2 units of packed red cells and 500 ml of crystalloid. His primary survey findings are as follows:

Airway: Intact, talking.
Breathing: Respiratory rate: 28/min, SpO$_2$ 100% on 15 l oxygen; no signs of thoracic injury
Circulation: Heart rate: 135 bpm, blood pressure: 70/56 mmHg, cool peripheries; seatbelt sign; slight abdominal tenderness
Disability: Glasgow Coma Score 15/15; pupils size 4, reactive; in pain
Exposure: No signs of bony or spinal injuries

An ultrasound (focused assessment with sonography in trauma [FAST] scan) of his abdomen is performed, which shows significant intraperitoneal fluid.

What is the next step of his management?

A. Immediate transfer to theatre for laparotomy
B. Further boluses of IV crystalloid
C. Computed tomography (CT) scan of chest, abdomen and pelvis
D. Diagnostic peritoneal lavage (DPL)
E. Plain abdominal radiograph

Answer: A

Short explanation

This patient is haemodynamically unstable despite fluids and has intraperitoneal fluid on FAST scan. He is highly likely to have intra-abdominal bleeding which requires assessment and control in theatre in the form of a 'damage control laparotomy.' DPL is not required if a FAST scan is positive; further crystalloid may worsen trauma coagulopathy, and he is not sufficiently stable for CT transfer. Plain abdominal radiographs have no place in trauma assessment.

Long explanation

Trauma is a major cause of death in people under age 45. Abdominal injuries are a major cause of morbidity and mortality and occur in both blunt and penetrating trauma. Blunt abdominal trauma is a source of considerable diagnostic uncertainty. Unlike penetrating trauma, major blood loss may be hidden from view and have little in the way of positive examination findings until patients are in extremis. Patients with major intra-abdominal injuries may have a soft abdomen or little in the way of tenderness, particularly in the context of distracting injuries, sedation, neurological injuries or intoxication. Bedside investigations have a crucial role.

The decision of whether to intervene (emergency laparotomy or, increasingly, interventional radiology) or manage conservatively depends on many factors, but chief among them is the haemodynamic stability of the patient. Tachycardia, tachypnoea and cool skin are early signs of shock that can easily be falsely attributed to pain or exposure. Likewise, narrowed pulse pressure or reduced urine output can be missed and reduced conscious level attributed to head injury or masked by sedation. Hypotension (systolic blood pressure <90 mmHg) is a late sign (particularly in the younger patient), signifying a loss of >30% of circulating volume. This patient falls into this category.

Symptoms and signs that are most associated (likelihood ratio >3.5) with intra-abdominal injury are the seat-belt sign (bruising from a seatbelt on the abdomen), rebound tenderness, hypotension (systolic blood pressure <90 mmHg), abdominal distension and guarding. The most sensitive signs (sensitivity >50%) are abdominal tenderness to palpation, abdominal pain and the seatbelt sign; the most specific signs (specificity >90%) are rebound tenderness, hypotension, distension, femoral fractures, the seatbelt sign and guarding.

Haemodynamically stable patients may be transferred for a trauma CT to elucidate any injuries or sources of bleeding. Haemodynamically unstable patients should have bedside testing for intraperitoneal fluid. Commonly, ultrasound is used (FAST scan), although diagnostic peritoneal aspiration or lavage (DPL) may be used in situations without access to ultrasound or the expertise to use it. If one of these tests is positive and/or the patient's condition is unstable, then prompt emergency intervention may be required, usually in the form of an emergency 'damage-control' laparotomy. This is a quick operation to identify and treat sources of bleeding (e.g. splenectomy, packing of liver lacerations) and prevent contamination to allow continuing resuscitation. Prolonged operations for extensive repair of multiple, complex or non-life-threatening injuries are not appropriate in unstable patients.

References

Al-Mudhaffar M, Hormbrey P. Abdominal trauma. *BMedJ* 2014;348:g1140.

Hoff WS, Holevar M, Nagy KK, et al. Practice management guidelines for the evaluation of blunt abdominal trauma: the East Practice Management Guidelines Work Group. *J Trauma*. 2002;53:602–615.

Simel DL. Does this adult patient have a blunt intra-abdominal injury? *JAMA*. 2012;307:1517.

Exam G: Questions

G1. A 28-year-old man suffered a traumatic transection of his spinal cord at the level of T4. Six weeks post-injury, he remains in the intensive care unit undergoing a respiratory wean via his tracheostomy. He suddenly complains of a headache and looks flushed.

Which of the following is LEAST accurate about this patient's condition?

A. Bladder distension is one of the commonest triggers.
B. Oral prazosin should be used to treat this patient.
C. Atropine may be required to treat associated bradycardia.
D. Vasodilation above the level T4 may be seen.
E. Pulmonary oedema is a rare complication.

G2. A 63-year-old woman is admitted to intensive care with suspected viral encephalitis. She has been treated with empirical aciclovir therapy before microbiology results. Cerebrospinal fluid (CSF) polymerase chain reaction results demonstrate the presence of herpes simplex (HSV) DNA.

In view of the presence of HSV, which of the following is the most appropriate therapy?

A. Ganciclovir
B. Continue aciclovir
C. Foscarnet
D. Zanamivir
E. Ribavirin

G3. A 64-year-old woman has been in the intensive care unit (ICU) for 12 days. She was initially admitted from theatre after an emergency laparotomy for colonic perforation and faecal peritonitis. She is currently sedated, ventilated and requires cardiovascular support and haemofiltration. Results of a blood culture from 2 days ago become available, showing a Gram-negative bacillus.

Which of the following newer antibacterial agents is likely to be the most effective against a Gram negative bacillus?

A. Teicoplanin
B. Ertapenem
C. Tigecycline
D. Daptomycin
E. Linezolid

G4. Which of these five patients is most likely to benefit from a reservoir ('non-rebreathe') oxygen mask at 15 l/min?

A. 56-year-old presenting with suspected myocardial infarction (MI) and ongoing chest pain with oxygen saturations of 95%

B. 78-year-old presenting with suspected stroke awaiting thrombolysis with oxygen saturations of 94%

C. 78-year-old following cardiac arrest, now stable with an unsupported circulation and spontaneously ventilating with oxygen saturations of 94%

D. 65-year-old lifelong smoker with possible pneumonia and oxygen saturations of 89%

E. 30-year-old trauma patient who was trapped in a smoke-filled car with oxygen saturations of 99%

G5. A 46-year-old man with no previous medical problems is admitted to the emergency department with respiratory failure secondary to severe bilateral pneumonia. He is intubated and ventilated and transferred to the intensive care unit. Initial investigations demonstrate bilateral infiltrates on chest radiography and a PaO_2/FiO_2 (P/F) ratio of 85 mmHg.

Which intervention is likely to have the biggest impact in respect of reducing his mortality risk?

A. Cisatracurium infusion
B. High positive end expiratory pressure (PEEP) strategy
C. Prone positioning
D. Ventilation at 6 ml/kg predicted body weight
E. Referral for extracorporeal membrane oxygenation (ECMO)

G6. You are called urgently by the nurses on intensive care. A patient who had a laryngectomy 2 days previously has suddenly desaturated. As you attend, you notice the tracheostomy appears to be several centimetres out from where it should sit. You administer oxygen both via the face and via the tracheostomy. You remove the inner cannula but cannot pass a suction catheter. The patient is continuing to desaturate, so you remove the tracheostomy. The patient is NOT breathing. You call the cardiac arrest team.

What is the most appropriate next step?

A. Reinsert the tracheostomy tube and ventilate through it
B. Pass an introducer (bougie) through the laryngectomy stoma and insert a 6.0 cuffed endotracheal tube over the introducer
C. Pass an introducer (bougie) through the laryngectomy stoma and insert a smaller size tracheostomy tube over the introducer
D. Commence bag-mask ventilation via the mouth
E. Ventilate via an LMA applied to the laryngectomy stoma

G7. Which of the following is the LEAST appropriate indication to perform flexible bronchoscopy (FB) on an intensive care unit (ICU) patient?

A. Obtaining a sample to identify an organism in ventilator-associated pneumonia (VAP) where noninvasive approaches have failed
B. To identify a cause for haemoptysis that cannot be identified on computed tomography (CT)
C. For the relief of atelectasis in a ventilated patient when physiotherapy has failed
D. For the prevention of atelectasis in a post-operative lobectomy patient
E. For the diagnosis of Acute Respiratory Distress Syndrome

G8. A previously healthy 3-month-old infant has been on the paediatric intensive care unit for 6 days after an admission with bronchiolitis. He is now ready for discharge to the ward, and the nurses ask you to prescribe his fluid regime. He weighs 3 kg and has normal renal function and electrolytes. He is tolerating six oral feeds per day of expressed breast milk but only taking about 30 ml each time.

Which of the following is the most appropriate prescription?

A. 3 ml/hr Hartmann's solution
B. 300 ml/day 0.45% saline + 5% dextrose
C. 5 ml/hr 0.45% saline + 5% dextrose
D. 7.5 ml/hr 0.45% saline + 10% dextrose
E. 12.5 ml/hr 0.18% saline + 20 mmol/l KCl

G9. Which of the following analgesics has the largest volume of distribution?

A. Alfentanil
B. Fentanyl
C. Remifentanil
D. Morphine
E. Diamorphine

G10. A previously healthy 70-year-old woman presents to the emergency department with significant lower GI blood loss. She is adequately resuscitated with blood and is now haemodynamically stable. The blood loss was described as dark altered blood, mixed with fresh red blood. It was estimated at 1 l over the previous 24 hours.

What is the most likely finding on endoscopy?

A. A left-sided colonic malignancy
B. A right-sided colonic malignancy
C. Diverticulitis
D. Angiodysplasia
E. Haemorrhoids

G11. Which of the following best describes the possible bias shown in the following statement?

Researchers carried out a trial to compare the effects of a new drug on blood pressure. They recruited 1000 volunteers and randomized them into two equal groups of 500 subjects using a random sequence generator. They compared each group for baseline characteristics. The volunteers and the researchers were blinded to which group they were assigned. One group received the drug and the other a placebo. By the end of the 6-month trial, a number of subjects had dropped out of the study groups. The researchers compared the 480 remaining patients in the placebo group to the 450 remaining in the test group using a Student t test. They found a significant difference and reported this, along with the baseline characteristics and study design.

A. Selection bias
B. Attrition bias
C. Reporting bias
D. Detection bias
E. Performance bias

G12. A 35-year-old woman presents with sudden-onset palpitations. She does not complain of any chest pain. Her heart rate is 240, blood pressure is 98/64 and her respiratory rate is 24 with oxygen saturations of 96% in air. A 12-lead electrocardio-gram (ECG) reveals a supraventricular tachycardia (SVT) at a rate of 240.

Which of the following management options should be performed FIRST?

A. Synchronized DC cardioversion (100 J)
B. Perform carotid sinus massage
C. 300-mg intravenous amiodarone infusion
D. 12-mg intravenous adenosine bolus
E. Arrange an urgent cardiology review

G13. When prescribing for a patient with acute kidney injury and a glomerular filtration rate (GFR) <30 ml/min/1.73 m², which of the following recommendations for each drug is LEAST appropriate?

A. Warfarin – reduce dose and monitor more closely
B. Beta-blockers – reduce dose by 50%
C. Fentanyl – avoid use
D. Low-molecular-weight heparins – halve the dose
E. Metformin – avoid if possible

G14. A 64-year-old man has been admitted with a intracerebral bleed. He has been anticoagulated with warfarin for atrial fibrillation (AF), and his international normal-ized ratio (INR) on admission is 6.4.

Which of the following would be the BEST treatment regime to reverse the effects his warfarin?

A. Vitamin K
B. Prothrombin complex concentrates (PCC)
C. Vitamin K and PCC
D. Tranexamic acid
E. Fresh frozen plasma (FFP)

G15. A previously well 65-year-old woman is admitted to the intensive care unit (ICU) from the emergency department. She has a history of breathlessness and pro-ductive cough but no contact with animals or history of recent travel. She is confused, her respiratory rate is 28 and blood pressure is 94/51. She is pyrexic with a white cell count of 17×10^9/l, a C-reactive protein of 147 and a urea of 8.2 mmol/l. Chest X-ray demonstrates right basal consolidation.

What is MOST LIKELY causative organism responsible?

A. *Staphylococcus aureus*
B. *Streptococcus pneumoniae*
C. *Legionella pneumophila*
D. *Influenza A*
E. *Pseudomonas aeruginosa*

G16. You attend the emergency department to review a patient who is unconscious after being trapped in a house fire.

Which of the following is the most likely cause of death after significant smoke inhalation in patients who survive to arrival at hospital?

A. Thermal injury below the vocal cords
B. Cyanide poisoning
C. Asphyxiation
D. Inhalation of particulate matter
E. Airway obstruction

G17. You have just admitted a 50-year-old man into the intensive care unit who is intubated and ventilated after a cardiac arrest. You commence sedation with propofol and alfentanil.

Which of the following is the LEAST likely consequence of prolonged sedation?

A. Prolonged ventilation
B. Hypotension
C. Hypercatabolism
D. Gastrointestinal ileus
E. Increased delirium

G18. A 64-year-old man is in the intensive care unit (ICU) with multiorgan failure secondary to pancreatitis. He has received broad-spectrum antibiotics including 48 hours of gentamicin, 5 days of steroids for septic shock and 40 mmol magnesium replacement during a bout of atrial fibrillation. After 3 weeks, he is being weaned from the ventilator and is noted to be profoundly weak. On examination, he exhibits flaccid paralysis and hyporeflexia in all four limbs, but there is sparing of the facial muscles.

Which is the most LIKELY cause of this patient's weakness?

A. Steroid treatment
B. Aminoglycoside treatment
C. Magnesium treatment
D. Cerebrovascular accident
E. Intensive Care Unit acquired weakness

G19. You have admitted a 40-year-old man to the intensive care unit who underwent elective laparoscopic removal of a phaeochromocytoma.

Which of the following is the most likely complication of this procedure?

A. Persistent hypotension
B. Hypoglycaemia
C. Hypocalcaemia
D. Persistent hypertension
E. Hypokalaemia

G20. You are performing a rapid sequence induction (RSI) for a patient with a severe pneumonia who remains hypoxic despite maximal medical therapy. Your attempts at direct laryngoscopy have failed, so you attempt to site a supra-glottic airway device (SAD). You are unable to ventilate the patient and they continue to desaturate.

   Which of the following is the most appropriate next step?

A. Perform an emergency cricothyroidotomy.
B. Release the cricoid pressure and attempt intubation once more.
C. Attempt face-mask ventilation.
D. Insert a different SAD and attempt ventilation with cricoid pressure maintained.
E. Wake the patient.

G21. You are asked to perform a manual noninvasive blood pressure reading on a hypotensive patient you have just admitted to the intensive care unit.
   Which of the following statements is LEAST accurate?

A. The blood pressure cuff width should be 20% larger than the width of the arm
B. The cuff should be inflated above systolic pressure, then deflated at 2 to 3 mmHg/sec.
C. If using a mercury manometer, it must have a functioning air vent at the top of the column of mercury.
D. The first Korotkoff phase represents systolic blood pressure.
E. The fourth Korotkoff phase represents diastolic blood pressure.

G22. Which of the following patients is MOST likely to benefit from parenteral nutrition (TPN)?

A. A 20-year-old man with a body mass index (BMI) of 19, with unintended weight loss of 3% in the last 3 months, with an unsafe swallow
B. A 20-year-old man with a BMI of 18, with severe Crohn disease, who has been nil-by-mouth for five days
C. A 20-year-old man with a BMI of 20, who has eaten nothing for 5 days, who finds the nasogastric tube intolerable
D. A 20-year-old man with a BMI of 21, who is unlikely to eat anything for the next 3 days, and is not absorbing nasogastric feed
E. A 20-year-old man with a BMI of 22, with unintended weight loss of 8% in the last 6 months, with inadequate enteral nutritional intake

G23. You are called to the emergency department at midnight to help with the ongoing management of a 16-year-old girl who took an overdose of paracetamol. She is well known to the psychiatric services with depression. She is fully conscious, has capacity and is refusing any further treatment. She has a markedly elevated paracetamol level and would benefit from treatment with N-acetyl cysteine.
   Which is the following is the most appropriate way to manage this patient?

A. Treat her against her wishes under the Mental Capacity Act
B. Treat her against her wishes under the Mental Health Act
C. Seek consent from her parents to treat her against her wishes
D. Accept her autonomy and discharge her home
E. Wait for legal advice

G24. A 42-year-old female patient with gallstone pancreatitis was ventilated on admission to ICU for respiratory failure. She deteriorates after a period of stability on day 5, with increasing FiO_2 requirements 0.4 to 0.7 and positive end expiratory pressure 8 to 12 cmH$_2$O, which continues for the next 48 hours. She has increased sputum production (no growth on culture), a temperature of 38.6°C, a white cell count of 18×10^9/l and new left basal consolidation on chest X-ray. Five days of antibiotics are prescribed as per local hospital guidelines.

Which of the following is the BEST term to describe her respiratory deterioration?

A. Probable ventilator-associated pneumonia
B. Ventilator-associated condition
C. Acute respiratory distress syndrome (ARDS)
D. Infection-related ventilator-associated complication
E. Possible ventilator-associated pneumonia

G25. A previously well 36-year-old female is brought to hospital after a seizure secondary to hyponatraemia. Investigations results include the following:

- Serum sodium concentration 113 mmol/l
- Serum osmolality 274 mOsm/kg
- Urine sodium concentration 10 mmol/l
- Urine osmolality 80 mOsm/kg

What is the MOST likely cause of her hyponatraemia?

A. Syndrome of inappropriate ADH secretion (SIADH)
B. Water intoxication
C. Renal failure
D. Cerebral salt wasting syndrome
E. Hypothyroidism

G26. A 60-year-old man has been admitted to the emergency department with shortness of breath. He smokes 40 cigarettes per day and has recently been prescribed a salbutamol inhaler by his general practitioner. His chest radiograph shows a right-sided pneumothorax, which measures 3 cm at the level of the hilum. There is no sign of tension pneumothorax. There is no history of trauma.

What is the best initial management?

A. Conservative: admit, treat with oxygen and observe
B. Aspiration with a 16-G cannula
C. Size 10-Fr Seldinger intercostal drain
D. Size 18-Fr open intercostal drain
E. Size 32-Fr open intercostal drain

G27. A 71-year-old man is admitted with acute decompensated cardiac failure. He is distressed and diaphoretic with a respiratory rate of 40, oxygen saturation 94% in air, heart rate of 100 bpm and blood pressure (BP) 90/65. Chest radiography shows signs of pulmonary oedema.

What is the most appropriate first therapy?

A. Glycerin trinitrate (GTN) 2 sprays sublingual
B. Furosemide 40 mg IV intravenous (IV)
C. Oxygen 6–10 l by face mask
D. Diamorphine 1–2.5 mg IV
E. Dobutamine infusion at 2–20 μg/kg/min

G28. A 26-year-old woman is admitted with fever, confusion, agitation and seizures. Following a normal CT scan of her brain, a lumbar puncture is performed. Her blood glucose is 6 mmol/l.

Which of the following sets of results would suggest viral meningoencephalitis?

A. Protein 100 mg/dl; glucose 4 mmol/l; white blood cell count (WBC) 50×10^9/l; predominant lymphocytes
B. Protein 250 mg/dl; glucose 1.5 mmol/l; WBC 500×10^9/l; predominant lymphocytes
C. Protein 500 mg/dl; glucose 0.5 mmol/l; WBC 5000×10^9/l; predominant neutrophils
D. Protein 50 mg/dl; glucose 4 mmol/l; WBC 1×10^9/l; predominant lymphocytes
E. Protein 500 mg/dl; glucose 6 mmol/l; WBC 50×10^9/l; predominant neutrophils

G29. You are called to the emergency department as part of the trauma team. A 23-year-old man has been involved in a road traffic collision. He was an unrestrained driver of a car that collided with a tree at high speed. His vital signs are as follows: heart rate 130, blood pressure (BP) 60/35, SpO$_2$ unrecordable, respiratory rate 35. He has no signs of external haemorrhage, pneumothorax or tamponade. His abdomen is soft, his pelvis is stable and he has no signs of long bone fractures. A focused assessment with sonography in trauma (FAST) scan is negative for blood in the peritoneum or pericardium. A chest radiograph is performed, showing a widened mediastinum. You suspect a traumatic aortic injury.

Where is the most likely site of aortic injury?

A. Aortic root
B. Ascending aorta
C. Aortic arch
D. Proximal descending aorta
E. Distal descending aorta

G30. Which of the following is the LEAST important risk factor in the development of delirium in adult patients in the intensive care unit (ICU)?

A. Pre-existing cognitive impairment
B. Presence of coma
C. Orientation to time with no visible clock present
D. High SAPS II score
E. Alcohol intake: 2 pints beer/day >3 units/day

Exam G: Answers

G1. A 28-year-old man suffered a traumatic transection of his spinal cord at the level of T4. Six weeks post-injury, he remains in the intensive care unit undergoing a respiratory wean via his tracheostomy. He suddenly complains of a headache and looks flushed.

Which of the following is LEAST accurate about this patient's condition?

A. Bladder distension is one of the commonest triggers.
B. Oral prazosin should be used to treat this patient.
C. Atropine may be required to treat associated bradycardia.
D. Vasodilation above the level T4 may be seen.
E. Pulmonary oedema is a rare complication.

Answer: B

Short explanation

This patient has autonomic dysreflexia, and increased blood pressure (BP) is a common sign. Non-pharmacological treatments should be instigated first, but if the patient's systolic BP remains >150 mmHg, pharmacological treatments may be required. Although prazocin is an antihypertensive and is used as prophylactic treatment for this condition, its onset of action is too slow once the condition is established.

Long explanation

Autonomic dysreflexia is a condition occurring in patients who have suffered spinal cord injuries, particularly at or above the T6 level. The incidence and severity is greater in those with higher injuries as well as those with complete rather than incomplete spinal cord injuries.

It results from a loss of the co-ordinated autonomic response to a stimulus below the level of injury. An uninhibited sympathetic response causes profound vasoconstriction below the level of injury, resulting in hypertension. There is associated vasodilatation above the level of injury, and bradycardia occurs as a result of the baroreceptor response.

The commonest triggers are distension or irritation of the bladder and bowel; however, any stimulus below the level of the injury can be implicated.

The severity varies from asymptomatic to fatal. Common symptoms and signs include severe headache, nasal obstruction, a sensation of anxiety, blurred vision and nausea. Flushed skin, diaphoresis and piloerection occur above the level of the injury can occur compared with pale and dry skin below. Hypertension may be associated with a reflex brachycardia.

Complications, although rare, can also occur as a result of the severe hypertensive crisis. These include myocardial infarction, cardiac dysrhythmias and arrest, intracranial haemorrhage, stroke, seizures, retinal haemorrhage, pulmonary oedema, apnoea and acute kidney injury.

Preventative treatment includes ensuring good bladder and bowel care and good pressure area and skin care. Oral nifedipine and prazosin can be used prophylactically to prevent attacks once patients identify early symptoms.

Treatment of established attacks includes monitoring of the BP. Patients should be immediately sat upright to aid lowering of BP. The trigger should be identified and treated. This includes catheterizing a patient in retention or ensuring the catheter is patent, performing a rectal examination to exclude/treat faecal impaction and removing tight-fitting garments. Prompt reduction of BP should be achieved if systolic BP remains >150 mmHg by titrating rapid onset and short-acting agents such as nitrates, nifedipine, sublingual captopril, intravenous hydrazine, labetalol or phentolamine. Propofol can also be used in patients already sedated. Treatment for complications may be required such as atropine or glycopyrrolate for bradycardia.

References
Bycroft J, Shergill I, Choong E, et al. Autonomic dysreflexia: a medical emergency. *Postgrad Med J*. 2005;81(954):232–235.
Krassioukov A, Warburton DE, Teasell R, Eng JJ; Spinal Cord Injury Rehabilitation Evidence Research Team. A systematic review of the management of autonomic dysreflexia following spinal cord injury. *Arch Phys Med Rehabil*. 2009;90(4):682–695.

G2. A 63-year-old woman is admitted to intensive care with suspected viral encephalitis. She has been treated with empirical aciclovir therapy before microbiology results. Cerebrospinal fluid (CSF) polymerase chain reaction results demonstrate the presence of herpes simplex (HSV) DNA.

In view of the presence of HSV, which of the following is the most appropriate therapy?

A. Ganciclovir
B. Continue aciclovir
C. Foscarnet
D. Zanamivir
E. Ribavirin

Answer: B

Short explanation
Aciclovir is first-line therapy for HSV encephalitis. Ganciclovir is used in cytomegaloviral infections. Foscarnet is active against HSV, but should only be used in cases resistant to aciclovir owing to its toxicity. Zanamivir is active against influenza A and B viruses. Ribavirin is used to treat respiratory syncytial viral infections and hepatitis C.

Long explanation
Encephalitis is a rare condition but one that can be disabling if not recognized or treated promptly. It may be infective or noninfective. The commonest pathogens responsible for encephalitis are viruses, although it may be caused by bacteria, fungi and parasites. Noninfective encephalitis is generally autoimmune in nature. Herpes

simplex virus (HSV) is the most commonly diagnosed cause of encephalitis, with up to 90% being due to HSV-1 and 10% HSV-2.

Viral encephalitis typically presents as a febrile illness with altered cognition, behaviour, personality or conscious level. New seizures or focal neurological signs in a patient with fever should also prompt investigations for viral encephalitis. An urgent lumbar puncture (LP) should be performed unless clinically contraindicated. Typical features of viral encephalitis include a normal or raised opening pressure and clear CSF with slightly raised white cell count (predominantly lymphocytes), normal glucose and normal or raised protein. If performing an LP is contraindicated, then magnetic resonance imaging is the investigation of choice.

Aciclovir is the antiviral of choice for the treatment of HSV encephalitis. It is a nucleoside analogue which acts to disrupt HSV DNA replication by replacing guanosine, leading to termination of the copied DNA strand, preventing viral replication. It should be started empirically within 6 hours of presentation if LP is contraindicated or CSF findings suggest viral encephalitis. Aciclovir should also be started if viral encephalitis is still suspected despite normal or equivocal CSF results. If bacterial meningitis is suspected, then ceftriaxone should also be commenced as soon as possible.

References

Joint Formulary Committee. *British National Formulary*. 69th ed. London: BMJ Group and Pharmaceutical Press; 2015.

Solomon T, Michael BD, Smith PE, et al. Management of suspected viral encephalitis in adults – Association of British Neurologists and British Infection Association National Guidelines. *J Infect*. 2012;64(4):347–373.

G3. A 64-year-old woman has been in the intensive care unit (ICU) for 12 days. She was initially admitted from theatre after an emergency laparotomy for colonic perforation and faecal peritonitis. She is currently sedated, ventilated and requires cardiovascular support and haemofiltration. Results of a blood culture from 2 days ago become available, showing a Gram-negative bacillus.

Which of the following newer antibacterial agents is likely to be the most effective against a Gram-negative bacillus?

A. Teicoplanin
B. Ertapenem
C. Tigecycline
D. Daptomycin
E. Linezolid

Answer: B

Short explanation

Ertapenem is a carbapenem, a class of drugs that has good activity against Gram-negative organisms. All of the other agents are newer antibacterials with predominantly Gram-positive activity.

Long explanation

Antibiotic resistance is an increasing threat to global public health. There are many factors implicated in its rise, principally related to antibiotic stewardship and drug development. The development of resistance by organisms is a natural occurrence and as such cannot be prevented. However, there are environmental conditions which encourage the development of resistance, such as inadequate dosing levels or antibiotic course duration, thus selecting out resistant organisms to thrive at the expense of nonresistant organisms.

Antibiotic stewardship is the choice of the right drug at the right dose for the right length of time. The overuse of antibiotics when not necessary is an important contributor to the development of resistance. This includes the use of antibiotics by industry (e.g. in animal feeds), over-the-counter purchasing or use of broad-spectrum antibiotics when a narrow-spectrum drug would work equally well. The problem is compounded by the huge reduction in new antibiotics being developed. Numbers of new agents have fallen from 17 per 5 years in the 1900s to 4 per 5 years in the 2000s. Appropriate use of the antimicrobial agents that we have is therefore of crucial importance.

The terms 'Gram-positive' and 'Gram-negative' are used to describe the retention of a violet stain used before microscopy. Gram-positive organisms have a thick peptidoglycan wall that retains the stain after washing, whereas Gram-negative bacteria have a much thinner peptidoglycan wall surrounded by a lipopolysaccharide-rich outer membrane. This difference in surface structures between the Gram-positive and -negative bacteria underlies the difference in activity between antimicrobial agents. Many drugs are particularly active against one or other of the Gram stain classes.

Gram-negative organisms are common causes of infection on the ICU, many of which are resistant to various common antibiotics. In particular, *Acinetobacter baumannii*, *Pseudomonas aeruginosa*, and *Enterobacter* species are widely drug resistant, forming part of the 'ESKAPE' group of pathogens (also including *Enterococcus faecium*, *Staphylococcus aureus* and *Klebsiella pneumonia*). The ESKAPE pathogens have been identified as one of the main current challenges in the battle against multi-drug-resistant bacteria.

Newer drugs that have been licensed and widely used in the past 15 years include linezolid, ertapenem, daptomycin, tigecycline and doripenem. The majority of these agents have a particular activity against Gram-positive organisms, possibly in response to the rise of methicillin-resistant *Staphylococcus aureus* over the same time period. The carbapenems are a class of drugs with good activity against both Gram-positive and Gram-negative organisms, although resistance to carbapenems is on the increase. More recent agents include telavancin and ceftaroline fosamil along with combinations of antibacterial drugs with beta-lactamase inhibitors. Most of the new drugs are in the same class as existing agents with similar actions and a risk of cross-resistance; drugs with novel mechanisms of action are few and far between.

References

Aksoy DY, Unal S. New antimicrobial agents for the treatment of Gram-positive bacterial infections. *Clin Microbiol Infect*. 2008;14(5):411–420.

Bassetti M, Merelli M, Temperoni C, Astilean A. New antibiotics for bad bugs: where are we? *Ann Clin Microbiol Antimicrob*. 2013;12(1):22.

Boucher HW, Talbot GH, Benjamin DK, et al. 10 × 20 Progress – development of new drugs active against Gram-negative bacilli: an update from the Infectious Diseases Society of America. *Clin Infect Dis*. 2013;56:1685–2694.

G4. Which of these five patients is most likely to benefit from a reservoir ('non-rebreathe') oxygen mask at 15 l/min?

A. 56-year-old presenting with suspected myocardial infarction (MI) and ongoing chest pain with oxygen saturations of 95%
B. 78-year-old presenting with suspected stroke awaiting thrombolysis with oxygen saturations of 94%
C. 78-year-old following cardiac arrest, now stable with an unsupported circulation and spontaneously ventilating with oxygen saturations of 94%
D. 65-year-old life-long smoker with possible pneumonia and oxygen saturations of 89%
E. 30-year-old trauma patient who was trapped in a smoke filled car with oxygen saturations of 99%

Answer: E

Short explanation

There is no evidence of benefit in stroke or MI unless there is concomitant hypoxaemia. Post–cardiac arrest patients should be treated with high-flow oxygen until stable. Patients with likely chronic obstructive pulmonary disease should be treated as such (target saturations 88–92%) until a forced expiratory volume in 1 second (FEV1) can be obtained. Patients exposed to carbon monoxide should receive high-flow oxygen until an arterial blood gas (ABG) can be obtained. Oxygen saturations may be falsely high in this group because of carboxyhaemoglobin.

Long explanation

Oxygen is a drug and should be prescribed and administered to patients only for the treatment of hypoxaemia. Administration should always target a saturation level or range, and this should be set to achieve a normal level for the patient and titrated to effect. The 2008 British Thoracic Society guidelines recommend the majority of patients should be targeted to an oxygen saturation of ≥94%. Those with COPD or chronic lung disease who retain CO_2 should be targeted to 88 to 92%. Chronic smokers should be assumed to have an underlying diagnosis of COPD and treated as such until formal testing can be arranged. The use of 15-l non-rebreathe mask oxygen should be reserved for critically ill unstable patients, in an emergency, until an oxygen saturation (SaO_2) level can be obtained and oxygen therapy titrated accordingly. There is no evidence that hyperoxia is helpful following stroke or cardiac arrest and in fact may be harmful.

Exposure to a smoke-filled environment risks inhalation of carbon monoxide (CO). CO binds much more avidly to haemoglobin than either oxygen or carbon dioxide, rendering the patient unable to transport oxygen to the tissues, despite adequate respiration and cardiac output. It also interferes with clinical assessment as it makes the patient appear 'cherry red' despite tissue hypoxia. Oxygen saturation meters are unreliable because the sensors are unable to distinguish between the absorption spectra for carbon monoxide bound haemoglobin and oxygenated haemoglobin. Therefore the SaO_2 will read as the total of both. An ABG result will be able to distinguish COHb as a fraction of total haemoglobin. However, the dissolved portion of oxygen in the blood will also be normal, therefore the PaO_2 on the blood gas result will also be deceptively high in the face of ongoing tissue hypoxia.

Once exposure to CO is stopped (i.e. the patient is removed from the smoky environment), the levels of CO in the blood will fall. However, this will occur slowly because of the strength of the CO bond to haemoglobin. The rate of CO removal can be increased by delivering a high inspired oxygen concentration or by placing the patient in a hyperbaric chamber.

Reference

O'Driscoll BR, Howard LS, Davison AG. BTS guideline for emergency oxygen use in
   adult patients. *Thorax*. 2008;63(Suppl VI):vi1–vi68.

G5. A 46-year-old man with no previous medical problems is admitted to the emergency department with respiratory failure secondary to severe bilateral pneumonia. He is intubated and ventilated and transferred to the intensive care unit. Initial
investigations demonstrate bilateral infiltrates on chest radiography and a PaO_2 / FiO_2
(P/F) ratio of 89 mmHg.

   Which intervention is likely to have the biggest impact in respect of reducing his
mortality risk?

A. Cisatracurium infusion
B. High positive end expiratory pressure (PEEP) strategy
C. Prone positioning
D. Ventilation at 6 ml/kg predicted body weight
E. Referral for extracorporeal membrane oxygenation (ECMO)

Answer: D

Short explanation

Although all of these interventions have been shown to improve oxygenation and/or
reduce mortality in acute respiratory distress syndrome (ARDS), lung-protective ventilation at 6 ml/kg predicted body weight has consistently been demonstrated to
greatly reduce mortality.

Long explanation

This patient fits the 2012 Berlin definition of severe ARDS, a condition with a mortality rate of around 40 to 45%. Attempts to find measures that will reduce mortality from ARDS include ventilatory strategies, positioning and pharmacological
treatments. Although many interventions have been shown to improve measures of
oxygenation or to reduce mortality in small studies, strong, replicable evidence of
improved outcomes in large multicentre trials is thinner on the ground.

   Cisatracurium infusion has been shown to reduce mortality in patients with moderate to severe ARDS (P/F ratio <150 mmHg). In a French multicentre RCT, 339
patients were randomized to receive cisatracurium or placebo for 48 hours in early
ARDS. No improvement in raw mortality was found, but a statistically significant
reduction in mortality was seen in the intervention group following adjustment for
severity of illness.

   PEEP should be employed in ARDS to reduce atelectrauma and maintain recruit-
ment; however, some advocate a 'high PEEP' strategy. This is also known as the 'open
lung' strategy and uses PEEP levels of >15 cmH_2O along with low tidal volumes.
Several studies have compared high and standard PEEP in ARDS, some showing no
benefit and others showing improved oxygenation. A recent meta-analysis demonstrated improvements in oxygenation but no mortality benefit.

   The use of lower tidal volumes (6 ml/kg predicted body weight) and restricted
plateau pressure (<30 cmH_2O) has been repeatedly shown to reduce mortality in
ARDS in multiple studies since 1998. It has more recently been shown to reduce mortality and measures of lung injury in patients without ARDS. All the other interventions described here include lower tidal volume ventilation as part of treatment.

   Prone positioning has been known to improve measures of oxygenation in moderate to severe ARDS for some years, but evidence of improved outcomes has only
recently been elucidated. A French multicentre randomized controlled trial compared prone positioning for 16 hours per day for up to 5 days in 466 patients with

moderate-severe ARDS (P/F ratio <150 mmHg). A significant reduction in mortality was demonstrated, but concerns remain about the safety of prone positioning in inexperienced ICUs.

Referral to an ECMO centre was found to reduce mortality in the CESAR trial, although not all patients in the intervention group received ECMO. Further studies are ongoing in this growing field.

References

Guérin C, Reignier J, Richard J-C, et al. Prone positioning in severe acute respiratory distress syndrome. *N Engl J Med*. 2013:368:2159–2168.

Papazian F, Forel JM, Gacouin A, et al. Neuromuscular blockers in early acute respiratory distress syndrome. *N Engl J Med*. 2010;363:1107–1116.

Peek GJ, Mugford M, Tiruvoipati R, et al. Efficacy and economic assessment of conventional ventilatory support versus extracorporeal membrane oxygenation for severe adult respiratory failure (CESAR): a multicentre randomised controlled trial. *Lancet*. 2009;374(9698):1351–1363. doi:10.1016/S0140–6736(09)61069–2.

Ranieri VM, Rubenfeld GD, Thompson BT, et al. Acute respiratory distress syndrome: the Berlin Definition. *JAMA*. 2012;307(23):2526–33. doi:10.1001/jama.2012.5669.

Santa Cruz R, Rojas J, Nervi R, et al. High versus low positive end-expiratory pressure (PEEP) levels for mechanically ventilated adult patients with acute lung injury and acute respiratory distress syndrome [review]. *Cochrane Database Syst Rev*. 2013;(6).

Slutsky AS, Ranieri VM. Ventilator-induced lung injury. *N Engl J Med*. 2013;369(22): 2126–2136.

G6. You are called urgently by the nurses on intensive care. A patient who had a laryngectomy 2 days previously has suddenly desaturated. As you attend, you notice the tracheostomy appears to be several centimetres out from where it should sit. You administer oxygen both via the face and via the tracheostomy. You remove the inner cannula but cannot pass a suction catheter. The patient is continuing to desaturate, so you remove the tracheostomy. The patient is NOT breathing. You call the cardiac arrest team.

What is the most appropriate next step?

A. Reinsert the tracheostomy tube and ventilate through it
B. Pass an introducer (bougie) through the laryngectomy stoma and insert a 6.0 cuffed endotracheal tube over the introducer
C. Pass an introducer (bougie) through the laryngectomy stoma and insert a smaller size tracheostomy tube over the introducer
D. Commence bag-mask ventilation via the mouth
E. Ventilate via an laryngeal mask airway (LMA) applied to the laryngectomy stoma

Answer: E

Short explanation

The national tracheostomy safety project has an algorithm for emergency tracheostomy management *without* a patent upper airway. It states that in a patient who is not breathing and in whom you have removed the tracheal tube, the resuscitation team should be called and then oxygenation attempted via the laryngectomy stoma with either a paediatric facemask or an LMA. If this fails, then intubation should be attempted of the laryngectomy stoma.

Long explanation

The national confidential enquiry into patient outcomes and death in 2014 reviewed the care received by patients who underwent a tracheostomy. One of the principal recommendations of this report was that all staff must be able to manage complications of tracheostomy use, including obstruction or displacement, in accordance with the algorithms produced by the national tracheostomy safety project.

In post-laryngectomy patients, there are fewer options available in an emergency situation of this kind. As soon as a displaced or obstructed tracheostomy is suspected, expert help should be called and the airway assessed by looking, listening and feeling for signs of breathing. Capnography should be used to assist this assessment. If the patient is breathing, high-flow oxygen should be applied to the tracheostomy site. For any patients who are not breathing, the resuscitation team should be called and cardiopulmonary resuscitation started if there is no pulse.

The patency of the laryngectomy stoma should be assessed immediately. Attempting to bag ventilate a displaced tracheostomy can quickly result in catastrophic surgical emphysema, so caution is advised. Any speaking valves and inner tubes (rarely present) should be removed and a suction catheter passed down the tracheostomy. If patent, suction should be performed and the stoma may be cautiously used for ventilation. If unsuccessful, any tracheostomy cuff should be deflated (not always present) and breathing reassessed.

If the patient is not breathing, the tracheostomy should be removed. Initial oxygenation attempts should be using a mask over the stoma site. A paediatric mask is preferred, but if not immediately available, a supraglottic airway with an inflated cuff is a widely available alternative. Reintubation of the laryngectomy stoma should be attempted by an expert, which may require the use of a fibreoptic bronchoscope or other airway equipment.

References

McGrath BA, Bates L, Atkinson D, Moore JA. Multidisciplinary guidelines for the management of tracheostomy and laryngectomy airway emergencies. *Anaesthesia*. 2012;67(9):1025–1041.

Wilkinson K, Martin I, Freeth H, et al. National Confidential Enquiry into Patient Outcome and Death. On the Right Trach? A review of the care received by patients who underwent a tracheostomy. 2014. http://www.ncepod.org.uk/2014 report1/downloads/On%20the%20Right%20Trach_FullReport.pdf (accessed January 2016).

G7. Which of the following is the LEAST appropriate indication to perform flexible bronchoscopy (FB) on an intensive care unit (ICU) patient?

A. Obtaining a sample to identify an organism in ventilator-associated pneumonia (VAP) where noninvasive approaches have failed

B. To identify a cause for haemoptysis which cannot be identified on computed tomography (CT)

C. For the relief of atelectasis in a ventilated patient when physiotherapy has failed

D. For the prevention of atelectasis in a post-operative lobectomy patient

E. For the diagnosis of acute respiratory distress syndrome

Answer: D

Short explanation

Although commonly performed, there is limited evidence for the use of FB on ICU. There is good evidence for attempting other, less invasive methods initially, such as directed catheter sampling in VAP and CT in haemoptysis. There is some evidence for the use of FB to treat atelectasis in certain circumstances where physiotherapy and other treatments have failed. There is, however, good evidence not to use FB in the prevention of atelectasis in post-lobectomy patients.

Long explanation

Flexible bronchoscopy is a commonly performed and vital diagnostic and therapeutic tool. The majority of FB is carried out by respiratory physicians in an outpatient setting, and those performed on critically ill patients on ICU should always be considered high risk. The risks of FB are increased for patients with hypoxaemia, cardiac abnormalities (particularly dysrhythmias or recent myocardial infarction), clotting abnormalities and immunocompromise. Risks include bleeding, arrhythmias, seizures, myocardial infarction and pneumothorax. For these reasons, FB is often only performed after less invasive measures have failed such as directed catheter sampling, CT or physiotherapy interventions.

Patients undergoing FB should be appropriately sedated. Local anaesthetic can be used to numb the airway, but it may be more appropriate in critical care patients to use sedation and muscle relaxants to prevent coughing. Ventilation should be adjusted to take into account leaks from the circuit and the significant obstruction to flow caused by the bronchoscope within the endotracheal tube. It is advised to pre-oxygenate with 100% oxygen and use appropriate positive end-expiratory pressure and 100% oxygen to avoid desaturation both during and following the procedure. Clinicians performing FB should be familiar with the British Thoracic Society guidelines.

Reference

Du Rand IA., Blaikley J, Booton R, et al. British Thoracic Society guideline for diagnostic flexible bronchoscopy in adults: accredited by NICE. *Thorax*. 2013;68:i1–44.

G8. A previously healthy 3-month-old infant has been on the paediatric intensive care unit for 6 days after an admission with bronchiolitis. He is now ready for discharge to the ward, and the nurses ask you to prescribe his fluid regime. He weighs 3 kg and has normal renal function and electrolytes. He is tolerating six oral feeds per day of expressed breast milk but only taking about 30 ml each time.

Which of the following is the most appropriate prescription?

A. 3 ml/hr Hartmann's solution
B. 300 ml/day 0.45% saline + 5% dextrose
C. 5 ml/hr 0.45% saline + 5% dextrose
D. 7.5 ml/hr 0.45% saline + 10% dextrose
E. 12.5 ml/hr 0.18% saline + 20 mmol/l KCl

Answer: C

Short explanation

His maintenance can be calculated as 4 ml/kg/hr or approximately 100 ml/kg/day equalling 300 ml/day. Six feeds of 30 ml comprise 180 ml, leaving a 120-ml deficit to be met by intravenous fluids per day or 5 ml/hr. This should be met with a dextrose containing fluid which is hypotonic so as not to cause hypernatraemia.

Long explanation

Bronchiolitis is a common presentation in this age group and often leads to problems with feeding. During the initial resuscitation phase, children should be resuscitated with fluid boluses and their deficit calculated based on haemodynamic parameters and clinical signs. The deficit should be replaced over the initial stabilization phase and corrected within the first 24 to 48 hours of the admission, ensuring ongoing losses are accounted for.

The Holliday and Segar formula recommends daily maintenance be calculated as

4 ml/kg/hr for the first 10 kg
2 ml/kg/hr for the second 10 kg
1 ml/kg/hr thereafter

Resuscitation and replacement fluids plus ongoing losses are added on top of this if needed.

When unwell, children often have an exaggerated anti-diuretic hormone response, leading to water retention. Children with diarrhoeal or vomiting illnesses rapidly lose sodium and water. Therefore it is important to avoid hypotonic solutions during the resuscitation phase. Much like adult guidelines, initial resuscitation fluids should always contain sodium in the range of 130 to 154 mmol/l. However, once stabilized, care should be given not to administer hypertonic solutions to children to avoid hypernatraemia. Neonates and those with metabolic disorders often require higher dextrose infusions but most centres recommend 0.45% Saline and 5% Dextrose as a maintenance fluid for children to balance energy and sodium requirements. Electrolytes should continue to be checked regularly for children on maintenance IV fluid and potassium and other electrolytes replaced as necessary.

When unwell, especially with respiratory or cardiac conditions, most centres will advocate limiting fluid intake to two-thirds of maintenance requirements during the acute illness. Once the child is improving, this often is increased to 100% maintenance. For children with nonrespiratory and noncardiac conditions and the very young, maintenance is often increased further to 120 to 150% to account for the further replacement and growth requirements.

Where possible, as much of the daily fluid volume requirement should be given via the oral route to avoid large volumes of intravenous fluids and to enable homeostatic electrolyte control to return to normal. In infants, the best fluid is expressed breast milk or a milk feed solution because this also contains the calories required for growth and development.

Reference

Association of Paediatric Anaesthetists of Great Britain and Ireland. APA Consensus guideline on perioperative fluid management in children, v1.1 2007. http://www.apagbi.org.uk/publications/apa-guidelines (accessed August 2015).

G9. Which of the following analgesics has the largest volume of distribution?

A. Alfentanil
B. Fentanyl
C. Remifentanil
D. Morphine
E. Diamorphine

Answer: E

Short explanation

Diamorphine has a volume of distribution of 5 l/kg compared to fentanyl with 4 l/kg, morphine with 3.5 l/kg, alfentanil with 0.6 l/kg and remifentanil with 0.3 l/kg.

Long explanation

The volume of distribution (Vd) of a drug is the theoretical volume of plasma into which the drug would need to be dissolved in order to obtain the plasma concentration which is achieved on administration of a bolus dose. Volume of distribution is measured in litres and is calculated as:

$$Vd = dose/plasma\ concentration\ of\ free\ drug\ at\ time\ 0$$

Knowledge of the Vd of a drug is helpful in understanding its pharmacokinetics, enabling modelling of drug characteristics etc. Understanding that propofol has a large Vd predicts that following a bolus dose the plasma levels will fall rapidly as it redistributes. The preceding equation can be rearranged to enable us to calculate the dose to give to a patient to achieve a specific plasma concentration.

As an example, a highly fat-soluble drug will rapidly disperse into other tissues such that perhaps only 1% of it remains in the plasma with the remaining 99% distributed around other body tissues. Now imagine we give a 4-mg dose to an 80-kg man. We measure the initial concentration to be 0.01 mg/l. We can then calculate his apparent volume of plasma (the Vd) to be 400 l, or 5 l/kg.

It therefore follows that drugs with a large Vd must disperse widely outside of the plasma, or sequester and accumulate in those tissues much more easily than in the plasma. Vd will also increase when drugs bind to molecules within the plasma and are therefore not measurable as 'free drug'.

Drugs such as muscle relaxants are charged molecules that struggle to cross phospholipid membranes and are not particularly fat soluble. They have a very small volume of distribution as most of the drug remains in the plasma. Highly lipid soluble drugs such as propofol and diamorphine will sit much more easily in fatty tissues than in plasma and therefore have a very large volume of distribution.

Reference
Peck TE, Hill SA, M Williams M. *Pharmacology for Anaesthesia and Intensive Care.* 3rd edition. Cambridge: Cambridge University Press.

G10. A previously healthy 70-year-old woman presents to the emergency department with significant lower gastrointestinal blood loss. She is adequately resuscitated with blood and is now haemodynamically stable. The blood loss was described as dark altered blood, mixed with fresh red blood. It was estimated at 1 l over the previous 24 hours.

What is the most likely finding on endoscopy?

A. A left-sided colonic malignancy
B. A right-sided colonic malignancy
C. Diverticulitis
D. Angiodysplasia
E. Haemorrhoids

Answer: C

Short explanation
The commonest cause of massive lower gastrointestinal bleeding (LGIB) is diverticulitis, followed by angiodysplasia. As diverticulitis is much more common in older patients, some sources quote angiodysplasia as the commonest cause of large LGIB in young patients. Haemorrhoids are the commonest cause of LGIB in younger patients but rarely present with significant bleeding. Malignancy classically presents with occult bleeding.

Long explanation
LGIB is defined as that which occurs distal to the ligament of Trietz within the GI tract. However, 15% of acute LGIB with altered blood will have a more proximal source,

therefore defined as an upper GI bleed. Haematochezia is the passage of fresh, red blood and melaena is the passage of old or altered blood. Melaena may be associated with a rise in blood urea levels as proteins from the partially digested blood are reabsorbed by the small bowel.

Acute GI bleeding carries a significant mortality. Patients admitted with lower GI bleeding have a mortality of 3.6%, whereas those who develop GI blood loss during their hospital stay secondary to another condition have a mortality of 23%. Known risk factors for severity and death include age, haemodynamic instability, the presence of comorbidity and the use of aspirin or NSAIDs. Those who bleed due to diverticular disease have a 10% recurrence rate at 1 year and 25% at 4 years.

Management should include resuscitation of the patient, initially with crystalloid and then blood products. Early activation of the hospital 'major haemorrhage protocol' may be appropriate. Alongside resuscitation, efforts should be made to identify and control the blood loss. Around 80–85% of cases of bleeding will stop without intervention.

The identification of the bleeding source may be with either computed tomography (CT) angiography or colonoscopy. The benefit of colonoscopy is the possibility of targeted investigations such as biopsy of lesions as well as therapy for the bleeding. CT-guided angiographic embolization is also an effective method of treatment. If one option fails, it is possible to try the other, with surgical resection kept as a last resort. Surgery for uncontrolled GI haemorrhage carries a 33% mortality rate.

Reference
Scottish Intercollegiate Guidelines Network (SIGN). *Management of acute upper and lower gastrointestinal bleeding. A national clinical guideline* (SIGN publication no. 105). Edinburgh: SIGN; 2008.

G11. Which of the following best describes the possible bias shown in the following statement?

Researchers carried out a trial to compare the effects of a new drug on blood pressure. They recruited 1000 volunteers and randomized them into two equal groups of 500 subjects using a random sequence generator. They compared each group for baseline characteristics. The volunteers and the researchers were blinded to which group they were assigned. One group received the drug and the other a placebo. By the end of the 6-month trial, a number of subjects had dropped out of the study groups. The researchers compared the 480 remaining patients in the placebo group to the 450 remaining in the test group using a Student *t* test. They found a significant difference and reported this, along with the baseline characteristics and study design.

A. Selection bias
B. Attrition bias
C. Reporting bias
D. Detection bias
E. Performance bias

Answer: B

Short explanation
The Cochrane group describes five types of bias commonly applied to medical research, as listed above. Attrition bias describes a failure to account for the withdrawals from the study groups. To avoid this, data should be analyzed on an 'intention to treat' basis.

Long explanation

Withdrawals from a study are classified as either exclusions – subjects not included in the analysis despite data being available (e.g. those who withdraw consent) or as attrition – those for whom no data are available (e.g. did not return for follow-up). The numbers of patients excluded from each group should be shown in the CONSORT diagram and the reasons published as part of the study design. Those lost to attrition should, as far as possible, be included in the study outcomes. For example, it may be that subjects withdraw from the treatment arm due to unpleasant side effects of the experimental drug. It is important to know this and for their outcomes to be included in the treatment group.

Selection bias is best avoided by using a blinded random sequencer to allocate subjects to the groups and can be demonstrated in differences in the baseline characteristics of the groups.

Performance bias includes performance of volunteers and researchers and their actions during the trial. For example, if patients suspect they are in the treatment arm, they may undertake extra measures to lower their blood pressure to improve the outcomes. This is best avoided by blinding patients and staff to the groups.

Detection bias is similar but comes into play during the collection of data. For example, if those researchers checking the blood pressure of the participants at the end of the study period are aware of the groupings, this may affect the results. Therefore those collecting and analyzing data should also be blinded.

Reporting bias occurs within the write-up, by authors placing more weight on positive findings within a study and seeking to 'explain away' negative findings. It may also occur on a publication level with editors more likely to select positive studies for publication. It is hoped that the pre-trial acceptance of larger studies for publication will reduce this effect because it is believed to be one of the most significant biases in medical research.

Reference

Higgins JPT, Altman DG. Chapter 8: Assessing risk of bias in included studies. In *Cochrane Handbook for Systematic Reviews of Interventions: Cochrane Book Series.* Higgins JPT, Green S. Chichester, UK: The Cochrane Collaboration. J Wiley, 2008.

G12. A 35-year-old woman presents with sudden-onset palpitations. She does not complain of any chest pain. Her heart rate is 240, blood pressure is 98/64 and her respiratory rate is 24 with oxygen saturations of 96% in air. A 12 lead electrocardiogram (ECG) reveals a supraventricular tachycardia (SVT) at a rate of 240.

Which of the following management options should be performed FIRST?

A. Synchronized DC cardioversion
B. Perform carotid sinus massage
C. 300-mg intravenous amiodarone infusion
D. 12-mg intravenous adenosine bolus
E. Arrange an urgent cardiology review

Answer: B

Short explanation

This patient has no adverse features from her SVT; therefore treatment with amiodarone or synchronized DC cardioversion is not warranted. Initial treatment should include vagal stimulation such as carotid sinus massage or Valsalva manoeuvres. The initial dose of adenosine is 6 mg and can be increased to 12 mg if the initial dose is ineffective. Seeking cardiology input is advised to guide further treatments if the SVT does not resolve with the above treatments.

Long explanation

The Resuscitation Council (UK) has produced guidelines for the management of SVT. It recommends the initial assessment to be conducted using an ABCDE approach, administering oxygen as required, obtaining intravenous (IV) access and assessing and treating reversible causes such as electrolyte abnormalities. If adverse features are present (the presence of shock, syncope, myocardial ischaemia or heart failure), patients should receive synchronized DC cardioversion. Patients will need appropriate sedation and airway support for this. The Resuscitation Guidelines recommend initial energy settings to be 70 to 120 J (biphasic). If electrical cardioversion is not achieved with three shocks the patient should receive amiodarone (300 mg IV over 10 to 20 minutes) before further attempts at synchronized DC cardioversion can be delivered, in conjunction with an ongoing amiodarone infusion.

If no adverse features are present, initial management should include vagal manoeuvres such as carotid sinus massage (avoid in the presence of carotid bruit or history of cerebrovascular disease) or getting the patient to perform the Valsalva manoeuvre. Adenosine, which slows conduction through the atrioventricular node thus interrupting the reentry pathways, is the drug treatment of choice. It is given at a dose of 6 and 12 mg for initial and subsequent two doses, respectively, and a rapid intravenous bolus followed by a large flush should be administered to ensuring an adequate concentration reaches the heart. Contraindications include second- or third-degree heart block or sick sinus syndrome with symptomatic bradycardias (unless a pacemaker has previously been inserted) or in patients susceptible to bronchospasm, such as asthmatics. The effects of adenosine are potentiated by methylxanthines such as theophylline and are antagonized by dipyridamole.

Most SVTs will respond and revert with the above treatments. If not, other treatment options include calcium channel blockers such as verapamil. If there is evidence of atrial flutter, a beta-blocker may be more appropriate, but a cardiology opinion should be sought at this point to help guide ongoing management.

Reference

Resuscitation Council (UK). Chapter 8: Peri-arrest arrhythmias. *Advanced Life Support Manual*. 6th edition. 2011. https://www.resus.org.uk/pages/GL2010.pdf (accessed October 2014).

G13. When prescribing for a patient with acute kidney injury and a glomerular filtration rate (GFR) <30 ml/min/1.73 m^2, which of the following recommendations for each drug is LEAST appropriate?

A. Warfarin – reduce dose and monitor more closely
B. Beta-blockers – reduce dose by 50%
C. Fentanyl – avoid use
D. Low-molecular-weight heparins – halve the dose
E. Metformin – avoid if possible

Answer: C

Short explanation

Opioids should be used with caution in those with an estimated GFR (eGFR) <60 ml/min/1.73 m^2 because doses can accumulate; however, their use should not be avoided entirely and short-acting opioids such as fentanyl may be preferable to agents such as morphine.

Long explanation

Guidance on prescribing for patients with a low eGFR can be found in both the 2013 KDIGO guidelines on chronic kidney disease and the renal drug database or British National Formulary. Some drugs simply require a dose adjustment in renal disease, whereas others should be avoided because they can worsen kidney function.

Physicians should always consider the eGFR when prescribing. Many drugs will need to be adjusted in renal disease, and if in doubt, consult one of the sources listed above. The following general considerations should be followed:

- Those drugs that are directly nephrotoxic such as non-steroidal anti-inflammatory drugs and angiotensin-converting enzyme (ACE) inhibitors should be stopped after an acute kidney injury and only prescribed under specific specialist guidance in chronic kidney injury.
- Drugs which may accumulate such as opioids or antibiotics should have their dose reduced to compensate.
- Herbal remedies or untested drugs should be avoided in renal disease.
- Drugs that interact with kidney functions, such as altering electrolyte levels or affecting blood pressure, should be closely monitored, alongside sodium, potassium and calcium levels.
- Drugs that are filtered by the kidney may also be filtered by haemodialysis, but clinicians should be cautious to review the pore size of the filtration membrane in relation to the size of the molecule. Patients on haemodialysis should be assumed to have significantly reduced kidney function and doses adjusted accordingly.

Warfarin doses may need adjusting, and its use should be monitored closely because there in an increased risk of bleeding when the eGFR falls below 30 ml/min/1.73 m^2. Similarly low-molecular-weight heparin doses should be halved at this eGFR, and there should be consideration of switching to unfractionated heparin to avoid accumulation.

Beta-blocker doses should be reduced by 50% with an eGFR of 30 ml/min/1.73 m^2. Other anti-hypertensives should be reviewed and ACE inhibitors stopped in acute kidney injury. However, ACE inhibitors may be nephroprotective in chronic kidney disease, and their use should be considered under specialist care.

Metformin should also be used cautiously in chronic kidney disease and avoided once the eGFR falls below this level. It should be stopped in acute kidney injury due to the risk of accumulation and metformin-induced lactic acidosis.

Reference

Kidney Disease: Improving Global Outcomes. KDIGO 2012 clinical practice guideline for the evaluation and management of chronic kidney disease. *Kidney Int.* 2013;3(1):103.

G14. A 64-year-old man has been admitted with a intracerebral haemorrhage. He has been anticoagulated with warfarin for atrial fibrillation (AF), and his international normalized ratio (INR) on admission is 6.4.

Which of the following would be the BEST treatment regime to reverse the effects of his warfarin?

A. Vitamin K
B. Prothrombin complex concentrates (PCC)
C. Vitamin K and PCC
D. Tranexamic acid
E. Fresh frozen plasma (FFP)

Answer: C

Short explanation

This patient requires urgent reversal of his prophylactic warfarin treatment. Tranexamic acid will not reverse the effects of warfarin. Vitamin K will work too slowly if used in isolation. Both PCC and FFP will reverse the effects of warfarin, but PCC does not have the associated risks of blood product transfusion. The effects of PCC and FFP wear off so after about 6 hours, so co-administration with vitamin K is required.

Long explanation

Patients who bleed on anticoagulants require:

- Supportive treatment
  - Fluid resuscitation with intravenous fluids and blood and appropriate associated blood products as required
  - Cardiovascular support with inotropes and vasopressors if necessary
- Antifibrinolytic – tranexamic acid
- Withholding of any further anticoagulants
- Specific reversal agents
- Investigation and treatment to stop the source of the bleeding

Drugs used to reverse the effects of anticoagulants:

- Vitamin K: the synthesis of clotting factors II, VII, IX and X (along with protein C and protein S) are dependent on vitamin K. The actions of vitamin K antagonists such as warfarin, a synthetic coumarin derivative, can be reversed with the administration of vitamin K. However, high doses of vitamin K have a long duration of action so can make re-anticoagulation difficult if required. Vitamin K can be administered intravenously or orally and takes 6 or 24 hours respectively to take effect.
- Protamine sulphate neutralizes unfractionated heparin (UFH) at a dose of 1 mg for every 100 units of UFH. It also has a partial effect against low molecular weight heparin. Excess doses have anticoagulant properties.
- Platelet transfusion is indicated for patients who are thrombocytopenic, on antiplatelet treatment or who are receiving massive transfusion.
- Fresh frozen plasma reverses anticoagulants that result in inhibition of coagulation factors. It takes 20 to 30 minutes to thaw before use and has the associated risk factors of transfusion of blood products.
- PCC contains clotting factors II, VII, IX, X, protein C and protein S. This provides rapid correction of INR in patients with raised INR on warfarin treatment and has also been recommended for patients who are anticoagulated with rivaroxaban.
- Activated PCC contains activated factor VII but inactivated forms of other clotting factors. There is some evidence that this can be used for the reversal of the new oral anticoagulants (NOAC), such as dabigatran.
- Recombinant factor VIIa (rFVIIa) has been used in life-threatening bleeding as a rescue treatment for patients on anticoagulants.
- Haemodialysis can be used for the removal of dabigatran in patients with life-threatening bleeding.
- There are specific reversal agents for NOAC drugs that are currently undergoing development.

References

Cushman M, Lim W, Zakai NA. American Society of Hematology. 2014 Clinical practice guide on antithrombotic drug dosing and management of antithrombotic-associated bleeding complications in adults. http://www.hematology.org (accessed April 2015).

Siegal DM, Garcia DA, Crowther MA. How I treat target-specific oral anticoagulant-associated bleeding. *Blood*. 2014;123(8):1152–1158.

G15. A previously well 65-year-old woman is admitted to the intensive care unit (ICU) from the emergency department. She has a history of breathlessness and productive cough but no contact with animals or history of recent travel. She is confused, her respiratory rate is 28 and blood pressure is 94/51. She is pyrexic with a white cell count of 17×10^9/l, a C-reactive protein of 147 and a urea of 8.2 mmol/l. Chest X-ray demonstrates right basal consolidation.

What is MOST LIKELY causative organism responsible?

A. *Staphylococcus aureus*
B. *Streptococcus pneumoniae*
C. *Legionella pneumophila*
D. *Influenza A*
E. *Pseudomonas aeruginosa*

Answer: B

Short explanation
This patient has a severe community-acquired pneumonia (CAP) with a CURB65 score of 4. *Streptococcus pneumoniae* is the commonest cause of CAP. *Staphylococcus aureus, Legionella pneumophila* and influenzae A can all cause CAP. *Pseudomonas aeruginosa* is often a cause of hospital-acquired pneumonia (HAP).

Long explanation
This patient has a severe CAP with a CURB65 score of 4. The CURB65 score grades the severity of CAP with patients scoring one point for each of the following:

- Confusion
- Urea >7 mmol/l
- Respiratory rate ≥30/min
- Blood pressure (systolic <90 mmHg or diastolic ≤60 mmHg)
- Age ≥65 years

Scores of 0–1, 2 and 3–5 equate to mortality risks of <3%, ~9% and 15 to 40%, respectively. Admission to critical care should be considered for those patients who score 4 to 5.

Streptococcus pneumoniae is the commonest cause of CAP whether treated in the community, in hospital or in the intensive care unit (ICU), although it is less common in those patients who require intensive care treatment.

Other causes include:

- *Legionella* spp (occurs with increased incidence in those patients who required treatment for CAP on ICU)
- *Staphylococcus aureus*
- *Mycoplasma pneumoniae*
- *Haemophilus influenza*
- *Moraxella catarrhalis*
- Viruses, most commonly influenza A and B
- Gram-negative enteric bacilli
- *Chlamydophila pneumoniae*
- *Chlamydophila psittaci*
- *Coxiella burnetii*

In about one-third of patients presenting to ICU with CAP, no causative organism is ever identified.

Individual patient history can give clues as to the aetiology of CAP. Important features include the following:

- Past medical history (e.g. COPD, immunosuppression, sickle cell disease, intravenous drug use)
- Risk of aspiration
- Residency in a long term care facility
- Immigration from areas with high prevalence of tuberculosis
- Travel history
- Contact with animals (e.g. farm animals and birds)

HAP develops >48 hours post-admission or is present at admission in those patients who live in long-term care facilities or have been admitted to an acute care facility for ≥2 days within the preceding 90 days. The causative organisms vary according to timing of onset of HAP. Disease onset ≤4 days of admission is usually associated with community-acquired pathogens, whereas disease onset ≥5 days is associated with hospital-acquired pathogens. These include methicillin-resistant *Staphylococcus aureus*, *Pseudomonas aeruginosa*, *Acinetobacter* and *Coliform* species (*Enterobacter, Klebsiella, Serratia, Escherichia coli* and *Citrobacter*), along with other multi-drug-resistant organisms. Viruses and fungi are less common unless the patient is immunocompromised. Exact rates differ between institutions due to variations in the local flora. Microbiological surveillance can identify the prevalence of organisms within an institution as well as resistance patterns that can be used to guide antimicrobial treatment.

References
Kieninger AN, Lipsett PA. Hospital-acquired pneumonia: pathophysiology, diagnosis and treatment. *Surg Clin North Am.* 2009;89(2):439–461.
Lim WS, Baudouin SV, George RC, et al. BTS guidelines for the management for community acquired pneumonia in adults: update 2009. *Thorax.* 2009;64(Suppl III):iii1–iii55.

G16. You attend the emergency department to review a patient who is unconscious after being trapped in a house fire.

Which of the following is the most likely cause of death after significant smoke inhalation in patients who survive to arrival at hospital?

A. Thermal injury below the vocal cords
B. Cyanide poisoning
C. Asphyxiation
D. Inhalation of particulate matter
E. Airway obstruction

Answer: D

Short explanation
Proximal burns are common, but burns below the vocal cords are rare. The main cause of pulmonary damage is inhalation of particulate matter, which in addition to causing obstruction and increased work of breathing leads to an inflammatory cascade causing acute respiratory distress syndrome (ARDS). Airway obstruction is either gradual and anticipated in hospital or so rapidly fatal that patients do not survive to hospital.

Long explanation

Burn injuries, especially facial, are strongly associated with smoke inhalation, but the smoke inhalation may not always be immediately apparent, despite the fact that that it has a profound impact on survival. The risk of death is increased nearly four times in the presence of an inhalational injury. Inhalation injury occurs through three main mechanisms: direct heat damage, irritation due to inhaled particulate matter and biochemical toxicity.

Proximal burns of the airway are common, but they almost exclusively occur above the vocal cords. This is because of the large surface areas available, which ensure such effective humidification of the inspiratory gases in health, also ensure transfer of heat from the inhaled hot gases to the mucosa. This means there is little heat damage below the cords, with the exception of superheated steam, which can penetrate further. Thermal injury to, and below, the cords, is usually rapidly fatal due to oedema of these structures. Clinical features of upper airway injury include swelling, redness and blistering. Shortness of breath and hoarseness may develop later on, followed by stridor. Oedema of the airway and face continues to develop in the days following injury and can be extreme. Early intubation with an uncut endotracheal tube is advised.

Inhaled soot contains toxic chemicals. The particles become deposited in the airways and alveoli, leading to an inflammatory cascade, which ultimately causes ARDS. In addition, particulates can cause mechanical obstruction, leading to atelectasis and reduced compliance.

Systemic biochemical dysfunction can either be hypoxic hypoxia, due to consumption of ambient oxygen by the fire itself, or impairment of tissue oxygen delivery as a result of toxic substances; one of the leading causes of death is carbon monoxide (CO) poisoning. CO has a much higher affinity for haemoglobin than oxygen does, thereby reducing oxygen carrying capacity. There is also a left-shift of the oxyhaemoglobin dissociation curve, which impairs oxygen offloading in the tissues. Hydrogen cyanide is many times more toxic than CO but much less frequently encountered. It has a high affinity for cytochrome oxidase in the mitochondria, which subsequently severely impairs aerobic metabolism.

Reference

Gill P, Martin R. Smoke inhalation injury. *Contin Educ Anaesth Crit Care Pain*. May 2014. doi:10.1093/bjaceaccp/mku017

G17. You have just admitted a 50-year-old man into the intensive care unit who is intubated and ventilated after a cardiac arrest. You commence sedation with propofol and alfentanil.

Which of the following is the LEAST likely consequence of prolonged sedation?

A. Prolonged ventilation
B. Hypotension
C. Hypercatabolism
D. Gastrointestinal ileus
E. Increased delirium

Answer: C

Short explanation

Prolonged sedation may cause all of the conditions listed with the exception of hypercatabolism, which would be rarely seen with effective sedation and is more commonly encountered with under-sedation, along with generalized discomfort, sympathetic activity and myocardial ischaemia.

Long explanation

Sedation is an essential component of critical care, but requirements vary widely at different stages of an illness and also show huge patient-to-patient variability. It is never a replacement for effective analgesia. The target should be to use the lowest dose of sedation necessary to achieve the appropriate level of sedation, accepting that there will be times when deeper sedation will be required. Use of deeper levels of sedation should be for the minimum time possible.

The Intensive Care Society, has published a review of best practice for analgesia and sedation in critical care. This lists the indications for the use of sedative drugs in the ICU and includes use to relieve pain and agitation and to help facilitate otherwise distressing treatments.

Sedation itself comes with a number of risks, and getting the balance of sedation right can be challenging. It is tempting to err on the side of over-sedation because this seems kinder. However, over-sedation may lead to prolonged ventilation, prolonged recovery, delirium, myopathy, muscle wasting and hypotension. On the other hand, under-sedation can increase catabolism, cause cardiovascular events from sympathetic activity and cause long-term psychological trauma. Sedation holds are key to reassessing and calibrating sedation levels and should occur daily unless specifically contraindicated.

Non-pharmacological methods should be used to help reduce stress and anxiety, with the aim of creating an environment that reduces stress. For example, regular orientation, low noise levels, and appropriate lighting levels at night can also reduce stress, anxiety and reduce delirium. Maintenance of normal sleep-wake cycles are important in reducing the risk of delirium, and music therapy has growing evidence of benefit.

Reference

Whitehouse T, Snelson C, Grounds M, on behalf of the Intensive Care Society. Intensive Care Society Review of best practice for analgesia and sedation in the critical care. June 2014. http://www.ics.ac.uk/ics-homepage/guidelines-and-standards (accessed April 2015).

G18. A 64-year-old man is in the intensive care unit (ICU) with multiorgan failure secondary to pancreatitis. He has received broad-spectrum antibiotics including 48 hours of gentamicin, 5 days of steroids for septic shock and 40 mmol magnesium replacement during a bout of atrial fibrillation. After 3 weeks he is being weaned from the ventilator and is noted to be profoundly weak. On examination, he exhibits flaccid paralysis and hyporeflexia in all four limbs, but there is sparing of the facial muscles.

Which is the most LIKELY cause of this patient's weakness?

A. Steroid treatment
B. Aminoglycoside treatment
C. Magnesium treatment
D. Cerebrovascular accident
E. ICU-acquired weakness

Answer: E

Short explanation

All of the factors listed can contribute to weakness in patients on the ICU. Cerebrovascular accidents would classically present with lateralizing signs. The doses of steroids, aminoglycosides and magnesium used in this patient are unlikely to be the sole cause of their weakness in the doses described in this case. They are also risk factors for the development of critical illness polyneuropathy, myopathy or neuro-myopathy.

Long explanation

ICU-acquired weakness occurs in approximately 46% of patients with severe sepsis, multiple organ failure or prolonged mechanical ventilation. ICU-acquired weakness is defined as clinically detectable weakness in critically ill patients where no other aetiology apart from critical illness is identified as its cause. It can be classified as critical illness polyneuropathy (CIP), critical illness myopathy (CIM) or critical illness neuromyopathy (CINM). Their clinical presentations are similar and cannot reliably be differentiated without further investigations. They present with symmetrical, flaccid paralysis and decreased reflexes in all four limbs. There is sparing of facial muscles and the autonomic nervous system.

There are a number of proposed risk factors:

- Severe sepsis and septic shock
- Increasing duration of systemic inflammatory response syndrome
- Multi-organ failure (increased duration increases risk)
- Catabolic state
- Prolonged duration of mechanical ventilation
- Prolonged immobility and bed rest
- Hyperglycaemia
- Drugs: corticosteroids, neuromuscular blocking agents, aminoglycosides, parenteral nutrition, vasopressors
- Female sex
- Severity of illness on admission and the admission APACHE II score
- Hypoalbuminaemia
- Renal replacement therapy
- Hyperosmolality

Other causes of weakness in ICU include:

- CNS pathology
  - o Stroke
  - o Intracerebral haemorrhage
  - o Subarachnoid haemorrhage
  - o Epilepsy (post-ictal)
  - o Multiple sclerosis
  - o Hemiplegic migraine
  - o Motor neurone disease
- Peripheral nerve disease
  - o Peripheral neuropathies
  - o Guillain-Barré syndrome
  - o Motor neurone disease
- Diseases effecting the neuromuscular junction:
  - o Myasthenia gravis
  - o Botulism
- Diseases effecting the muscles
  - o Myopathy or myositis
  - o Disuse atrophy
  - o Rhabdomyolysis
  - o Compartment syndrome
  - o Polymyalgia rheumatica
- Other causes:
  - o Drugs: magnesium, corticosteroids, gentamicin

- ○ Familial periodic paralysis
- ○ Generalized weakness secondary to anaemia, hypoglycaemia.

References
Appleton R, Kinsella J. Intensive care unit–acquired weakness. *Contin Educ Anaesth Crit Care Pain*. 2012;12(2):62–66.

Kress JP, Hall JB. ICU-acquired weakness and recovery from critical illness. *N Engl J Med*. 2014;370(17):1626–1635.

G19. You have admitted a 40-year-old man to the intensive care unit who underwent elective laparoscopic removal of a phaeochromocytoma. Which of the following is the most likely complication of this procedure?

A. Persistent hypotension
B. Hypoglycaemia
C. Hypocalcaemia
D. Persistent hypertension
E. Hypokalaemia

Answer: A

Short explanation
Patients with phaeochromocytoma are established on long-acting alpha-antagonists such as phenoxybenzamine before elective surgery. The result of this is marked hypotension once the phaeochromocytoma is removed, which can be resistant to treatment with alpha-agonists such as noradrenaline. Post-operative hypoglycaemia is also seen, while significant electrolyte disturbances are unusual.

Long explanation
Phaeochromocytoma is a tumour of the adrenal medulla that secretes catecholamines, leading to profound hypertension and symptoms such as palpitations, sweating and anxiety. Preoperative treatment is with alpha-blockers, usually the nonselective blocker phenoxybenzamine. If there are residual excessive beta-adrenergic effects then beta-blockade can be added, once phenoxybenzamine therapy is established. Beta-blockade should not be initiated first because this can result in unopposed alpha effects causing excessive hypertension and cardiac failure.

Intraoperative treatment of acute severe hypertension is with short-acting alpha antagonists, such as phentolamine, to try and treat the acute catecholamine surges while minimizing the risk of residual hypotension. Magnesium sulphate can also be used for hypertensive crises, as can sodium nitroprusside, glyceryl trinitrate and calcium channel blockers such as nicardipine.

Once the venous drainage of the phaeochromocytoma is clamped, there is a sudden fall in plasma catecholamine concentration, reflecting the short half-life of the catecholamines. This, combined with residual alpha blockade from phenoxybenzamine and hypovolaemia from blood loss, can lead to a profound hypotension. Fluid boluses should be given initially, but vasopressors are often needed. Alpha-agonists such as noradrenaline and phenylephrine may have little effect, and resistant cases may need treatment with vasopressin.

Other post-operative complication include hypoglycaemia, but this occurs much less frequently than hypotension. It occurs because catecholamine levels fall, leading to reduced glycogenolysis.

References
Myburgh J. Vasodilators and antihypertensives. In Bersten AD, Soni N. *Oh's Intensive Care Manual*. 6th edition. Oxford: Butterworth Heinemann Elsevier, 2009, Chapter 83.

Woodrum D, Kheterpal S. Anaesthetic management of phaeochromocytoma. *World J Endocr Surg*. 2010;2(3):111–117.

G20. You are performing a rapid sequence induction (RSI) for a patient with a severe pneumonia who remains hypoxic despite maximal medical therapy. Your attempts at direct laryngoscopy have failed, so you attempt to site a supra-glottic airway device (SAD). You are unable to ventilate the patient and they continue to desaturate.

Which of the following is the most appropriate next step?

A. Perform an emergency cricothyroidotomy.
B. Release the cricoid pressure and attempt intubation once more.
C. Attempt face-mask ventilation.
D. Insert a different SAD and attempt ventilation with cricoid pressure maintained.
E. Wake the patient.

Answer: C

Short explanation

The Difficult Airway Society (DAS) publishes guidelines for the management of a difficult tracheal intubation in an RSI scenario. In the event of failed tracheal intubation, and failed oxygenation with a face mask, the next step is to insert an LMA, and attempt ventilation with cricoid pressure maintained. If this fails, then cricoid pressure can be released and ventilation again attempted through the LMA before emergency cricothyroidotomy is required.

Long explanation

The Royal College of Anaesthetists' 4th National Audit Project (NAP4) found that 25% of airway-related adverse events occurred outside the operating theatre (in intensive care units or emergency departments), with a lack of thorough preparation for difficult airways and sub-standard use and interpretation of capnography. The number of airway events occurring in these locations was disproportionately high.

Anticipation of potential difficulty, and verbalization of the plan if difficulty is encountered before commencement of induction is recommended. The DAS guidelines for the management of difficult intubation are as follows.

Plan A: standard tracheal intubation. There should be a maximum of three attempts at tracheal intubation, ensuring good head and neck position. Anaesthesia must be maintained, as should cricoid pressure and oxygenation between attempts with a facemask. Options to improve the view include reducing the cricoid pressure, external laryngeal manipulation, or use of a bougie.

Plan B: maintenance of oxygenation. Oxygenation should maintained with a second-generation SAD. If successful, the SAD may be used as sole airway device (not applicable to ICU), intubation through the SAD may be attempted, the patient woken (not suitable in this case) or tracheostomy performed. Cricoid pressure can be reduced if ventilation remains difficult.

Plan C: if oxygenation fails via a SAD, then face-mask ventilation should be attempted to oxygenate and ventilate the patient. Two people may be required, along with airway adjuncts. Cricoid pressure may need to be reduced or removed.

Plan D: can't intubate, can't oxygenate. This life-threatening scenario requires prompt emergency management with surgical cricothyroidotomy.

References

Cook T, Woodall N, Frerk C. *4th National Audit Project (NAP4) of The Royal College of Anaesthetists and The Difficult Airway Society: Major complications of airway management in the United Kingdom. Report and Findings.* London: Royal College of Anaesthetists, March 2011.

Difficult Airway Society. Intubation Guidelines. http://www.das.uk.com/guidelines (accessed January 2016).

G21. You are asked to perform a manual noninvasive blood pressure reading on a hypotensive patient you have just admitted to the intensive care unit.

Which of the following statements is LEAST accurate?

A. The blood pressure cuff width should be 20% larger than the width of the arm.
B. The cuff should be inflated above systolic pressure, then deflated at 2 to 3 mmHg/sec.
C. If using a mercury manometer, it must have a functioning air vent at the top of the column of mercury.
D. The first Korotkoff phase represents systolic blood pressure.
E. The fourth Korotkoff phase represents diastolic blood pressure.

Answer: E

Short explanation

Although the fourth Korotkoff phase (an abrupt decrease and muffling in the level of the sound) may be taken to represent diastolic blood pressure, it is more widely accepted that it is the fifth Korotkoff phase (the final disappearance of all sound) that represents diastolic blood pressure, making this the least accurate answer. The others are all correct.

Long explanation

Blood pressure is a measure of the pressure within the circulation. Hydrostatic effects mean that it varies with the site of measurement and patient position. A reference point is therefore needed, otherwise all blood pressure readings would need the site and position recorded as well. In the standing position, for example, the blood pressure will be much higher in the leg than in the head by around 140 to 170 cmH$_2$O, depending on the patient's height. This would equate to 103 to 125 mmHg. The accepted reference point is the level of the right atrium.

The simplest way to measure blood pressure is using an inflatable cuff connected to either an aneroid gauge or a manometer. The auscultation of Korotkoff phases or sounds in an artery distal to the cuff indicates systolic and diastolic blood pressure. This simple process is also the default when an automated technique fails (typically due to hypotension or obesity).

Appropriate cuff selection and positioning are imperative for accurate blood pressure recording. First, the width of the cuff will affect accuracy of the readings. Ideally, it will be 20% wider than the diameter of the arm, with too small a cuff resulting in a falsely high reading, and too large a cuff producing a falsely low reading. Second, the cuff needs to be positioned with its centre over the brachial artery.

If using a mercury manometer, it should be used in the upright position with all tubes and air vents unobstructed. If using an aneroid gauge, it will need regular calibration.

The cuff is inflated above the systolic pressure, then slowly deflated (roughly 2 mmHg per second). The systolic pressure can be detected by manual palpation of the radial pulse, or both systolic and diastolic pressures can be identified by auscultation over the brachial artery. The first Korotkoff phase correlates with systolic blood pressure and is generated when blood begins to flow through the brachial artery again. The different phases of Korotkoff sounds (phase 2–4) represent the stages as the sounds get quieter, louder and then quieter again, but it is the pressure at which there is loss of all sound, or Korotkoff phase 5, which is generally accepted as the diastolic blood pressure.

Reference
Blood pressure measurement. In Davis P, Kenny G. *Basic Physics and Measurement in Anaesthesia*. 5th edition. London: Elsevier, 2003, pp 187–198.

G22. Which of the following patients is MOST likely to benefit from parenteral nutrition (TPN)?

A. A 20-year-old man with a body mass index (BMI) of 19, with unintended weight loss of 3% in the past 3 months, with an unsafe swallow
B. A 20-year-old man with a BMI of 18, with severe Crohn disease, who has been nil-by-mouth for five days
C. A 20-year-old man with a BMI of 20, who has eaten nothing for 5 days, who finds the nasogastric tube intolerable
D. A 20-year-old man with a BMI of 21, who is unlikely to eat anything for the next 3 days, and is not absorbing nasogastric feed
E. A 20-year-old man with a BMI of 22, with unintended weight loss of 8% in the past 6 months, with inadequate enteral nutritional intake

Answer: B

Short explanation
Following NICE guidance, the indications for parenteral nutrition require people to be malnourished or at risk of malnutrition (B and C fulfil this), and either have inadequate or unsafe oral and/or enteral nutritional intake, or have a non-functioning, inaccessible or perforated gastrointestinal tract.

Long explanation
The NICE guidance for nutritional support in adults (CG32) is clear in the indications for the parenteral nutrition. First, the patient must fulfil the criteria for nutrition support, being malnourished or at risk of malnutrition. Malnutrition is defined by BMI (under 18.5); weight loss (10% over 6 months); or a combination of the two (a BMI of <20 with 5% loss of weight over 6 months). Risk of malnutrition is defined by grossly inadequate oral intake for 5 or more days, poor absorption, hypermetabolism or increased losses of nutrients.

If a patient is malnourished or at risk; feeding support should be instituted. Parenteral feeding should only be considered if enteral feeding is not possible, for reasons of safety or efficacy.

Parenteral nutrition should always be started slowly, with a target of 50% for the first 24 to 48 hours, via a dedicated line. Those at risk of refeeding syndrome should have this started at a much lower rate (starting at 5–10 kcal/kg/day), and this should be gradually increased to full requirements by day 7.

Daily nutritional replacement should include roughly 30 kcal/kg/day, 30 ml water/kg/day, and around 1 g/kg protein (roughly 0.2 g/kg nitrogen). In addition, there should be adequate electrolyte replacement, which are roughly 1 to 2 mmol/kg for Na^+ and 0.7 to 1 mmol/kg for K^+.

Sodium and potassium must be measured daily in all patients in whom feeding is being commenced until levels are stable. Magnesium and phosphate need only be measured three times a week initially if patients are not at risk of refeeding syndrome, or daily if they are.

Reference

National Institute for Health and Clinical Excellence (NICE). *Nutrition support in adults: Oral nutrition support, enteral tube feeding and parenteral nutrition* (CG32). London: NICE, 2006.

G23. You are called to the emergency department at midnight to help with the ongoing management of a 16-year-old girl who took an overdose of paracetamol. She is well known to the psychiatric services with depression. She is fully conscious, has capacity, and is refusing any further treatment. She has a markedly elevated paracetamol level and would benefit from treatment with N-acetyl cysteine.

Which is the following is the most appropriate way to manage this patient?

A. Treat her against her wishes under the Mental Capacity Act
B. Treat her against her wishes under the Mental Health Act
C. Seek consent from her parents to treat her against her wishes
D. Accept her autonomy and discharge her home
E. Wait for legal advice

Answer: E

Short explanation

The Mental Health Act provides a legal means of treating a *mental health condition* without the patient's consent, regardless of capacity. The Mental Capacity Act (MCA) allows the emergency treatment of a patient, in the interests of preservation of life, when the patient has a lack of capacity. The above scenario is complex, and although early treatment will benefit her medically, it is not so urgent that legal advice cannot be sought first.

Long explanation

The Mental Health Act allows treatment of mental disorder, and some aspects of medical care may be done without consent under the act, although whether treatment of a paracetamol overdose can be included under this act remains contentious, and court orders have often been sought in such a scenario. There are short-term sections of this allowing emergency detention of patients (usually for 72 hours) until full assessment can be carried out.

The MCA provides a structure for the management of patients who may lack capacity. Patients should be assumed to have capacity, but this can be assessed by ensuring the patient can understand the information, retain it, process the information to make a decision and communicate that decision. Capacity is decision specific, and so if a patient lacks capacity for one decision, he or she should not be assumed to lack it for a different decision. If the patient has capacity and makes a decision, then that must be respected, however unwise it is felt to be. The deprivation of liberty safeguards, which were an amendment to the MCA in 2005, must be considered if treatment is being instigated that will deprive a patient of his or her liberty and provides an extra safety net for patients.

This scenario is complicated by the patient's age. The General Medical Council states that from age 16, a young person can be presumed to have capacity to consent. Parents could potentially override a young person's competent refusal of treatment, but the law on this is complicated. The GMC recommends seeking legal advice if you think treatment is in the best interests of a competent young person who refuses.

References

Doy R, Burroughs D, Scott J. Mental health: consent, the law and depression-management in emergency settings. *Emerg Med J.* 2005;22(4):279–285.

General Medical Council. Good Medical Practice 2013. http://www.gmc-uk.org/guidance/ethical_guidance/children_guidance_24_26_assessing_capacity.asp (accessed January 2015).

Hull A, Haut F. Managing patients with deliberate self harm who refuse treatment in accident and emergency departments : Advice and procedure require correction. *BMJ*. 1999;319(7214):916 (and the letters thereafter). http://www.ncbi.nlm.nih.gov/pmc/articles/PMC1116734/pdf/916.pdf. (accessed July 2015).

United Kingdom Clinical Ethics Network. Education resources: Mental Capacity Act and Mental Health Act. http://www.ukcen.net/index.php/education_resources/mental_capacity/mca_and_the_mental_health_act (accessed May 2015).

G24. A 42-year-old female patient with gallstone pancreatitis was ventilated on admission to ICU for respiratory failure. She deteriorates after a period of stability on day 5, with increasing FiO_2 requirements 0.4 to 0.7 and positive end expiratory pressure 8 to 12 cmH_2O, which continues for the next 48 hours. She has increased sputum production (no growth on culture), a temperature of 38.6°C, a white cell count of 18×10^9/l and new left basal consolidation on chest X-ray. Five days of antibiotics are prescribed as per local hospital guidelines.

Which of the following is the BEST term to describe her respiratory deterioration?

A. Probable ventilator-associated pneumonia (VAP)
B. Ventilator-associated condition
C. Acute respiratory distress syndrome (ARDS)
D. Infection-related ventilator-associated complication
E. Possible ventilator-associated pneumonia

Answer: D

Short explanation

The US Centers for Disease Control and Prevention (CDC) has produced guidelines to describe ventilator-associated events. Although this patient is likely to have had ARDS initially, it is unlikely to be the cause of this deterioration. Probable and possible VAP cannot be confirmed because no positive cultures have been obtained. Infection-related ventilator-associated complication is the best description because there is evidence of significant respiratory deterioration along with evidence of infection.

Long explanation

There is no universal standardized definition of VAP. Many existing scores take account of a mixture of clinical, biochemical and laboratory indices as well as radiological changes. Some elements are subjective, making a standardized definition difficult and therefore making it harder to compare VAP rates between different institutions or use it as a quality indicator. The CDC has designed an algorithm describing a range of ventilator-associated events with strict parameters thus making the diagnosis more objective.

- The CDC's guidelines require the patient to have a period of stability or improvement in their daily minimum FiO_2 or PEEP for ≥2 days after initiation of ventilation.
- If the patient subsequently deteriorates with an increase in his or her daily minimum FiO_2 of ≥0.2 or PEEP ≥3 that is sustained for ≥2 days, they are described as having a ventilator-associated condition.
- This is further described as an infection-related ventilator-associated complication if on or after day 3 of ventilation and within 2 days of the

worsening of oxygenation, the patient has evidence of infection (temperature >38°C or <36°C or a white cell count ≥12 × 10⁹/l or ≤4 × 10⁹/l) and has been started on new antimicrobials which are continued for ≥4 days.

- This can be further categorized as possible or probable VAP on the basis of microbiological samples from day 3 of ventilation onward and ≤2 days before or after worsening oxygenation.
  - Possible VAP requires the presence of purulent respiratory secretions with ≥25 neutrophils and ≤10 squamous epithelial cells per low power field or a positive culture from respiratory secretions (except normal or mixed respiratory/oral flora, candida, coagulase-negative *Staphylococcus* or *enterococcus*) or lung tissue.
  - Probable ventilator-associated pneumonia requires purulent secretions as defined above with the presence of specific number of organisms per colony forming unit. The number required for diagnosis is dependent on the sample type or the presence of other positive microbiological samples from pleural fluid analysis, lung histopathology or other positive tests such as viral PCR or *legionella* antigens.

Reference
The United States Centers for Disease Control and Prevention. Ventilator-associated event (VAE). 2014 http://www.cdc.gov/nhsn/pdfs/pscManual/10-VAE_FINAL.pdf (accessed July 2014).

G25. A previously well 36-year-old woman is brought to hospital after a seizure secondary to hyponatraemia. Investigations results include the following:

- Serum sodium concentration 113 mmol/l
- Serum osmolality 274 mOsm/kg
- Urine sodium concentration 10 mmol/l
- Urine osmolality 80 mOsm/kg

What is the MOST likely cause of her hyponatraemia?

A. Syndrome of inappropriate ADH secretion (SIADH)
B. Water intoxication
C. Renal failure
D. Cerebral salt wasting syndrome
E. Hypothyroidism

Answer: B

Short explanation
All the factors listed are causes of hyponatraemia. This patient has hypotonic hyponatraemia with a normal extracellular fluid volume which suggests water intoxication to be the likely cause.

Long explanation
Hyponatraemia occurs in up to 20% of hospital admissions. Recent guidelines have been produced by the European Society of Intensive Care Medicine regarding its diagnosis and treatment. It defines hyponatraemia as a serum sodium concentration as <135 mmol/l. It further defines disease as mild, moderate and profound with sodium concentrations of 130 to 135, 125 to 139 and <125 mmol/l, respectively. Acute hyponatraemia is defined as that occurring within <48 hours, and chronic disease is hyponatraemia present for at least 48 hours. If the time frame is unknown, it is assumed to be chronic unless there is evidence to the contrary.

Moderately symptomatic hyponatraemia is hyponatraemia with the presence of moderately severe symptoms such as nausea without vomiting, confusion or headache. Severely symptomatic hyponatraemia is disease that occurs with the presence of severe symptoms such as vomiting, cardio-respiratory distress, and central nervous system disturbance (abnormal and deep somnolence, seizures or coma). Symptoms may occur at a higher sodium concentrations if hyponatraemia is of rapid onset.

Elevated glucose levels can give a false impression of hyponatraemia. Sodium levels can be corrected by using the following formula:

$$\text{Corrected serum } [\text{Na}^+] = \text{measured } [\text{Na}^+]$$
$$+ 2.4 \times [\text{glucose (mmol/l)} - 5.5]$$

Causes of hyponatraemia:

- Hyposmolar hyponatraemia (due to increased ADH)
  - Hypovolaemia
    - Diuretics
    - Dehydration
    - Bleeding
  - Hypotension
    - Cardiac failure
    - Cirrhotic liver disease
  - Syndrome of inappropriate ADH
  - Others
    - Hypothyroidism
    - MDMA use
    - Pregnancy
    - Adrenal insufficiency
- Hyperosmolar hyponatraemia
  - Hyperglycaemia
- Normosmolar hyponatraemia
  - Hyperlipidaemia
  - Hyperproteinaemia
- Hyponatraemia with low ADH levels
  - Primary polydipsia
  - Renal failure
  - Low dietary solute intake

Reference
Spasovski G, Vanholder R, Allolio B, et al. Clinical practice guideline on diagnosis and treatment of hyponatraemia. *Intensive Care Med*. 2014;40:320–333.

G26. A 60-year-old man has been admitted to the emergency department with shortness of breath. He smokes 40 cigarettes per day and has recently been prescribed a salbutamol inhaler by his general practitioner. His chest radiograph shows a right-sided pneumothorax, which measures 3 cm at the level of the hilum. There is no sign of tension pneumothorax. There is no history of trauma.

What is the best initial management?

A. Conservative: admit, treat with oxygen and observe
B. Aspiration with a 16-G cannula
C. Size 10-Fr Seldinger intercostal drain
D. Size 18-Fr open intercostal drain
E. Size 32-Fr open intercostal drain

Answer: C

Short explanation

This patient has a secondary spontaneous pneumothorax (SSP) and conservative management is inappropriate. Small (<2 cm), asymptomatic SSPs may be aspirated with a cannula, but this is neither small nor asymptomatic. An intercostal drain should be sited. The British Thoracic Society recommend the use of small-bore intercostal drains (8–14 Fr) as large-bore 'open' intercostal drains cause more pain without evidence of increased efficacy.

Long explanation

A pneumothorax is air in the pleural cavity. Spontaneous pneumothoraces occur in the absence of trauma, iatrogenic or otherwise. Primary spontaneous pneumothoraces (PSP) occur in patients with no history of lung disease; commonly in young male smokers. PSP may be managed conservatively in asymptomatic patients or aspirated with a cannula in cases with dyspnoea, hypoxia or large PSP. If aspiration fails, an intercostal drain should be inserted.

Pneumothoraces in patients with underlying lung disease or those older than 50 with a significant smoking history are classified as secondary spontaneous pneumothorax (SSP). These are less well tolerated and much less likely to resolve spontaneously, so conservative management is not appropriate. Patients with SSP should be admitted to hospital and given high flow oxygen, even if the SSP is <1 cm. One to two centimetre SSP may be treated with cannula aspiration, although this is less likely to be effective than in PSP. Larger SSP or failed cannula aspiration should prompt the insertion of an intercostal drain.

Choice of intercostal drain can be a controversial subject, although most UK practitioners would agree that the use of trocars is outdated and dangerous. The British Thoracic Society guidelines stipulate that there is no advantage to the use of large-bore intercostal drains for pneumothoraces but evidence that they cause a great deal more pain than small-bore intercostal drains. There is a place for the use of large-bore intercostal drains in patients with chest trauma or following thoracic surgery, but even empyemas may be effectively drained using small-bore drains, providing that regular flushing occurs.

Reference
MacDuff A, Arnold A, Harvey J. Management of spontaneous pneumothorax: British Thoracic Society Pleural Disease Guideline 2010. *Thorax*. 2010;65(Suppl 2 D):ii18–31.

G27. A 71-year-old man is admitted with acute decompensated cardiac failure. He is distressed and diaphoretic with a respiratory rate of 40, oxygen saturation 94% in air, heart rate of 100 bpm and blood pressure (BP) 90/65. Chest radiography shows signs of pulmonary oedema.

What is the most appropriate first therapy?

A. Glycerin trinitrate (GTN) 2 sprays sublingual
B. Furosemide 40 mg intravenous (IV)
C. Oxygen 6–10 l by face mask
D. Diamorphine 1–2.5 mg IV
E. Dobutamine infusion at 2–20 μg/kg/min

Answer: B

Short explanation

Oxygen should only be used to treat hypoxaemia (SpO$_2$ <90%). Vasodilators are not appropriate in patients with systolic BP <110 mmHg. Dobutamine should be used in cardiogenic shock (SBP <85 mmHg). Diamorphine acts as an anxiolytic and venodilator but may cause nausea, requiring antiemetics. Diuretics have been shown to improve symptoms through both vasodilation and fluid removal but should be used with careful renal function monitoring.

Long explanation

Acute heart failure (AHF) is a rapid onset of heart failure requiring urgent medical attention. It is a life-threatening condition that usually occurs in the context of chronic heart failure but may also be the primary presentation of cardiovascular disease. Causes include ischaemia, arrhythmias, tamponade, emboli, aortic disease and cardiomyopathy. AHF requires immediate treatment, usually before the exact trigger is elucidated.

Oxygen should be given in the presence of hypoxaemia (SpO$_2$ <90%) but should not be given to patients with AHF without hypoxaemia because excessive oxygen is associated with vasoconstriction and worsening of AHF. Noninvasive ventilation (e.g. continuous positive airway pressure) should be considered in hypoxic patients with a respiratory rate of >20/min, but hypotension may be exacerbated.

Intravenous loop diuretics (e.g. furosemide) provide rapid relief of symptoms, both from a short-lived venodilatation effect and then from diuresis. Electrolytes and renal function should be closely monitored. Resistant cases may require the addition of a thiazide diuretic for a short period of time.

Opiates (e.g. diamorphine) are commonly used in AHF because they provide anxiolysis, reduce catecholamine levels and have a venodilator effect. However, respiratory depression is a significant problem, as is nausea, which often requires treatment with vasoconstrictive antiemetics.

Vasodilators (e.g. glyceryl trinitrate) improve stroke volume by reducing afterload and preload (which can help a failing ventricle). However, significant reductions in blood pressure are common, and so the use of vasodilators should be restricted to patients with SBP >110 mmHg without significant valvular lesions.

Inotropes (e.g. dobutamine) improve contractility and stroke volume but cause tachycardia and increased myocardial work and can lead to arrhythmias or ischaemia, worsening the clinical situation. Their use should be restricted to patients with cardiogenic shock (SBP <85 mmHg or signs of hypoperfusion).

Vasoconstrictors (e.g. noradrenaline) improve blood pressure but increase afterload and myocardial work. They should not be used in AHF except in cases of cardiogenic shock after the introduction of an inotrope.

Once symptomatic relief has been achieved, management can focus on treatment of the underlying cause, for example, coronary revascularization, cardioversion for arrhythmias, thrombolysis of pulmonary emboli etc.

Reference

McMurray JJV, Adamopoulos S, Anker SD, et al. ESC Guidelines for the diagnosis and treatment of acute and chronic heart failure 2012: The Task Force for the Diagnosis and Treatment of Acute and Chronic Heart Failure 2012 of the European Society of Cardiology. *Eur Heart J.* 2012;33(14):1787–1847.

G28. A 26-year-old woman is admitted with fever, confusion, agitation and seizures. Following a normal CT scan of her brain, a lumbar puncture is performed. Her blood glucose is 6 mmol/l.

Which of the following sets of results would suggest viral meningoencephalitis?

A. Protein 100 mg/dl; glucose 4 mmol/l; white blood cell count (WBC) 50×10^9/l; predominant lymphocytes
B. Protein 250 mg/dl; glucose 1.5 mmol/l; WBC 500×10^9/l; predominant lymphocytes
C. Protein 500 mg/dl; glucose 0.5 mmol/l; WBC 5000×10^9/l; predominant neutrophils
D. Protein 50 mg/dl; glucose 4 mmol/l; WBC 1×10^9/l; predominant lymphocytes
E. Protein 500 mg/dl; glucose 6 mmol/l; WBC 50×10^9/l; predominant neutrophils

Answer: A

Short explanation

Viral meningoencephalitis typically has cerebrospinal fluid (CSF) findings of clear CSF, slightly increased WBCs (mainly lymphocytes), normal glucose (~60% of blood glucose) and normal to high protein. Answer B would suggest tuberculosis meningitis, answer C bacterial meningitis and answer D is normal CSF. Answer E does not fit any clinical pattern.

Long explanation

Encephalitis is inflammation of the brain parenchyma and may be caused by viruses, bacteria, protozoa or autoimmune processes. The commonest causes of viral encephalitis in the developed world are herpes viruses, varicella, cytomegalovirus, Epstein-Barr virus, enteroviruses and paramyxoviruses. There are many other pathogens responsible for viral encephalitis in the developing world, such as dengue fever, rabies, tickborne encephalitis and West Nile fever.

Patients who present with a fever and a change in behaviour, seizures or new focal neurology should be suspected as having viral encephalitis. Any features of meningism should prompt treatment for bacterial meningitis until proven otherwise. In the absence of meningism, a computed tomography brain scan should be performed followed by lumbar puncture if there is no suggestion of raised intracranial pressure. Different infectious agents cause different patterns of CSF abnormalities, as presented in the following table. It is important to be aware that not all patients and diseases will fit these patterns completely. CSF should be sent for Gram stain, culture and microscopy and viral polymerase chain reaction.

Finding	Normal	Viral	Bacterial	Fungal	TB
Opening pressure	<20 cmH$_2$O	Normal/high	High	Very high	High
Colour	Clear	Clear	Cloudy	Clear/cloudy	Yellow
Glucose	⅔ blood glucose	Normal	Low	Low-normal	Very low
Protein	35 mg/dl	50–100	>100	20–50	100–500
WBCs	0–3/μl	5–1000	100–50,000	0–1000	25–500
Predominant type	N/A	Lymphocytes	Neutrophils	Lymphocytes	Lymphocytes

Empirical treatment of suspected viral encephalitis is aciclovir 10 mg/kg TDS. If there is any suspicion of bacterial meningitis, then antibacterials should be commenced: ceftriaxone or cefotaxime 2g QDS with ampicillin for suspected listeria. Dexamethasone should be started at the same time as the antibacterials (as per NICE guidelines), particularly if pneumococcal meningitis is suspected.

Reference
Solomon T, Hart IJ, Beeching NJ. Viral encephalitis: a clinician's guide. *Pract Neurol.* 2007;7:288–305.

G29. You are called to the emergency department as part of the trauma team. A 23-year-old man has been involved in a road traffic collision. He was the unrestrained driver of a car that collided with a tree at high speed. His vital signs are as follows: heart rate 130, blood pressure (BP) 60/35, SpO$_2$ unrecordable, respiratory rate 35. He has no signs of external haemorrhage, pneumothorax or tamponade. His abdomen is soft, his pelvis is stable and he has no signs of long bone fractures. A focused assessment with sonography in trauma (FAST) scan is negative for blood in the peritoneum or pericardium. A chest radiograph is performed, showing a widened mediastinum. You suspect a traumatic aortic injury.

Where is the most likely site of aortic injury?

A. Aortic root
B. Ascending aorta
C. Aortic arch
D. Proximal descending aorta
E. Distal descending aorta

Answer: D

Short explanation
The vast majority of traumatic aortic injury in blunt trauma is in the proximal descending aorta, at the site of the ligamentum arteriosum, just distal to the origin of the left subclavian artery. The aorta is fixed to the chest wall here, whereas the mediastinum is relatively free to move. Rapid deceleration causes tears at this transition point.

Long explanation
Traumatic aortic injury (TAI) from blunt trauma is usually fatal. Approximately 80% of TAIs die at the scene, and the mortality rate for patients who make it to hospital is 10 to 30%. The incidence of other associated injuries is high, but death may

be attributed to the aortic injury in nearly two-thirds of cases. TAI may be graded according to severity as determined with imaging:

- Grade I – Intimal tear
- Grade II – Intramural haematoma
- Grade III – Pseudo aneurysm
- Grade IV – Rupture

There are various theorized mechanisms of injury associated with TAI. Rapid deceleration and high-impact collisions are implicated. Shearing forces occur at the proximal descending aorta because this is a transition point between the relatively fixed descending aorta and the mobile mediastinum. Approximately 90% of all TAI occurs at this point. Other mechanisms include a 'water hammer' effect in which abdominal compression occludes the abdominal aorta, causing huge increases in intra-aortic pressure and intimal rupture. An 'osseous pinch' mechanism is sometimes thought to occur as the aortic arch is crushed between the sternum and vertebral column.

Patients typically complain of chest pain, sometimes radiating to the back, with dyspnoea or dysphagia. Suspicion should be raised in patients with a significant mechanism of injury, patients are commonly unrestrained drivers in high-velocity collisions. New murmurs or BP discrepancies may also suggest TAI, but imaging is required for a diagnosis. Chest radiograph signs associated with TAI include a widened mediastinum, abnormal aortic arch silhouette, left sided haemothorax (particularly at the left apex) and displacement of mediastinal structures to the right. Diagnosis may be made with a transoesophageal echocardiogram or computed tomography (CT) scanning. To effectively rule out TAI, an ECG-gated contrast CT is required to eliminate cardiac movement artifact.

The management of TAI is surgical, either with an open repair or intravascular stenting. In the interim, BP and heart rate control should be instituted, to reduce the risk of the aortic injury worsening. Labetalol or esmolol infusions are typically used, with nitrates or calcium channel blockers as second- and third-line therapies.

References

American College of Surgeons (ACS) Committee on Trauma. Advanced Trauma Life Support Student Course Manual. 9th edition. Chicago, IL: ACS, 2012.

Goarin JP, Riou B. Traumatic aortic injury. *Curr Opin Crit Care*. 1998;4(6):417–423.

Lee WA, Matsumura JS, Mitchell RS, et al. Endovascular repair of traumatic thoracic aortic injury: clinical practice guidelines of the Society for Vascular Surgery. *J Vasc Surg*; 2011;53(1):187–192.

G30. Which of the following is the LEAST important risk factor in the development of delirium in adult patients in the intensive care unit (ICU)?

A. Pre-existing cognitive impairment
B. Presence of coma
C. Orientation to time with no visible clock present
D. High SAPS II score
E. Alcohol intake: >3 units/day

Answer: C

Short explanation

Pre-existing cognitive impairment, the presence of coma, a high severity of illness, such as with a high SAPS II score and high alcohol intake are all risk factors that have been demonstrated to be associated with the development of delirium in ICU patients and have been specifically identified in the Critical Care Medicine Society guidelines.

Although orientating patients is thought to aid prevention of delirium, the absence of a visible clock has not been found to be a risk factor for its development.

Long explanation

Delirium is a disturbance of consciousness with inattention, accompanied by cognitive change or perceptual disturbance. It is acute in onset and its severity fluctuates with time. Delirium is thought to occur in up to 80% of mechanically ventilated patients on ICU. Its presentation varies according to the form the disease takes. Only 10% have hyperactive disease and they present with a heightened state of arousal, restlessness, agitation and aggressive paranoid behaviour. Hypoactive disease occurs in 45% of patients and is harder to detect. Patients are often quiet, withdrawn, sleepy and inattentive to their surroundings. The remaining 45% of patients present with mixed disease (features of hypoactive and hyperactive delirium).

A number of risk factors have been suggested to be associated with the development of delirium in patients on the ICU. These include the following:

Predisposing factors – pre-existing patient related features:

- Pre-existing cognitive impairment
- History of excess alcohol intake, especially >3 units/day
- Smoking history, especially >10 cigarettes/day
- History of drug misuse or withdrawal from psychoactive drugs
- Visual or auditory impairments
- Pre-existing malnutrition, vitamin B_{12}/thiamine deficiencies
- Co-morbidities: history of hypertension, congestive cardiac failure, cerebrovascular accidents, epilepsy, depression, HIV, renal impairment or hepatic impairment/hyperbilirubinaemia
- Age >70 years

Precipitating factors:

- Acute disease and physiological related factors:
- Increased severity of illness
- Coma/central nervous system pathology
- Metabolic derangement
- Thyroid function/glycaemic control/electrolyte derangement/acidosis
- Acute kidney injury
- Fever
- Sepsis
- Hypoxia
- Hypotension
- Anaemia
- Anxiety
- Uncontrolled pain
- Urinary retention
- Faecal impaction

Procedures:

- Presence of indwelling rectal or urethral catheters, central venous cannulas, or arterial lines
- Surgery, especially involving extracorporeal circulation

Environmental:

- Physical restraints
- Sleep deprivation/disturbance
- No visible daylight

Medications:

- Benzodiazepines
- Propofol
- Anticholinergics
- Opiates (although some of the data are conflicting)

The Critical Care Medicine Society guidelines identify the presence of coma, the use of benzodiazepines and four baseline risk factors; pre-existing dementia, a history of hypertension, a history of alcoholism and a high severity of illness at admission as being specifically associated with the development of delirium in the ICU.

References

Barr J, Fraser GL, Puntillo K, et al. Clinical practice guidelines for the management of pain, agitation, and delirium in adult patients in the intensive care unit. *Crit Care Med*. 2013;41(1):263–306.

Reade MC, Finfer S. Sedation and delirium in the intensive care unit. *N Engl J Med*. 2014;370(5):444–454.

Van Romcompaey B, Elseviers MM, Schuurmans MJ, et al. Risk factors for delirium in intensive care patients: a prospective cohort study. *Crit Care*. 2009;13:R77.

Exam H: Questions

H1. A 70-kg patient is 'nil by mouth'. The nursing staff has asked you to prescribe some maintenance fluids for the night. He is clinically euvolaemic, and his electrolyte levels are all within normal range.

Which prescription is the most appropriate for intravenous (IV) maintenance fluids?

A. 1 l 0.18% NaCl with 4% glucose and 27 mmol/l potassium over 12 hours
B. 1 l 0.9% NaCl with 40 mmol/l potassium over 12 hours
C. 1 l compound sodium lactate (Hartmann's solution) over 12 hours
D. 1 l 5% dextrose over 12 hours
E. 1 l 0.18% NaCl with 4% glucose over 12 hours

H2. A 37-year-old patient in the intensive care unit (ICU) who is intubated and ventilated is about to undergo a change of dressings. He sustained a thermal injury to his leg and abdomen 3 days ago and has been sedated with propofol and morphine since that time. You are aware that the procedure will be painful and decide to increase his sedation to ensure he is both comfortable and not aware of the dressing change. To target his sedation you attach a bispectral index (BIS) monitor.

Which is the most appropriate value to target before commencing the dressing change?

A. 10
B. 25
C. 35
D. 50
E. 75

H3. A 46-year-old woman is admitted with suspected tuberculosis (TB) meningitis. She requires a lumbar puncture for diagnosis.

Which of the following is the best way to identify the L4/5 intervertebral space for the procedure?

A. Identify the body of the T12 vertebra from the 12th rib, and count down
B. Midpoint of a line drawn between the iliac crests
C. Measure 10 cm superiorly from the inferior pole of the coccyx
D. Identify the vertebra prominens (C7 spinous process) and count down
E. Midpoint of a line drawn between the posterior superior iliac spines (PSIS)

H4. Which of the following terms best describes the precision to which a collection of samples represents the population from which they are drawn?

A. Standard error of the mean
B. Standard deviation
C. Variance
D. Mean
E. Central tendency

H5. You are call to see a 35-year-old woman in the emergency department who has a severe metabolic acidosis. Her arterial blood gas result is as follows:

pH	7.15	Na^+	137 mmol/l
pO_2	13.6 kPa	K^+	4.8 mmol/l
pCO_2	2.5 kPa	Cl^-	105 mmol/l
BE^-	−18	Salicylates	Undetectable
HCO_3^-	9.8 mmol/l	Glucose	4.2 mmol/l
Lactate	1.6 mmol/l		

Her body mass index is 18. The only other medical history of note is a recent case of sinusitis which was treated with 2 weeks of flucloxacillin and chronic back pain for which she take regular paracetamol. There is no history of other drug misuse.
What is the most likely cause of her metabolic acidosis?

A. Hyperchloraemic acidosis
B. Pyroglutamic acidosis
C. Renal tubular acidosis
D. Diabetic ketoacidosis
E. Aspirin overdose

H6. Which of the following molecules has the highest affinity for oxygen?

A. Haemoglobin in the presence of a high partial pressure of CO_2 (6.1 kPa)
B. Haemoglobin in the presence of a low partial pressure of CO_2 (3.2 kPa)
C. Fetal haemoglobin (HbF)
D. Myoglobin
E. Haemoglobin in a hypothermic patient (core temp 33°C)

H7. Which of the following drugs will remain active for the longest, following cessation of a 12-hour infusion aimed at maintaining a constant plasma concentration?

A. Fentanyl
B. Alfentanil
C. Remifentanil
D. Thiopentone
E. Propofol

H8. You are admitting a 32-year-old patient with diabetic ketoacidosis (DKA), and you have decided to site a radial arterial line to facilitate regular blood gas analysis.
Which of the following is the most likely complication of this procedure?

A. Arterial line site infection
B. Air embolism
C. Inadvertent intra-arterial injection
D. Permanent ischaemic damage
E. Arterial occlusion

H9. A 50-kg 67-year-old woman has presented to the emergency department with cellulitis. Her initial observations demonstrate a sinus tachycardia of 116 bpm, a blood pressure of 74/36, a respiratory rate of 28 and oxygen saturations (SpO_2) of 95% in air.

Which of the following interventions should be performed FIRST?

A. Inspired oxygen therapy
B. Perform peripheral blood cultures
C. Broad spectrum antibiotics
D. Administer a 2-l crystalloid fluid bolus
E. Central venous catheter

H10. A 70-kg man has been intubated on the ward and requires transfer to the intensive care unit (ICU). He is making no respiratory effort and requires hand ventilation. You wish to put together the equipment to form a Water's circuit. Which of the following arrangements of components best describes the Water's circuit, commonly used to ventilate patients at a resuscitation event?

(APL = adjustable pressure limiting valve; Bag = 2-l collapsible bag; ETT = connection to patient's endotracheal tube)

A. Oxygen tubing – Bag – Tubing – APL – ETT
B. Oxygen tubing – APL – ETT – Tubing – Bag
C. Oxygen tubing – APL – ETT – Bag
D. Oxygen tubing – ETT – Tubing – APL – Bag
E. Oxygen tubing – ETT – Tubing – Open-ended bag

H11. A 23-year-old, 36-week pregnant woman is admitted to labour ward with headaches and visual disturbance. She has a blood pressure of 168/112 and proteinuria (3+ of protein on urine dipstick). She is commenced on oral labetalol, but her blood pressure remains high at 156/104. Whilst on labour ward, she suffers a tonic clonic seizure, which self terminates after 90 seconds. A post-fit cardiotocography (CTG) is normal.

Which of the following interventions should be performed FIRST?

A. A magnesium infusion (1 g/hr)
B. Delivery of the fetus
C. A 4-g bolus of magnesium
D. Intravenous hydralazine
E. An intravenous labetalol infusion

H12. A 46-year-old woman is admitted to your intensive care unit with a subarachnoid haemorrhage (SAH). Before a reduction in her Glasgow Coma Score (GCS), she complained of a sudden onset severe headache.

What is the MOST likely cause of her SAH?

A. A coagulation disorder
B. An aneurysm
C. Head trauma
D. Cocaine use
E. Polycystic kidney disease

H13. A previously fit and well 32-year-old primigravida woman, who is 34 weeks pregnant, has presented with hypertension, proteinuria and abdominal pain. She is otherwise feeling well. Investigations reveal an increased alanine transaminase (ALT), bilirubin and lactate dehydrogenase as well as thrombocytopenia. She has no coagulopathy or raised bile acids. She has been taking paracetamol 1 g QDS for right upper quadrant pain for 3 days.

What is the MOST likely cause of her symptoms?

A. HELLP syndrome
B. Acute liver failure secondary to paracetamol
C. Acute fatty liver disease of pregnancy (AFLP)
D. Intrahepatic cholestasis of pregnancy
E. Hyperemesis gravidarum

H14. A 34-year-old patient with a background of Crohn disease has been admitted to the medical ward for the past 4 days due to an exacerbation in symptoms. He was started on parenteral nutrition 2 days ago. In the past few hours, he has developed generalized weakness and a respiratory alkalosis. He appears mildly confused and complains of double vision.

What is the likely electrolyte abnormality responsible?

A. Hypomagnesaemia
B. Hyperkalaemia
C. Hyponatraemia
D. –Hypophosphatemia
E. Hypercalcaemia

H15. A 24-year-old intravenous drug user presents with a rapidly progressive skin infection of the right arm. His skin is erythematous with purpuric areas, and he is in extreme pain. Systemically, he demonstrates signs of sepsis and on imaging there is evidence of gas in the soft tissues.

Which is the MOST important treatment?

A. Broad-spectrum antibiotics
B. Intravenous immunoglobulins
C. Hyperbaric oxygen therapy
D. Early surgical debridement
E. Supportive treatment in the intensive care unit

H16. A 34-year-old female patient suffered an isolated unsurvivable head injury. She does not meet brain-stem death criteria, but after discussions with family, withdrawal of active treatment and a plan for organ donation is made.

Which of the following organs has the LONGEST warm ischaemic time?

A: Lungs
B: Liver
C: Heart
D: Pancreas
E: Kidney

H17. Regarding aortic dissection which of the following is MOST accurate?

A. Aortic valve and root replacement is the treatment of choice for repair of all type A dissections

B. Transoesophageal echocardiography (TOE) is the diagnostic investigation of choice

C. Sodium nitroprusside is first line antihypertensive for uncomplicated type B dissections

D. Back pain is the commonest cause of presentation of type A and B aortic dissections

E. Cerebrospinal fluid drainage can treat lower limb neurology developing post aortic dissection

H18. A 28-year-old is intubated and ventilated in the intensive care unit with an isolated traumatic brain injury. His intracranial pressure (ICP) is being monitored with an intraparenchymal monitor, and he is receiving management in accordance with guidelines for the management of traumatic brain injury. Despite this, he suffers a spike in his ICP to 35 mmHg for 15 minutes.

Out of the following which is the most appropriate FIRST step to instigate in his ongoing management?

A. Perform a repeat computed tomography (CT) scan

B. Perform a decompressive craniectomy

C. Vasopressors to maintain a MAP of 80–90 mmHg

D. Administer a 100-ml bolus of 5% saline

E. Insertion of an extraventricular drain

H19. In which condition is treatment with corticosteroids LEAST useful?

A. Meningococcal meningitis

B. Thyrotoxic crisis

C. Acute severe asthma

D. Septic shock

E. Alcoholic hepatitis

H20. A 76-year-old man undergoes an emergency repair of his ruptured abdominal aortic aneurysm. On post-operative day 4, he remains sedated and ventilated and has a distended abdomen with gastric aspirates of around 200–300 ml every 4 hours. When measured, his abdominal pressures are 16 and 18 mmHg 4 hours apart.

Which of the following interventions should be implemented FIRST to reduce his intraabdominal pressure?

A. Administration of enemas per rectum

B. Continuous infusion of muscle relaxants

C. Reduction in rate of enteral nutrition

D. Administration of gastric prokinetics

E. Haemodialysis for fluid management

H21. You have been asked to prescribe enteral nutrition for a 26-year-old patient. The nurses ask you what normal daily nutritional requirements are.

Which of the following is the most accurate daily nutritional requirements?

A. Na$^+$ 2 mmol/kg, K$^+$ 1 mmol/kg, water 25 ml/kg, energy 20 kcal/kg, protein 1 g/kg

B. Na$^+$ 1 mmol/kg, K$^+$ 1 mmol/kg, water 25 ml/kg, energy 20 kcal /kg, protein 0.2 g/kg

C. Na$^+$ 1 mmol/kg, K$^+$ 1 mmol/kg, water 25 ml/kg, energy 20 kcal/kg, protein 0.2 g/kg

D. Na$^+$ 2 mmol/kg, K$^+$ 1 mmol/kg, water 25 ml/kg, energy 30 kcal/kg, protein 1 g/kg

E. Na$^+$ 2 mmol/kg, K$^+$ 1 mmol/kg, water 35 ml/kg, energy 30 kcal/kg, protein 0.2 g/kg

H22. A 48-year-old alcoholic presents with acute severe pancreatitis. He has evidence of a systemic inflammatory response syndrome, has evidence of pancreatic necrosis on computed tomography (CT) on day 3 and is improving with conservative treatment.

Which of the following is the MOST appropriate antimicrobial regime for this patient?

A. Ciprofloxacin
B. Imipenem
C. Imipenem and fluconazole
D. Metronidazole
E. None required

H23. You are performing a rapid sequence intubation (RSI) on a patient through an 18-gauge cannula sited in the right antecubital fossa. As you administer the thiopentone, the patient screams in pain, and his right hand goes white. You suspect an intra-arterial (IA) injection.

Which of the following is the most appropriate immediate action?

A. Remove the cannula, site a new one in the other arm, and continue the RSI
B. Urgently perform a stellate ganglion block
C. Request an urgent vascular review
D. Flush the cannula with saline
E. Inject papaverine through the cannula

H24. You are about to perform a lumbar puncture in a patient in the intensive care unit, but the nurses tell you he received an anticoagulant recently.

Which of the following patients is at highest risk of developing an epidural haematoma?

A. A 40-year-old male on an intravenous (IV) heparin infusion, stopped 4 hours ago
B. A 40-year-old male on treatment-dose subcutaneous low molecular weight heparin (LWMH), last dose 24 hours ago
C. A 40-year-old male on aspirin, taken 6 hours ago
D. A 40-year-old male on clopidogrel, stopped 4 days ago
E. A 40-year-old male on prophylactic-dose subcutaneous LMWH, last dose 12 hours ago

H25. You have admitted a patient with an acute exacerbation of chronic obstructive pulmonary disorder (COPD) for a trial of noninvasive ventilation (NIV).

Which of the following approaches, with reference to inspiratory positive airway pressure (IPAP) and expiratory positive airway pressure (EPAP), is the best way to commence this?

A. Commence an IPAP of 10 cmH$_2$O, and EPAP 5 cmH$_2$O, and titrate up the IPAP by 5 cmH2O every 10 minutes until a response is achieved.

B. Commence an IPAP of 10 cmH$_2$O, EPAP 5 cmH$_2$O, and titrate up the IPAP by 5 cmH$_2$O, and the EPAP by 2 cmH$_2$O every 10 minutes until a response is achieved.

C. Commence an IPAP of 14 cmH$_2$O, EPAP 4 cmH$_2$O, and titrate up both by 2 cmH$_2$O every 5 minutes until the IPAP is 20 cmH$_2$O and the EPAP is 10 cmH$_2$O.

D. Commence an IPAP of 14 cmH$_2$O, EPAP 4 cmH$_2$O, and titrate up both by 2 cmH$_2$O every 5 minutes until a response is achieved.

E. Commence an IPAP of 14 cmH$_2$O, EPAP 4 cmH$_2$O, and titrate up both by 2 cmH$_2$O every 10 minutes until a response is achieved, or the IPAP is 20 cmH$_2$O and the EPAP is 10 cmH$_2$O.

H26. You have just inserted a percutaneous tracheostomy into a 50-year-old patient to aid ventilatory weaning.

Which of the following statements has the strongest recommendation from the Intensive Care Society?

A. In mechanically ventilated patients, the inner tube should be changed every 8 hours.

B. Cuff pressure should not exceed 20 cmH$_2$O.

C. Humidification is essential for all patients.

D. Before decannulation, a fibreoptic inspection of the upper airway should be undertaken.

E. Fenestrated tracheostomy tubes should be the first choice tube type in ventilated patients.

H27. You are a member of the trauma team in a major trauma centre and are called to the emergency department (ED). A 30-year-old man has just been brought in after his lower limbs were run over by a truck. He has suffered a traumatic amputation of his left leg and an obvious open fracture to his right leg. He has no obvious head, chest, abdominal or pelvic injuries. He has a heart rate of 145, blood pressure of 90/60, a respiratory rate of 40 and is drowsy. The pre-hospital team have placed a tourniquet on his left thigh, and his right leg has been splinted and covered. Two large-bore cannulae have been sited, and he has received a total of 1000 ml of 0.9% saline.

Which of the following initial fluid management strategies would be most appropriate?

A. Warmed Hartmann's solution

B. 0.9% saline solution

C. O negative packed red cells, fresh frozen plasma (FFP) and platelets

D. Human albumin solution 4.5%

E. Cross-matched packed red cells, FFP and platelets

H28. A 64-year-old man with a history of angina and depression is brought in by ambulance after taking an overdose of atenolol. He is spontaneously ventilating, with SpO_2 99% (FiO_2 0.6) and Glasgow Coma Score of 15. Monitoring shows heart rate 30 bpm, blood pressure 85/37 mmHg, and a 12-lead electrocardiogram (ECG) shows complete heart block with a wide QRS complex. He has intravenous (IV) access in situ and has been given 500 µg of atropine with no response.

What is the most appropriate next step?

A. Commence transcutaneous pacing
B. Administer 600 µg glycopyrrolate
C. Arrange transvenous pacing
D. Administer glucagon IV
E. Administer IV adrenaline 2 to 10 µg/min

H29. You are asked to review a 27-year-old man in the emergency department with burns.

Which of the following features is LEAST likely to warrant referral to the regional burns unit?

A. A circumferential burn to the left forearm that is non-blanching, and 6% of total body surface area
B. A burn to the back that is 7% total body surface area (BSA) and is non-blanching
C. A full-thickness burn to the soles of both feet
D. A chemical burn to the right thigh
E. Partial-thickness burns of 12% total body surface area

H30. A 45-year-old 63-kg female is ventilated in the intensive care unit for community-acquired pneumonia and requires noradrenaline to maintain a mean arterial pressure ≥ 65 mmHg. On day 3, her urine output declines, with 150 ml urine in 6 hours. Her creatinine has risen to 148 µmol/l from a baseline value of 84 µmol/l on admission.

Which of the following is the LEAST important measure to implement in this patient's management?

A. Check and adjust drug doses
B. Active management of blood glucose to avoid hyperglycaemia
C. Renal tract ultrasound
D. Avoid all nephrotoxic drugs and contrast media where possible
E. Optimize patient haemodynamics using cardiac output monitoring

Exam H: Answers

H1. A 70-kg patient is 'nil by mouth'. The nursing staff have asked you to prescribe some maintenance fluids for the night. He is clinically euvolaemic and his electrolyte levels are all within normal range.

Which prescription is the most appropriate for IV maintenance fluids?

A. 1 l 0.18% NaCl with 4% glucose and 27 mmol/l potassium over 12 hours
B. 1 l 0.9% NaCl with 40 mmol/l potassium over 12 hours
C. 1 l compound sodium lactate (Hartmann's solution) over 12 hours
D. 1 l 5% dextrose over 12 hours
E. 1 l 0.18% NaCl with 4% glucose over 12 hours

Answer: A

Short explanation
This patient requires maintenance fluid only. Adults require 25–30 ml/kg/day of fluid with 1 mmol/kg/day of sodium, potassium and chloride and 50–100 g/day of glucose. Two litres of 0.18% NaCl with 4% glucose and 27 mmol/l potassium is the fluid recommended by NICE for the first 24 hours of maintenance fluids.

Long explanation
Fluid and electrolyte management is crucial for good patient care but is often left to the most inexperienced medical staff with deleterious consequences. Recent guidelines have been published by NICE to help to address this shortcoming. Patients' fluid and electrolyte needs should be assessed and managed on every ward round by skilled, trained professionals.

There are four steps involved: assessment, resuscitation, routine maintenance and replacement. The first step is assessment for hypovolaemia and the requirement for resuscitation. If resuscitation is not required, then patients should be assessed for their fluid and electrolyte requirements, and whether they have the capacity to fulfil these needs enterally. If not, then routine maintenance fluids should be prescribed. If patients have complex fluid and electrolyte needs, or ongoing losses that need to be replaced, then a plan should be developed taking this into account in addition to routine fluid requirements.

In the clinical scenario described here, the patient is not in need of fluid resuscitation and has no additional complex fluid or electrolyte requirements nor ongoing losses. Routine maintenance is therefore all that is required. Daily requirements are

25 to 30 ml/kg/day of fluid, with 1 mmol/kg/day of sodium, potassium and chloride. A small amount of glucose (50–100 g/day) is required to prevent ketosis, but this will not meet the nutritional requirements of the patient. A 70-kg patient therefore requires 1750 to 2100 ml/day of fluid with 70 mmol of sodium, potassium and chloride in addition to glucose. 0.18% NaCl and 4% glucose is isotonic, and contains 31 mmol of sodium and chloride and 40 g glucose. With the addition of potassium, 2 l of this fluid approaches daily requirements. After the first 24 hours, fluids and electrolyte requirements should be reassessed.

Reference
National Institute for Health and Care Excellence (NICE). *Intravenous fluid therapy in adults in hospital* (CG174). 2013. http://www.nice.org.uk/nicemedia/live/14330/66015/66015.pdf (accessed February 23, 2016).

H2. A 37-year-old patient in the intensive care unit (ICU), who is intubated and ventilated, is about to undergo a change of dressings. He sustained a thermal injury to his leg and abdomen 3 days ago and has been sedated with propofol and morphine since. You are aware that the procedure will be painful and decide to increase his sedation to ensure he is both comfortable and not aware of the dressing change. To target his sedation you attach a Bispectral index (BIS) monitor. Which is the most appropriate value to target before commencing the dressing change?

A. 10
B. 25
C. 35
D. 50
E. 75

Answer: D

Short explanation
For general anaesthesia and amnesia of events a target of 40 to 60 should be achieved. BIS values range from 0 to 100 where 0 is no brain activity and 100 is wide awake.

Long explanation
BIS monitoring is recommended by NICE during general anaesthesia for any patient undergoing total intravenous anaesthesia (TIVA). There have been no recommendations regarding its routine use in ICU. Targeting a BIS value of 40 to 60, alongside other clinical monitoring, is thought to reliably reduce the risk of awareness and induce amnesia of events. In routine ICU practice, this is not often desirable because sedation targets are set to avoid distress and agitation rather than induce deep anaesthesia. However, in this clinical case, its use is justified because the patient is effectively undergoing a painful procedure for which general anaesthesia would be indicated, although it is being delivered in an ICU setting.

BIS monitors contain proprietary algorithms which use electroencephalograph signals to calculate a level of brain activity from 0 to 100. The signal is detected using a four-point electrode attached to the patient's forehead and displayed as a number. Both the Narcotrend and E-Entropy systems use similar technologies to produce a similar range of numbers.

Reference
National Institute for Health and Care Excellence. *DG6: Depth of anaesthesia monitors – bispectral Index (BIS), E-Entropy and Narcotrend-Compact M*. London: NICE, 2012. www.nice.org.uk/guidance/dg6 (accessed May 2015).

H3. A 46-year-old woman is admitted with suspected TB meningitis. She requires a lumbar puncture for diagnosis.

Which of the following is the best way to identify the L4/5 intervertebral space for the procedure?

A. Identify the body of the T12 vertebra from the 12th rib, and count down
B. Midpoint of a line drawn between the iliac crests
C. Measure 10 cm superiorly from the inferior pole of the coccyx
D. Identify the vertebra prominens (C7 spinous process) and count down
E. Midpoint of a line drawn between the posterior superior iliac spines (PSIS)

Answer: B

Short explanation

Any method based on counting vertebral bodies or measuring distances from an alternative landmark is notoriously unreliable. Tuffier's line (a line connecting the iliac crests) has the greatest accuracy for identification of the L4/5 vertebral space. The line connecting the PSISs overlies the S1 to S2 spinous processes in most cases.

Long explanation

Lumbar puncture (LP) is an important diagnostic test in many clinical situations. The procedure involves advancing a fine (22- to 25-G) needle into the subarachnoid space of the lumbar spine to collect CSF. In doing so, the needle passes through the following structures: skin, subcutaneous tissue, supraspinous ligament, interspinous ligament, ligamentum flavum, extradural space, dura mater and arachnoid mater.

An LP attempt should be performed at the level of the L3/4 or L4/5 interspace. It is important to identify the correct intervertebral space at which to aim one's needle for several reasons. In adults, the termination of the spinal cord, the conus medullaris, usually lies at the level of the L1 vertebral body. In approximately 10 to 12% of patients, the spinal cord terminates at or below the L2 vertebral body, so attempts should be well below this to avoid cord damage. However, below the L4/5 interspace, performing an LP becomes progressively more challenging and uncomfortable.

Identification of the correct intervertebral space is notoriously difficult. Magnetic resonance imaging and other radiological studies have demonstrated that even experienced anaesthetists routinely misidentify lumbar vertebrae. Any method based on counting vertebral bodies from a landmark is highly inaccurate, as is any method dependant on measurement from a landmark. The line connecting the iliac crests (Tuffier's line or the intercristal line) is the most accurate means of identifying lumbar vertebrae. It most commonly crosses the L3/4 intervertebral space. The line connecting the PSISs most commonly runs through the S2 spinous process and so is not particularly useful for LP.

References

Broadbent CR, Maxwell WB, Ferrie R, et al. Ability of anaesthetists to identify a marked lumbar interspace. *Anaesthesia*. 2000;55;1122–1126.

Chakraverty R, Pynsent P, Isaacs K. Which spinal levels are identified by palpation of the iliac crests and the posterior superior iliac spines? *J Anat*. 2007;210:232–236.

H4. Which of the following terms best describes the precision to which a collection of samples represents the population from which they are drawn?

A. Standard error of the mean (SEM)
B. Standard deviation
C. Variance
D. Mean
E. Central tendency

Answer: A

Short explanation

Given a collection of samples from a population, each will have a mean that approximates but is unlikely to equal the true population mean. The larger the sample, the closer this approximation is likely to be. The precision to which the sample mean approximates the true population mean is described as the SEM.

Long explanation

The SEM is calculated as the standard deviation (SD) divided by the square root of the sample size. The variance is a measure of the dispersion of the data around the mean – that is, the spread of the frequency distribution curve. The standard deviation is the square root of the variance. The mean is a measure of the central tendency of a sample, or the average.

The SEM and the SD are both measures of spread and will therefore take on the same units as the data (e.g. if measuring heart rate in beats per minute, the SD and SEM should also be expressed in beats per min). However, SD and SEM demonstrate very different aspects of the data and are often confused. SD measures how scattered the data are; as sample sizes increase, SD will be more precise but may increase or decrease. In contrast, as a sample size increases, the spread of the means will decrease, and therefore the SEM will decrease.

Some graph data are shown with error bars demonstrating the SEM. This is often difficult to interpret and simply demonstrates how well the mean you measure approximates the population mean. It is often confused with confidence intervals or SD, which are both much more useful and easier to interpret on a graph. For normally distributed data, 68% of the population will fall within 1 SD of the mean and 95% will fall within 2 SDs.

Reference

Kirkwood BR, Sterne JAC. *Medical Statistics*. 2nd edition. Oxford: Blackwell Science Ltd.

H5. You are call to see a 35-year-old woman in the emergency department who has a severe metabolic acidosis. Her arterial blood gas result is as follows:

pH	7.15	Na$^+$	137 mmol/l
pO$_2$	13.6 kPa	K$^+$	4.8 mmol/l
pCO$_2$	2.5 kPa	Cl$^-$	105 mmol/l
BE$^-$	−18	Salicylates	Undetectable
HCO$_3^-$	9.8 mmol/l	Glucose	4.2 mmol/l
Lactate	1.6 mmol/l		

Her body mass index is 18. The only other medical history of note is a recent case of sinusitis which was treated with 2 weeks of flucloxacillin and chronic back pain for which she take regular paracetamol. There is no history of other drug misuse.

What is the most likely cause of her metabolic acidosis?

A. Hyperchloraemic acidosis
B. Pyroglutamic acidosis
C. Renal tubular acidosis
D. Diabetic ketoacidosis
E. Aspirin overdose

Answer: B

Short explanation

This patient has a high anion gap metabolic acidosis which excludes renal tubular acidosis. The glucose and salicylate levels are normal, so this excludes aspirin overdose or diabetic ketoacidosis. Her chloride level is within the normal range. She has a history of chronic paracetamol use, which can result in pyroglutamic acidosis, particularly when taken regularly and/or in combination with flucloxacillin

Long explanation

This patient has a high anion gap metabolic acidosis. The anion gap (AG) is the difference between unmeasured anions and cations and is calculated as:

$$AG = ([Na^+] + [K^+]) - ([Cl^-] + [HCO3^-]) = \text{typically 8–12 mEq/l}$$

Hypoalbuminaemia and hypophosphataemia can result in a falsely low anion gap so a corrected anion gap (AGc) can be used to correct for these abnormalities:

$$AGc = [([Na^+] + [K^+]) - ([Cl^-] + [HCO3^-])]$$
$$- [0.2[\text{alb (g/l)}] + 1.5[PO_4^{3-} (\text{mmol/l})]]$$

Causes of high anion gap metabolic acidoses include:

- Lactic acidosis
- Ketoacidosis
  - Diabetes mellitus (diabetic ketoacidosis)
  - Starvation (starvation ketoacidosis)
  - Ethanol (alcoholic ketoacidosis)
  - Genetic/metabolic disorders of amino acids and the urea cycle, e.g. methylmalonic acidaemia
- Other acids
  - Salicylates which impair oxidative phosphorylation
  - Methanol and glycols (ethylene glycol, diethylene glycol and propylene glycol): these are metabolized to organic acids; propylene glycol is used in a number of parenteral medications

- ○ 5-oxoproline (pyroglutamic acid): this is most commonly seen with chronic paracetamol, ingestion especially in malnourished women
- ○ Isoniazid
- ○ Cyanide
- ○ Paraldehyde and formaldehyde
- ○ Renal failure (acute or chronic): increased retention of hydrogen ions, sulphate and phosphate ions

Pyroglutamic acidosis is an uncommon but greatly under-recognised cause of a high anion gap metabolic acidosis. Increased levels of pyroglutamic acid (5-oxoproline) arise from the γ-glutamyl cycle in the liver, which is important for uptake and/or metabolism of amino acids and dipeptides. Glutathione plays a key role in this cycle. In circumstances of glutathione depletion, such as malnutrition, pregnancy, sepsis or paracetamol use, levels of pyroglutamic acid rise. Some drugs (particularly flucloxacillin) exacerbate this by inhibiting 5-oxoprolinase, the enzyme responsible for breakdown of pyroglutamic acid, leading to pyroglutamic acidosis.

Treatment of pyroglutamic acidosis involves removal of exacerbating factors (e.g. stopping paracetamol and flucloxacillin), supportive care and the administration of N-acetyl cysteine or methionine to replenish glutathione stores.

References

Kellum JA. Clinical review: reunification of acid-base physiology. *Crit Care.* 2005; 9:500–507.

Morris CG, Low J. Metabolic acidosis in the critically ill: part 2. *Causes Treat Anaesth.* 2008;63(4):396–411.

H6. Which of the following molecules has the highest affinity for oxygen?

A. Haemoglobin in the presence of a high partial pressure of CO_2 (6.1 kPa)
B. Haemoglobin in the presence of a low partial pressure of CO_2 (3.2 kPa)
C. Fetal haemoglobin (HbF)
D. Myoglobin
E. Haemoglobin in a hypothermic patient (core temp 33°C)

Answer: D

Short explanation

The normal oxygen-haemoglobin dissociation curve has a P50 (the partial pressure of oxygen at which it is 50% saturated) of 3.5kPa. The curve is shifted to the left (i.e. the P50 is reduced) in hypothermia and low CO_2 states, but only marginally compared with the HbF curve, which has a P50 of 2.5kPa. Myoglobin has the highest affinity for oxygen with a P50 of <0.4 kPa and binds oxygen in very low oxygen tensions.

Long explanation

The oxygen dissociation curve links the saturation of haemoglobin molecules (SaO_2; %) with the partial pressure of oxygen dissolved in the blood (mmHg or kPa). The curve for haemoglobin is sigmoid shaped owing to the interactions between the four binding sites for oxygen. The 'normal' curve is drawn with oxygen saturation on the y-axis and partial pressure on the x-axis. The curve passes through the origin and will have a saturation of 50% at 3.5 kPa, of 75% (venous) at 5.3 kPa and approach 100% (arterial) at 13.3 kPa.

The curve shifts to the right or left with various physiological variables, which therefore mean that haemoglobin is more or less likely to bind oxygen in various situations. The degree to which the curve shifts can be measured by the P50, which is the partial pressure of oxygen at which haemoglobin is 50% saturated. The curve will shift right in conditions of increased temperature, reduced pH and increased pCO_2 and the P50 will be higher. The curve shifts left in low temperatures, alkalosis and in the presence of decreased 2,3-diphosphoglycerate and so the P50 will be lower. In effect, the affinity of haemoglobin for oxygen is lower in conditions found in active muscles, and so oxygen is more likely to be released.

These shifts are small in comparison to the left shift seen when haemoglobin is replaced with fetal haemoglobin. This ensures that the fetus will always have a higher affinity for oxygen than the mother, and therefore oxygen will dissociate from the mother's blood and bind to the fetus' haemoglobin.

The curve for myoglobin is exponential, not sigmoidal, because there is only one oxygen binding site. This curve exists at the extreme left of the graph with a P50 of 0.4 kPa. This ensures that even in the lowest oxygen tensions, oxygen will dissociate from haemoglobin in the blood to bind to myoglobin in the muscle. In other words, the greater the gap between the curves for haemoglobin during exercise and myoglobin (difference between the P50s), the greater the concentration gradient for oxygen to move from the blood to the muscle.

Reference

Thomas C, Lumb AB. Physiology of haemoglobin. *Contin Educ Anaesth Crit Care Pain*. 2012. http://ceaccp.oxfordjournals.org/content/early/2012/05/15/bjaceaccp.mks025.full (accessed January 2016).

H7. Which of the following drugs will remain active for the longest, following cessation of a 12-hour infusion aimed at maintaining a constant plasma concentration?

A. Fentanyl
B. Alfentanil
C. Remifentanil
D. Thiopentone
E. Propofol

Answer: A

Short explanation

The context sensitive half time for fentanyl at 12 hours is approximately 300 minutes. Thiopentone is 100 minutes and the other drugs listed are below 50 minutes.

Long explanation

This question is referencing 'context sensitive half-time' (CSHT), or the time taken for the plasma concentration of a drug to fall to 50% following the cessation of an infusion aimed at maintaining a constant plasma concentration. The half-life given for a drug refers to the time taken for the plasma concentration to halve following a bolus dose. Once drugs have been given by infusion, their pharmacokinetic modelling alters because drugs accumulate in various compartments of the body. The time taken for drugs to wash out of these compartments will vary; therefore the half time of the drug varies according to the duration of the infusion.

Drugs that are highly lipid soluble, such as fentanyl and propofol, will accumulate in fatty tissues. The rate of elimination from these compartments depends on the time constants between each compartment and the central plasma compartment. In this case, fentanyl redistributes much faster than propofol, therefore the large fatty compartment is able to maintain much higher plasma concentrations of fentanyl for longer. Propofol may take longer to leave the body but will do so at very low plasma levels, much lower than are able to maintain drug activity. Half-times are only valid for a plasma concentration to fall from 100% to 50%. This should be distinguished from half-lives which are a constant duration for a drug concentration to fall by half – either from 100% to 50% or 80% to 40% or from 10% to 5%.

Drugs with low CSHTs, such as alfentanil and remifentanil, make excellent choices where a rapid wake-up following a long infusion is desired. Remifentanil is considered 'context insensitive' because its CSHT is so low that time to wake-up is almost the same irrespective of the duration of the infusion.

Reference
Peck TE, Hill SA, Williams M. *Pharmacology for Anaesthesia and Intensive Care*. 3rd edition. Cambridge. Cambridge University Press, 2008.

H8. You are admitting a 32-year-old patient with diabetic ketoacidosis (DKA), and you have decided to site a radial arterial line to facilitate regular blood gas analysis. Which of the following is the most likely complication of this procedure?

A. Arterial line site infection
B. Air embolism
C. Inadvertent intra-arterial injection
D. Permanent ischaemic damage
E. Arterial occlusion

Answer: E

Short explanation
Temporary radial artery occlusion (20%) and bleeding/haematoma formation (14%) are the two most likely complications of this procedure, followed by local line infection (roughly 1%). Inadvertent intra-arterial (as opposed to intra-venous) injection is much rarer.

Long explanation
Arterial cannulation is one of the most commonly performed procedures in ICUs, and although it has not received the same degree of safety interest nationally as central-line and nasogastric tube insertion, it has had renewed interest since National Patient Safety Agency guidance in 2008. This guidance led to changing of fluid solutions to 0.9% NaCl after patient deaths were reported with glucose-containing infusion sets. The deaths were a result of misleadingly high glucose readings on blood samples taken through the lines, leading to inappropriate insulin administration.

The radial artery is the most often used site because of its superficial anatomy and easy palpability. The frequency of complications is variably reported, but temporary radial artery occlusion and bleeding or haematoma formation are the most common. Line infection is much less common with arterial lines than with venous lines but is still site dependant, with the femoral site offering a higher rate of infection than the radial or brachial sites. Arterial lines do not suffer the same problems with air embolism that central lines do, given their higher pressure, although air can accidentally be injected into the line if the infusion line is not set up properly. Inadvertent intra-arterial injection remains a topical subject, particularly if using thiopentone for a rapid-sequence-induction, which can crystallize and cause end-artery occlusion.

References
National Patient Safety Agency. Rapid response report: problems with infusions and sampling from arterial lines. 2008. http://www.nrls.npsa.nhs.uk/resources/?entryid45=59891 (accessed March 2015).
Scheer B, Perel A, Pfeiffer UJ. Clinical review: complications and risk factors of peripheral arterial catheters used for haemodynamic monitoring in anaesthesia and intensive care medicine. *Crit Care*. 2002;6(3):199–204.

H9. A 50-kg 67-year-old woman has presented to the emergency department with cellulitis. Her initial observations demonstrate a sinus tachycardia of 116 bpm, a blood pressure of 74/36, a respiratory rate of 28 and oxygen saturations (SpO$_2$) of 95% in air.

Which of the following interventions should be performed FIRST?

A. Inspired oxygen therapy
B. Perform peripheral blood cultures
C. Broad spectrum antibiotics
D. Administer a 2L crystalloid fluid bolus
E. Central venous catheter

Answer: B

Short explanation

This patient is not hypoxic so does not require inspired oxygen therapy. In patients with severe sepsis, measuring blood lactate, performing blood cultures before administering broad-spectrum antibiotics and administering a 30 ml/kg fluid bolus (1.5 l here) are all part of the surviving sepsis campaign 3-hour management bundle. Insertion of a central line and measurement of CVP is included in the 6-hour bundle.

Long explanation

The Surviving Sepsis Committee (SSC) has produced guidelines regarding the management of severe sepsis and septic shock. It has produced both a 3-hour and a 6-hour bundle of interventions that should be completed within these time frames.

The 3-hour bundle includes the following:

- Measure lactate level.
- Obtain blood cultures (at least two sets) before commencing antimicrobials.
- Give broad-spectrum antimicrobials. Initial empiric broad spectrum anti-infective agents as per local hospital guidelines should be administered within an hour following the recognition of severe sepsis or septic shock.
- In the presence of hypotension or hyperlactataemia (lactate ≥4 mmol/l) a 30-mL/kg fluid bolus should be given. Fluid challenges should be continued whilst the patient continues to demonstrate fluid responsiveness.

The 6-hour bundle includes the following:

- Vasopressors should be used to treat hypotension unresponsive to fluids aiming for a mean arterial pressure (MAP) ≥65 mmHg.
- Patients presenting with an initial lactate level ≥4 mmol/l or with septic shock should have a central line inserted to measure:
  - Central venous pressure (CVP)
  - Central venous oxygen saturation (ScvO$_2$)
- Patients presenting with initially high lactate levels should have this re-measured with ongoing treatment to ensure its return to normal.

The SSC guidelines recommend the initial resuscitation of patients who are identified as having septic shock should be protocolized. Targets for the first 6 hours of resuscitation include achieving the following:

- A CVP 8–12 mmHg
- A MAP ≥65 mmHg
- A urine output ≥0.5 mL/kg/hr
- A central venous (superior vena cava) or mixed venous oxygen saturation of 70% (ScvO$_2$) or 65% (SvO$_2$), respectively
- Normalization of lactate

Following the results of the ProCESS and ARISE trials the SSC have stated that measurement of CVP and ScvO$_2$ in those patients who receive timely antibiotics and fluid resuscitation does not confer a survival benefit. However, it is not associated with adverse outcomes, and therefore it remains an element in the SSC 6-hour bundle.

The SSC guidelines also stress the importance of the prompt diagnosis the source of sepsis and timely source control if appropriate. Interventions to achieve this should be as minimally invasive as possible and should be undertaken with the first 12 hours after diagnosis of severe sepsis or septic shock is made. The exception to this rule is peripancreatic necrosis, for which delayed intervention is preferred.

References

Dellinger RP, Levy MM, Rhodes A, et al. Surviving Sepsis Campaign: international guidelines for management of severe sepsis and septic shock. 2012. *Intensive Care Med*. 2013;39(2):165–228.

Surviving Sepsis Campaign. Surviving Sepsis Campaign statement regarding hemodynamic and oximetric monitoring in response to ProCESS and ARISE Trials. 2014. http://www.survivingsepsis.org/SiteCollectionDocuments/ProCESS-ARISE.pdf (accessed October 14).

H10. A 70-kg man has been intubated on the ward and requires transfer to the ICU. He is making no respiratory effort and requires hand ventilation. You wish to put together the equipment to form a Water's circuit. Which of the following arrangements of components best describes the Water's circuit, commonly used to ventilate patients at a resuscitation event?

(APL = adjustable pressure limiting valve; Bag = 2-l collapsible bag; ETT = connection to patient's endotracheal tube)

A. Oxygen tubing – Bag – Tubing – APL – ETT
B. Oxygen tubing – APL – ETT – Tubing – Bag
C. Oxygen tubing – APL – ETT – Bag
D. Oxygen tubing – ETT – Tubing – APL – Bag
E. Oxygen tubing – ETT – Tubing – Open-ended bag

Answer: C

Short explanation

Answer C is the Mapleson C or Water's circuit. Answer D describes a Mapleson D or Bain circuit and Answer A is a Mapleson A or Magill circuit. These are sometimes used in anaesthetic practice. Answer B is a Mapleson B (rarely used), and answer E is a Jackson Rees modification of the Mapleson E, used in paediatric practice.

Long explanation

Breathing circuits are combinations of components connecting patients to gas supplies and ventilators. Circuits attached to modern ICU ventilators provide fresh gas in a controlled, pressurized flow with inspiratory and expiratory limbs meeting close to the patient. One-way valves and flow sensors ensure that there can be minimal rebreathing of expired gases. However, when transferring a patient or maintaining ventilation by hand in an emergency, a more simple circuit is required which can run from cylinder gas. These systems are described as semi-open and often include the components listed in the question.

The order in which the components are arranged will determine the efficiency of the circuit at reducing rebreathing of CO$_2$. The most common arrangement used in resuscitation is a Mapleson C or Water's circuit. This has the advantages of being light and compact; however, it will require up to 15 l/min to prevent rebreathing. In paediatric practice a Mapleson E circuit with an open-ended bag is commonly used, known as the Jackson-Rees modification. This is used in paediatric intensive care settings, particularly for children under 25 kg, to reduce the risks of barotrauma.

The semi-closed circuits were described by Professor Mapleson and labelled A through E. There are variations in their efficiency depending on whether the patient is spontaneously or mechanically ventilated. It is usually desirable to choose a circuit that requires the lowest fresh gas flow (FGF) whilst avoiding the patient re-breathing their expired gases. The APL valve can usually by adjusted from open (almost no resistance to expiration) to closed (up to 60 cmH$_2$O pressure). This allows control over ventilation pressures and the application of positive end-expiratory pressure.

Reference
Al-Sheik B, Stacey S. *Essentials of Anaesthetic Equipment*. 4th edition. London: Elsevier, 2013.

H11. A 23-year-old female 36-week pregnant woman is admitted to labour ward with headaches and visual disturbance. She has a blood pressure of 168/112 and proteinuria (3+ of protein on urine dipstick). She is commenced on oral labetalol, but her blood pressure remains high at 156/104. Whilst on labour ward, she suffers a tonic clonic seizure, which self terminates after 90 seconds. A post-seizurecardiotocogram (CTG) is normal.

Which of the following interventions should be performed FIRST?

A. A magnesium infusion (1 g/hr)
B. Delivery of the fetus
C. A 4-g bolus of magnesium
D. Intravenous hydralazine
E. An intravenous labetalol infusion

Answer: C

Short explanation
All the interventions are appropriate treatment options for this woman with eclampsia. Delivery of the fetus, however, should wait until she has stabilized. Intravenous hydrazine and labetalol are appropriate treatment options to control her blood pressure; however, a bolus dose of magnesium followed by a magnesium infusion should be administered first.

Long explanation
This patient presented with severe pre-eclampsia and then went on to develop eclampsia on labour ward.

NICE has produced guidelines regarding hypertensive disease in pregnancy. It defines pre-eclampsia as hypertension developing after 20 weeks' gestation with the presence of significant proteinuria (a urinary protein/creatinine ratio of >30 mg/mmol or >300 mg protein on 24-hour urine collection). Severe pre-eclampsia is pre-eclampsia with the presence any of the following:

- Severe hypertension
  - Systolic blood pressure (SBP) of ≥160 mmHg
  - Diastolic blood pressure (DBP) of ≥110 mmHg
- Symptoms (severe headache, visual disturbance, right upper quadrant, vomiting)
- Signs (papilloedema, clonus, liver tenderness)
- Biochemical derangement (increased liver enzymes, alanine transaminase or aspartate transaminase >70 IU/l)
- Haematological impairment (platelets <100 × 10^9/l).

Definitions of moderate and mild hypertension are a SBP of 150 to 159 and 140 to 149 or a DBP of 110 to 109 and 90 to 99, respectively. Eclampsia is present when a seizure occurs as a result of pre-eclampsia.

NICE recommend patients with moderate or severe hypertension to be treated with oral labetalol to control blood pressure to <150/80 to 100 mmHg. Labetalol should be used as first-line treatment but alternatives include methyldopa or nifedipine. Intravenous labetalol or hydralazine are alternatives in critical care patients.

NICE recommend intravenous magnesium sulphate to be given to women in the critical care setting if they have severe hypertension, severe pre-eclampsia or a previous eclamptic fit or to women with severe pre-eclampsia when delivery is planned within 24 hours. They recommend using the Collaborative Eclampsia Trial regimen of a 4-g bolus dose to be given over 5 minutes. An infusion at a rate of 1 g/hr should then be commenced for 24 hours, and any further seizures should be treated with a bolus dose of 2 g over 5 minutes. NICE also recommends not using diazepam or phenytoin as alternatives to magnesium.

The fetus should be delivered in these patients once the patient has been stabilized and after a course of steroids if appropriate. The mode of delivery (induction of labour or caesarean delivery) should be considered on an individual case-by-case basis.

Reference

National Institute for Health and Care Excellence. *Hypertension in pregnancy. The management of hypertensive disorders during pregnancy* (GC107). London: NICE, 2010. http://www.nice.org.uk/guidance/cg107 (accessed October 2014).

H12. A 46-year-old woman is admitted to your intensive care unit with a subarachnoid haemorrhage (SAH). Before a reduction in her Glasgow Coma Score (GCS), she complained of a sudden onset severe headache.

What is the MOST likely cause of her SAH?

A. A coagulation disorder
B. An aneurysm
C. Head trauma
D. Cocaine use
E. Polycystic kidney disease

Answer: B

Short explanation

Aneurysms cause approximately 85% of all SAHs. Head trauma and coagulation disorders are other causes, but this patient has no history of either. Polycystic kidney disease and the use of cocaine are risk factors for the development and rupture of aneurysms.

Long explanation

Subarachnoid haemorrhage (SAH) is the presence of blood in the subarachnoid space, lying between the pia and arachnoid dural membranes. Symptoms and signs include the classical 'thunderclap' headache, focal neurological signs, reduced GCS, signs of meningism, seizures, cardiovascular compromise/arrhythmias and pulmonary oedema.

Approximately 5% of all cerebrovascular events in the United Kingdom are the result of a SAH. The majority (approximately 85%) occur due to rupture of aneurysms. Other causes include trauma, arteriovenous malformations, bleeding diatheses/use of anticoagulants or idiopathic.

Risk factors for the development of aneurysms include:

- Positive family history
- Female sex
- Hypertension
- Amyloid angiopathy
- Vasculitides
- Inherited disorders:
  - Autosomal dominant polycystic kidney disease
  - Connective tissue disease (type IV Ehlers-Danlos syndrome, alpha1-antitrypsin deficiency)
  - Neurofibromatosis
- Modifiable risk factors:
  - Smoking
  - High alcohol intake
  - Use of sympathomimetic drugs (e.g. cocaine and amphetamines)
  - Very low body mass index

Most aneurysms arise from the Circle of Willis arteries

- 30 to 35% anterior communicating artery
- 30 to 35% internal carotid artery and posterior communicating artery
- 20% middle cerebral artery
- 5% bifurcation of the basilar artery
- 8% the remaining posterior circulation

The World Federation of NeuroSurgeons (WFNS) classification grades SAHs I–V according to the patient's GCS and the degree of neurology.

Grade	GCS	Motor Deficit
I	15	Absent
II	13–14	Absent
III	13–14	Present
IV	7–12	Present or absent
V	≤6	Present or absent

The outcome for these patients is often poor with more severe grades being associated with increased morbidity and mortality. The WFNS grade can therefore be used to help guide treatment.

References

Connolly ES, Rabinstein AA, Carhyapoma JR, et al. Guidelines for the management of aneurysmal subarachnoid haemorrhage. A guideline for healthcare professionals form the association/american stroke association. *Stroke*. 2012;43(6):1711–1137.

Gouch MJ, Goodwin APL, Shotton H, et al. NCEPOD report. Managing the flow? A review of the care received by patients who were diagnosed with an aneurysmal subarachnoid haemorrhage. National Confidential Enquiry into Patient Outcome and Death, 2013. http://www.ncepod.org.uk/2013sah.htm

Vega C, Kwoon JV, Lavine SD. Intracranial aneurysms: current evidence and clinical practice. *Am Fam Physician*. 2002;66(4):601–609.

H13. A previously fit and well 32-year-old primigravida woman, who is 34 weeks pregnant, has presented with hypertension, proteinuria and abdominal pain. She is otherwise feeling well. Investigations reveal an increased alanine transaminase (ALT), bilirubin and lactate dehydrogenase and thrombocytopenia. She has no coagulopathy or raised bile acids. She has been taking paracetamol 1 g QDS for right upper quadrant pain for 3 days.

What is the MOST likely cause of her symptoms?

A. HELLP syndrome
B. Acute liver failure secondary to paracetamol
C. Acute fatty liver disease of pregnancy (AFLP)
D. Intrahepatic cholestasis of pregnancy
E. Hyperemesis gravidarum

Answer: A

Short explanation

Abnormal liver function tests can be seen in all of these conditions. Hyperemesis gravidarum presents in early pregnancy with severe vomiting. Symptoms of intrahepatic cholestasis of pregnancy are related to increased bile acids, absent in this case. This patient has taken a short course of paracetamol at the standard dose, which is unlikely to be the causative factor. HELLP and AFLP can present similarly, but HELLP presents with increased LDH and patients with AFLP are generally unwell.

Long explanation

Deranged liver function tests (LFTs) in pregnancy can be the result of diseases specific to pregnancy and those that are not. Examples of conditions specific to pregnancy include the following:

- Intrahepatic cholestasis of pregnancy presents insidiously with pruritus, nausea, malaise and rarely jaundice in the third trimester. It is characterized by an increase in bile acids. It is associated with increased fetal complications including intrauterine death but no long-term physical sequelae for the mother.
- Hyperemesis gravidarum tends to occur in the first trimester of pregnancy with the main clinical symptom of vomiting, which can be intractable. Most clinical sequelae are a result of this.
- HELLP (Haemolysis, Elevated Liver enzymes, Low Platelets) syndrome often presents with gastrointestinal symptoms – right upper quadrant or epigastric pain, nausea and vomiting as well as nonspecific symptoms of being generally unwell. Neurological symptoms including headache and visual disturbance may also be present. Although not always present hypertension, proteinuria and generalized oedema are common.
- AFLP: symptoms and signs of AFLP can be similar to that of HELLP syndrome; however, they tend to be more severe. Nausea, vomiting, anorexia, abdominal pain, ascites, jaundice and lethargy often occur. About half of patients have symptoms and signs of pre-eclampsia. Acute renal failure and hepatic encephalopathy can also be present. AFLP occurs less commonly than HELLP.

	Acute liver failure	Hyperemesis gravidarum	Intrahepatic cholestasis of pregnancy	HELLP	AFLP
Onset	Any time in pregnancy, after trigger	Usually first trimester	Usually third trimester	Second half of pregnancy to postpartum	Second half of pregnancy to postpartum
Bile acids/salts	Normal	Normal	Raised	Normal	Normal
ALT	Raised	Raised	Raised	Raised	Raised
Bilirubin	Raised	Raised	Normal/raised	Raised	Raised
Lactate dehydrogenase	Normal	Normal	Normal		Normal
Glucose	Low	Normal	Normal	Normal	Low
Blood pressure	Normal/low	Normal	Normal	Raised	Raised
Proteinuria	No	No	No	Yes	Yes
Creatinine	Normal/raised	Normal/raised	Normal	Raised	Raised
Platelets	Normal/low	Normal	Normal	Low	Low
Coagulopathy	Yes	No	No	Yes/No	Yes

References

Walker I, Chappell LC, Williamson C. Abnormal liver function tests in pregnancy. *BMJ*. 2013;347:f6055.

Hepburn IS. Pregnancy-associated liver disorders. *Dig Dis Sci*. 2008;53(9):2334–2358.

H14. A 34-year-old patient with a background of Crohn's disease has been admitted to the medical ward for the past 4 days due to an exacerbation in symptoms. He was started on parenteral nutrition 2 days ago. In the past few hours, he has developed generalized weakness and a respiratory alkalosis. He appears mildly confused and complains of double vision. What is the likely electrolyte abnormality responsible?

A. Hypomagnesaemia
B. Hyperkalaemia
C. Hyponatraemia
D. Hypophosphatemia
E. Hypercalcaemia

Answer: D

Short explanation

Acute hypophosphatemia may be precipitated by refeeding a patient who has had a long-term reduced phosphate intake. It is characterized by global muscle weakness, which can include the extraocular muscles, causing double vision. Respiratory muscle weakness classically presents as tachypnoea with a respiratory alkalosis. Neurological complications include confusion and coma. It may be associated with rhabdomyolysis and haemolytic anaemia. Severe hypophosphataemia can precipitate cardiac failure.

Long explanation

Phosphate is a key mineral, required for almost all cellular functions. The majority of total body phosphate is stored in the bones and plasma phosphate levels are controlled by parathyroid hormone. Increased phosphate levels lead to the release of parathyroid hormone which in turn acts to increase phosphate losses by the kidneys. Parathyroid hormone also increases calcium reabsorption and promotes the activation of vitamin D, both of which are required for the sequestering of calcium phosphate into bone by osteoblasts.

Hyperphosphataemia is rare and often associated with renal disease, reflecting a reduction in excretion. It can also be seen in hypoparathyroidism or following tumour lysis syndrome. Hypophosphataemia is commonly a disorder of decreased intake such as chronic malnutrition, chronic inflammatory bowel diseases, alcoholism or long-term antacid use. Increased excretion causing hypophosphataemia is seen in hyperparathyroidism. Low plasma levels are commonly asymptomatic, but a sudden drop and hyposphosphataemic crisis may be precipitated by refeeding syndrome, caused by a sudden intracellular shift of phosphate. Conversely, the rapid replacement of phosphate through intravenous administration can precipitate hypocalcaemia.

Refeeding syndrome occurs following the reinstitution of carbohydrate as the main source of energy following a period of starvation. Patients are often relatively hypophosphataemic due to the starvation period and decreased intake. Intracellular phosphate stores are moved into the extracellular space in an attempt to maintain plasma levels. When carbohydrate is reintroduced the body switches from a starvation mode to utilization of the new energy source. This results in the release of insulin, which in turn activates the Na/K ATPase pumps and the production of proteins. Both of these processes facilitate the intracellular shift of phosphate, potassium and magnesium as co-factors, precipitating the sudden fall in their serum concentrations.

The symptoms of hypophosphataemia are widespread and generalized but include muscle weakness which can lead to respiratory, cardiac and gut failure, cell division problems leading to haematological dysfunction and central nervous system problems leading to coma and death.

References
Field MJ, Burnett L, Sullivan DR, Stewart P. Clinical biochemistry and metabolism. In *Davidson's Principles and Practice of Medicine*. 20th edition. Edinburgh: Churchill Livingstone Elsevier, 2006, pp 441–442.
Shiber JR, Mattu A. Serum phosphate abnormalities in the emergency department. *J Emerg Med*. 2002;23(4):395–400.

H15. A 24-year-old intravenous drug user presents with a rapidly progressive skin infection of the right arm. His skin is erythematous with purpuric areas, and he is in extreme pain. Systemically, he demonstrates signs of sepsis and on imaging there is evidence of gas in the soft tissues.

Which is the MOST important treatment?

A. Broad spectrum antibiotics
B. Intravenous immunoglobulins
C. Hyperbaric oxygen therapy
D. Early surgical debridement
E. Supportive treatment in the intensive care unit

Answer: D

Short explanation
This patient is presenting with necrotizing fasciitis, and all of these treatments are important components of management. The combination of early aggressive surgical debridement along with early broad-spectrum antibiotics and supportive intensive care management are needed to minimize the morbidity and mortality associated with this condition. However, few patients will survive without surgical debridement.

Long explanation
Necrotizing fasciitis is a severe, rapidly progressive, necrotizing soft tissue infection that involves the fascial planes and the tissues surrounding them. It is associated

with a high degree of morbidity and mortality and can be differentiated into types according to the causative organism(s).

- Type 1 – polymicrobial infections, which can be due to a mixture of anaerobic, aerobic and facultatively anaerobic Gram-positive and Gram-negative organisms
- Type 2 – usually monomicrobial and due to Gram-positive organisms, classically group A beta-haemolytic *Streptococcus* (*Streptococcus pyogenes*) but can be due to *Staphylococcus aureus* (especially community acquired methicillin-resistant *Staphylococcus aureus*)
- Type 3 – some guidelines include a third subtype of necrotizing fasciitis secondary to *Clostridium perfringens* (gas gangrene)

The combination of early clinical suspicion and prompt aggressive surgical debridement, along with the administration of timely broad-spectrum antibiotics and supportive intensive care management, are needed to minimize the morbidity and mortality associated with this condition.

Surgery for source control is time critical. Aggressive debridement needs to be performed early, and repeated surgery is required to ensure adequate margins have been achieved. Despite treatment with antibiotics and supportive care, if this condition is treated without surgical debridement, it is almost universally fatal.

Antibiotics need to be administered early. Antibiotic regimes should include broad-spectrum antimicrobials to cover for Gram-positive and -negative aerobic and anaerobic organisms, which can then be de-escalated when appropriate. An example of an appropriate initial regime would be meropenem, clindamycin (which suppresses toxin production in infections associated with Group A beta-haemolytic streptococcus and clostridium) and vancomycin or linezolid (as cover against MRSA).

As with all critical illnesses, good supportive management on critical care is vitally important to provide multiorgan support and treat the systemic complications that may arise.

There is limited evidence reviewing intravenous immunoglobulins (IVIGs) and hyperbaric oxygen (HBO) therapy; however, they may be useful adjuncts in selected groups. IVIGs have been used to neutralize the exotoxins associated with Group A beta-haemolytic *Streptococcus* (streptococcal super-antigens) and *Clostridium*. HBO treatment may improve outcomes, decrease mortality and decrease the amount of further surgical debridement that is required, especially with anaerobic infections.

References
Irwin K, The diagnosis and management of necrotising fasciitis. *Anaesth Tutorial Week*. 2013;298. http://www.aagbi.org/sites/default/files/298%20Diagnosis%20and%20Management%20of%20Necrotising%20Fasciitis.pdf (accessed October 2014).
Sultan HY, Boyle AA, Sheppard N. Necrotising fasciitis. *Brit Med J*. 2012;345:e4274.

H16. A 34-year-old woman suffered an isolated catastrophic head injury. She does not meet brain-stem death criteria, but after discussions with family, withdrawal of active treatment and a plan for organ donation is made.

Which of the following organs has the LONGEST warm ischaemic time?

A: Lungs
B: Liver
C: Heart
D: Pancreas
E: Kidney

Answer: E

Short explanation

The lungs, liver, pancreas and kidney can be transplanted post–cardiac death in the United Kingdom. Whilst heart transplants have occurred after cardiac death in the United States, they are not performed in the United Kingdom. The accepted warm ischaemic times are 30 minutes for liver and pancreas, 60 minutes for lungs and 120 minutes for the kidneys.

Long explanation

Donation of organs can occur after brain-stem death or cardiac death. Organs donated following cardiac death in the United Kingdom include the kidneys, liver, pancreas and lungs. Tissues such as bone, skin, corneas, heart valves and tendons can also be transplanted.

When considering donation after cardiac death, the warm ischaemic time needs to be considered. This is important because during this time the organs will suffer ischaemic damage that will affect future graft function. It occurs during such periods that the patient still has a circulation inadequate for organ perfusion, leading to adenosine triphosphate store depletion due to anaerobic metabolism.

The warm ischaemic time is defined as the duration of time following a sustained decrease in systolic blood pressure <50 mmHg for >2 minutes and until the organs have been cooled with cold perfusate in theatre. A reduction in the patient's oxygen saturation below 70% has previously been used to define warm ischaemia time. The British Transplantation Society UK guidelines recommend that the transplant teams should monitor this. However, although this measure may be used with other criteria to guide which organs are donated, it should not be used to define warm ischaemia time.

The quoted acceptable warm ischaemic times are:

- Liver: <20 minutes is preferable; however, up to 30 minutes will be accepted
- Pancreas: 30 minutes (although some organs have been successfully been donated after this time frame)
- Lungs: 60 minutes (this is the time from onset of warm ischaemia until the lungs have been inflated post mortem)
- Kidneys: 120 minutes (longer duration can be accepted on a case by case basis)

Although these are guides, the transplant team can put further restrictions on time frames.

Cardiac transplantation in the United Kingdom is only performed after confirmation of brain-stem death. However, it has been successfully performed in North America after a period of asystole. Tissues such as bone, skin, corneas, heart valves and tendons can also be donated after cardiac death without restriction on warm ischaemic time.

Reference

British Transplantation Society. United Kingdom Guidelines. Transplantation from donors after deceased circulatory death. 2013. http://www.bts.org.uk/Documents/2013–02–04%20DCD%20guidelines.pdf (accessed May 2015).

H17. Regarding aortic dissection which of the following is MOST accurate?

A. Aortic valve and root replacement is the treatment of choice for repair of all type A dissections.
B. Transoesophageal echocardiography (TOE) is the diagnostic investigation of choice.
C. Sodium nitroprusside is first line antihypertensive for uncomplicated type B dissections.
D. Back pain is the commonest cause of presentation of type A and B aortic dissections.
E. Cerebrospinal fluid drainage can treat lower limb neurology developing post aortic dissection.

Answer: E

Short explanation

Surgery is the treatment of choice for type A dissections, but replacement of the aortic valve is not always required. TOE is not the diagnostic investigation of choice but is useful in unstable patients. Type B dissections should be treated conservatively with antihypertensives – beta-blockers should be used first line. Chest pain is the commonest symptom of all aortic dissections; however, back pain occurs with increased frequency in type B dissections.

Long explanation

The Stanford classification categorizes aortic dissections according to their location. Type A involves the ascending aorta and type B is limited to affecting the descending aorta (occurring distal to the origin of the left subclavian artery).

Clinical presentations and complications of aortic dissections:

- Chest pain (commonest symptom)
- Back pain – increased with type B dissections compared with type A dissections
- Cardiac complications
  - Aortic regurgitation
  - Cardiac tamponade
  - Myocardial ischaemia or infarction
  - Heart failure
- Pulmonary complications:
  - Tachypnoea/breathlessness
  - Haemoptysis
  - Pleural effusion
- Distal hypoperfusion
  - Neurological sequelae
    - Syncope
    - Coma/cerebral vascular accident (CVA)
    - Spinal cord injury/ischaemia due to occlusion/hypoperfusion of spinal arteries
    - Ischaemic neuropathy
    - Hoarseness due to compression of left recurrent laryngeal nerve
  - Mesenteric ischaemia
  - Acute renal failure
  - Lower limb ischaemia

Electrocardiogram (ECG)-gated computed tomography (CT) angiogram is the investigation of choice. Magnetic resonance imaging provides good images, but for practical reasons, it is less useful. Although echocardiography is not as good as other techniques, it can be particularly useful for the evaluation of unstable patients.

Initial blood pressure (BP) and heart rate control using opiate analgesia and anti-hypertensives are important for all patients. Intravenous beta-blockers are the anti-hypertensives of choice as they reduce shear stress on the aortic wall. Additional agents such as sodium nitroprusside or calcium channel blockers can be added in if BP remains uncontrolled.

The recommended treatment for type A aortic dissection is urgent surgical repair with grafting of the ascending aorta ± the arch. The aortic valve may or may not need replacement depending on the extent of the dissection.

Patients with uncomplicated type B dissections may be treated with medical therapy alone. For those with complicated disease, treatment with thoracic endovascular aortic repair (TEVAR) is recommended, and open surgery may be indicated. Patients who develop neurological compromise due to spinal hypoperfusion require treatment to increase perfusion pressure. This can be achieved by increasing mean arterial pressure and decreasing the pressure exerted by cerebrospinal fluid (CSF) with the insertion of a lumbar drain to drain CSF. Prophylactic insertion of a drain should be considered in patients undergoing TEVAR or surgical repair of type B dissections.

References

Braverman AC. Acute aortic dissection: clinician update. *Circulation*. 2010;122(2):184–188.

Erbel R, Aboyans V, Boileau C. 2014 ESC guidelines on the diagnosis and treatment of aortic diseases: Document covering acute and chronic aortic diseases of the thoracic and abdominal aorta of the adult. The Task Force for the Diagnosis and Treatment of Aortic Diseases of the European Society of Cardiology (ESC). *Eur Heart J*. 2014;35(41):2873–2926.

H18. A 28-year-old is intubated and ventilated in the intensive care unit with an isolated traumatic brain injury. His intracranial pressure (ICP) is being monitored with an intraparenchymal monitor, and he is receiving management in accordance with guidelines for the management of traumatic brain injury. Despite this, he suffers a spike in his ICP to 35 mmHg for 15 minutes.

Out of the following which is the most appropriate FIRST step to instigate in his ongoing management?

A. Perform a repeat computed tomography (CT) scan
B. Perform a decompressive craniectomy
C. Vasopressors to maintain a MAP of 80 to 90 mmHg
D. Administer a 100-ml bolus of 5% saline
E. Insertion of an external ventricular drain

Answer: D

Short explanation

This patient's cerebral perfusion pressure (CPP) would be 55 mmHg with a MAP of 90, so his mean arterial pressure (MAP) should be increased to ensure the CPP is 60 to 70 mmHg. All of the other options are potential treatment options. This is a time-critical event, and urgent reduction in ICP is required. A bolus of hypertonic saline should be given initially whilst further treatment options are considered and arranged.

Long explanation

Management of severe traumatic brain injury pressure includes the following:

- Control of ventilation; keeping $PaCO_2$ 4.5–5 kPa and PaO_2 >10–13 kPa
- MAP 80 to 90 mmHg (if ICP unknown)

- Maintain CPP 60 to 70 mmHg (CPP = MAP-ICP)
- Maintain head-up tilt of at least 30 degrees
- Ensure unobstructed venous drainage:
  - Keep the neck in a neutral position
  - Removing C-spine collars as soon as possible
- Maintain glucose 6 to 10 mmol/l
- Keep Na^{++} 145 to 150 mmol/l
- Maintain normothermia; avoid pyrexia
- Keep Hb >10 g/dl
- Ensure adequate sedation
- Antiepileptics (short-term 7-day course of phenytoin or levetiracetam) – reduce the incidence of early seizures but do not prevent the later development of epilepsy

If ICP rises ≥20 mmHg further treatment strategies include the following:

- Sedation bolus ± paralyze
- Osmotherapy
  - Hypertonic saline, such as 100 ml 5% saline boluses, aiming to increase serum Na^+ but maintained ≤155 mmol/l
  - Mannitol 1 g/kg boluses, ensure osmolality remains <320 mOsm/l
- Furosemide can aid the effects of hypertonic saline and mannitol
- Further imaging with CT to exclude a surgically treatable lesion
- Insertion of external ventricular drain; or if present, drain off cerebrospinal fluid
- Temperature management with avoidance of pyrexia or therapeutic hypothermia
- Hyperventilation – increase minute ventilation to reduce $PaCO_2$ to 4 kPa and further if ICPs remain uncontrolled; once ICP has been controlled adjust ventilation to slowly return $PaCO_2$ to 4.5–5 kPa
- Consideration of decompressive craniotomy
- Consideration of thiopentone coma

Strategies to treat raised ICP should start with the simplest and quickest to achieve and then work down the list. CT scanning usually requires lying the patient flat, and so sedation and osmotic therapy need to have been instigated before this.

References

Brain Trauma Foundation. American Association of Neurological Surgeons, Congress of Neurological Surgeons. Guidelines for the Management of Severe Traumatic Brain Injury. 3rd edition. *J Neurotrauma*. 2007;24(Suppl 1):S1–106.

Haddad SH, Arabi YM. Critical care management of severe traumatic brain injury in adults. *Scand J Trauma Resusc Emerg Med*. 2012;20:12.

H19. In which condition is treatment with corticosteroids LEAST useful?

A. Meningococcal meningitis
B. Thyrotoxic crisis
C. Acute severe asthma
D. Septic shock
E. Alcoholic hepatitis

Answer: A

Short explanation

Corticosteroids can be used to treat thyrotoxic crisis, acute severe asthma, septic shock, alcoholic hepatitis and meningitis. Although not all patients with alcoholic hepatitis will improve with steroids, some will. The evidence of benefit in meningitis is for pneumococcal meningitis and not meningococcal disease.

Long explanation
The use of corticosteroids has a number of indications:

- Airway and respiratory
  - Airway swelling/oedema, e.g. croup, post-operative, anaphylaxis
  - Chronic obstructive pulmonary disease
  - Asthma: the British Thoracic Society guidelines recommend steroids for all acute asthma attacks in adults and children ≥2 years and should be considered in patients <2 years with severe asthma attacks
  - Organizing pneumonia
  - Pneumocystis jirovecii pneumonia
- Cardiovascular
  - Septic shock: the 2008 CORTICUS trial demonstrated increased resolution of shock, although this was not associated in a difference in 28-day mortality. The use of low-dose steroids (200-mg hydrocortisone/day) for patients who remain shocked despite appropriate fluids and vasopressors has been recommended by the Surviving Sepsis Campaign.
- Neurological
  - Central nervous system tumours – to decrease surrounding oedema
  - Meningitis: Corticosteroids have been used to decrease neurological sequelae and mortality when given before the first dose of antibiotics. They act by reducing the inflammatory response to cell lysis that is induced following treatment with antibiotics. In adults, the evidence for benefit is only for pneumococcal meningitis. However, in paediatrics, there is some evidence that steroids are beneficial in pneumococcal meningitis and meningitis secondary to *Haemophilus influenzae*. There is no evidence for use in patients with meningococcal meningitis unless there is co-existing evidence of septic shock.
- Gastroenterology
  - Alcoholic hepatitis – a steroid trial should be considered; however, it should be discontinued if the patient does not respond to treatment
  - Inflammatory bowel disease: Crohn disease, ulcerative colitis
- Haematology/oncology
  - Idiopathic thrombocytopenic
  - Lymphoma
  - Leukaemias
- Transplant patients – immune suppression to prevent rejection
- Anaphylaxis
- Endocrinological
  - Thyrotoxic crisis
  - Hypothyroid coma
  - Addison's disease
- Steroid cover for those patients who routinely receive >10 mg prednisolone/day or equivalent and are acutely unwell or undergoing surgery
- Autoimmune diseases such as vasculitides, rheumatoid arthritis, systemic lupus erythematosus, myasthenia gravis, multiple sclerosis and polymyalgia rheumatica

Steroids have previously been used to treat acute traumatic brain injuries; however, they are now no longer recommended. The use of steroids in acute respiratory distress syndrome is controversial but not currently recommended, although they may be useful in the fibrotic phase (>2 weeks after onset).

References

British Thoracic Society, Scottish Intercollegiate Guidelines Network. British guideline on the management of asthma. *Thorax*. 2014;69(Suppl 1):1–192.

Dellinger RP, Levy MM, Rhodes A, et al. Surviving Sepsis Campaign: international guidelines for management of severe sepsis and septic shock: 2012. *Crit Care Med*. 2013;41(3):580–637.

European Association for the Study of the Liver. EASL clinical practical guidelines: management of alcoholic liver disease. *J Hepatol*. 2012;57(2):399–420.

Hurlbert RJ, Hadley MN, Walters BC, et al. Pharmacological therapy for acute spinal cord injury. *Neurosurgery*. 2013;72(Suppl 2):93–105.

Roberts I, Yates D, Sandercock P, et al. Effect of intravenous corticosteroids on death within 14 days in 10 008 adults with clinically significant head injury (MRC CRASH trial): randomised placebo-controlled trial. *Lancet*. 2004;364(9442):1321–1328.

Sprung CL, Annane D, Keh D, et al. Hydrocortisone therapy for patients with septic shock. *N Engl J Med*. 2008;358(2):111–124.

Tunkel AR, Hartman BJ, Kaplan SL, et al. Practice guidelines for the management of bacterial meningitis. *Clin Infect Dis*. 2004;39(9):1267–1284.

H20. A 76-year-old man undergoes an emergency repair of his ruptured abdominal aortic aneurysm. On post-operative day 4, he remains sedated and ventilated and has a distended abdomen with gastric aspirates of approximately 200 to 300 ml every 4 hours. When measured, his abdominal pressures are 16 and 18 mmHg 4 hours apart.

Which of the following interventions should be implemented FIRST to reduce his intraabdominal pressure?

A. Administration of enemas per rectum
B. Continuous infusion of muscle relaxants
C. Reduction in rate of enteral nutrition
D. Administration of gastric prokinetics
E. Haemodialysis for fluid management

Answer: D

Short explanation

This patient has grade 2 intra-abdominal hypertension. All of the interventions are recommended as strategies to reduce intra-abdominal pressure. The administration of gastric prokinetics is first-line therapy, whereas the others are later in the World Society of the Abdominal Compartment Syndrome's suggested treatment algorithm.

Long explanation

The World Society of the Abdominal Compartment Syndrome (WSACS) defines intra-abdominal hypertension (IAH) and abdominal compartment syndrome (ACS) as pathological and sustained increases in intra-abdominal pressure (IAP).

It is classified as:

IAH	Grade 1	IAP 12–15 mmHg
	Grade 2	IAP 16–20 mmHg
	Grade 3	IAP 21–25 mmHg
	Grade 4	IAP >25 mmHg
ACS		IAP >20 mmHg with new organ dysfunction or failure

The WSACS has recommended that all critically ill patients who have any risk factors for the development of IAH/ACS should have their intra-abdominal pressure measured. Risk factors include conditions that result in decreased abdominal wall compliance, increased intra-luminal or intra-abdominal contents, or systemic pathology or physiological derangement (such as coagulopathy, hypothermia, acidosis, major trauma/burns or high volume fluid resuscitation).

The WSACS suggests a step wise approach to the treatment of IAH addressing a variety of factors involved in the aetiology of the condition. The approach to each factor has first-line, second-line and third-line interventions. Treatments should be used simultaneously to address these different aspects and should be tailored to the individual patient.

Treatments to improve abdominal wall compliance:

1. Adequate sedation and analgesia: ensure no external constriction e.g. dressings/eschars
2. Positioning
   o Avoid proning/head up >20°
   o Consider reverse Trendelenburg positioning
3. Neuromuscular blockade

Treatments to evacuate intra-luminal content:

1. Nasogastric/rectal decompression via aspiration/free drainage of nasogastric/rectal tubes; Administration of gastric/colonic prokinetics
2. Reduce enteral nutrition volume; enemas
3. Colonoscopic decompression: stop enteral feeding

Treatments to identify and evacuate intra-abdominal collections:

1. Ultrasound scan
2. Computed tomography scan
   Percutaneous drainage or paracentesis
3. Surgical evacuation

Treatments to optimize fluid balance:

1. Optimal, not excessive, fluid resuscitation; by day 3, aim to achieve a zero to negative fluid balance
2. Fluid resuscitation with hypertonic solutions and colloids; diuretics to drive negative fluid balance if haemodynamically stable
3. Renal replacement therapy

Treatments to optimize tissue perfusion to maintain an abdominal perfusion pressure (APP = Mean arterial pressure-IAP) ≥60 mmHg:

1. Use goal directed fluid resuscitation
2. Use haemodynamic monitoring to guide resuscitation
3. Using vasoactive drugs

If ACS is refractory to medical management the WSACS strongly suggests considering surgical management with abdominal decompression.

Reference

Kirkpatrick AW, Roberts DJ, De Waele J, et al. Intra-abdominal hypertension and the abdominal compartment syndrome: updated consensus definitions and clinical practice guidelines from the World Society of the Abdominal Compartment Syndrome. *Intensive Care Med.* 2013;39(7):1190–1206.

H21. You have been asked to prescribe enteral nutrition for a 26-year-old patient. The nurses ask you what normal daily nutritional requirements are.

Which of the following is the most accurate daily nutritional requirements?

A. Na^+ 2 mmol/kg, K^+ 1 mmol/kg, water 25 ml/kg, energy 20 kcal/kg, protein 1 g/kg

B. Na^+ 1 mmol/kg, K^+ 1 mmol/kg, water 25 ml/kg, energy 20 kcal /kg, protein 0.2 g/kg

C. Na^+ 2 mmol/kg, K^+ 2 mmol/kg, water 25 ml/kg, energy 20 kcal/kg, protein 0.2 g/kg

D. Na^+ 2 mmol/kg, K^+ 1 mmol/kg, water 25 ml/kg, energy 30 kcal/kg, protein 1 g/kg

E. Na^+ 2 mmol/kg, K^+ 1 mmol/kg, water 35 ml/kg, energy 30 kcal/kg, protein 0.2 g/kg

Answer: D

Short explanation

The NICE guidance (CG32) gives guidance on suggested nutritional prescriptions for total daily intake. This states it should include roughly 30 kcal/kg/day, 30 ml water/kg/day, and approximately 1 g/kg protein (roughly 0.2 g/kg nitrogen). In addition, there should be adequate electrolyte replacement. None of the answers match this guidance, but Answer D is the closest.

Long explanation

Nutritional requirements depend on a large number of factors, including pre-existing deficits, risk of refeeding syndrome, activity levels and underlying medical condition, such as pyrexia or hypercatabolic states. The route of nutrition will also be dependant on factors such as gastrointestinal tolerance, and anticipated longevity of nutritional support.

For patients who are not at risk of refeeding syndrome or who are not critically unwell, the nutritional replacement should include the preceding constituents. Daily electrolyte requirements are roughly 1 to 2 mmol/kg for Na^+ and 0.7–1 mmol/kg for K^+.

Those patients who are critically ill should have nutritional support commenced at less than 50% of their anticipated energy needs and slowly built up over 2 days. If patients are at risk of refeeding syndrome, however, it should be started at a much lower rate (roughly 20–30% of calculated needs) and slowly increased over several days with close monitoring of electrolytes, as well as replacement of vitamins. These supplements, which include thiamine, vitamin B complex, multivitamins and trace elements, should be given during the first 10 days of feeding.

Reference

National Institute for Health and Clinical Excellence (NICE). *Nutrition support in adults: Oral nutrition support, enteral tube feeding and parenteral nutrition* (CG 32). London: NICE, 2006.

H22. A 48-year-old alcoholic presents with acute severe pancreatitis. He has evidence of a systemic inflammatory response syndrome, has evidence of pancreatic necrosis on computed tomography (CT) on day 3 and is improving with conservative treatment.

Which of the following is the MOST appropriate antimicrobial regime for this patient?

A. Ciprofloxacin
B. Imipenem
C. Imipenem and fluconazole
D. Metronidazole
E. None required

Answer: E

Short explanation

This patient is improving with treatment so is unlikely to have untreated infected pancreatic necrosis. Both bacteria and fungi can be responsible for the development of infected pancreatic necrosis; however, prophylactic antibiotics or antifungals are not recommended in patients with acute severe pancreatitis or pancreatic necrosis.

Quinolones, carbapenems and metronidazole are all known to penetrate pancreatic necrosis so can be used in the treatment of infected necrosis.

Long explanation

The presence of infected pancreatic necrosis as well as extrapancreatic infections increase morbidity and mortality in patients with pancreatitis. However, it can often be difficult to differentiate clinically which symptoms and signs are caused by infection versus the systemic inflammatory response syndrome (SIRS) that commonly occurs due to the underlying pancreatitis. Markers such as procalcitonin can be used to help differentiate the two.

The routine use of prophylactic antibiotics for patients with pancreatitis is not recommended by the current evidence. The Cochrane review in 2010 did not support the use of prophylactic antibiotics. They found no significant reduction in mortality with the use of prophylactic antibiotics. Imipenem specifically was associated with a significant reduction in the rate of pancreatic infection, but this did not correlate with a significant survival advantage.

The 2013 American College of Gastroenterology guidelines on the management of acute pancreatitis do not recommend the use of prophylactic antibiotics or antifungals for patients with acute severe pancreatitis or in patients with sterile necrosis. They recommend considering the diagnosis of infected pancreatic necrosis in patients who fail to improve with treatment or deteriorate after 7 to 10 days of hospital treatment. Whilst they recommend starting antibiotics when infection (pancreatic or extrapancreatic) is suspected, they stress the importance of obtaining cultures including fine needle aspiration (FNA) of necrotic pancreatic tissue to confirm the presence of infection and guide treatment. If the cultures prove to be sterile and there is no other evidence of a source of infection, the antibiotics should be discontinued. Whilst antibiotic regimes should go in line with trust protocols, appropriate antibiotics include quinolones, carbapenems, high-dose cephalosporins, metronidazole and beta-lactams, although there is the most evidence for carbapenems.

Fungi can be responsible for infection of pancreatic necrosis so treatment with antifungals should be considered in patients with confirmed infection. However, prophylactic antifungals are not recommended.

References

Mofidi R, Suttie SA, Patil PV, et al. The value of procalcitonin at predicting the severity of acute pancreatitis and development of infected pancreatic necrosis: systematic review. *Surgery*. 2009;146:72–81.

Tenner S, Baillie J, DeWitt J, Swaroop S. American College of Gastroenterology Guideline: management of acute pancreatitis. *Am J Gastroenterol*. 2013;108(9):1400–1415.

Villatoro E Mulla M, Larvin M. Antibiotic therapy for prophylaxis against infection of pancreatic necrosis in acute pancreatitis. *Cochrane Database Syst Rev*. 2010;(5):CD002941.

H23. You are performing a rapid sequence intubation (RSI) on a patient through an 18-gauge cannula sited in the right antecubital fossa. As you administer the thiopentone, the patient screams in pain, and his right hand goes white. You suspect an intra-arterial (IA) injection.

Which of the following is the most appropriate immediate action:

A. Remove the cannula, site a new one in the other arm, and continue the RSI
B. Urgently perform a stellate ganglion block
C. Request an urgent vascular review
D. Flush the cannula with saline
E. Inject papaverine through the cannula

Answer: D

Short explanation

Immediate management of an inadvertent IA injection involves leaving the cannula in, dilution with saline, elevating the limb to improve drainage and analgesia. Papaverine, procaine, calcium channel blockers, iloprost and stellate ganglion blocks may all feature in the multimodal management but would not be the immediate priority. In this scenario, continuing the RSI may be necessary, depending on the clinical situation.

Long explanation

Inadvertent intra-arterial injection of drugs can have profound consequences, based on the properties of the drug. Thiopentone has classically been associated with damage when given via the intra-arterial route. This occurs by precipitation of crystals that form when the drug, which is alkaline in the syringe, reaches arterial pH. This appears not to occur with normal venous injection because dilution into the central circulation occurs, compared with injection into a narrowing vasculature when given via the IA route. Recent formulations of thiopentone have been specifically made to minimize this problem, but it still exists.

Some anatomical sites are particularly high risk for IA injection, with the antecubital fossa, forearm and hand being common sites. The brachial artery lies close to the veins in the antecubital fossa, whereas inter-patient variation means radial artery branches can lie along typical venous paths in the forearm and hand.

The first symptom of IA injection of drugs is severe pain, and this will be lost in sedated or unconscious patients, potentially removing a key early marker of IA injection. Without this warning, the next sign may be that the injected drugs fail to have any effect, followed by limb signs such as pallor and cyanosis, which can progress to marked oedema and necrosis.

Treatment of IA injection needs to be immediate. The general principle is to optimize distal perfusion. Saline should be injected through the cannula to dilute the drug and encourage more distal spread, and analgesia should be given urgently to minimize the sympathetic response, which will worsen vasoconstriction. Keeping the limb warm will encourage dilation of vessels, while elevation will improve venous drainage and oedema, which can otherwise exacerbate the problem. Anticoagulation with heparin can reduce the extent of ischaemic damage, while sympathetic blocks (e.g. stellate ganglion) will cause vascular dilation, but these are not risk-free interventions.

Other treatment options that may be useful include calcium channel antagonists, thromboxane inhibitors and prostacyclin analogues.

Reference
Lake C, Beecroft C. Extravasation injuries and accidental intra-arterial injection. Contin Educ Anaesth. *Crit Care Pain*. 2010;10(4):109–113.

H24. You are about to perform a lumbar puncture in a patient on the intensive care unit, but the nurses tell you he received an anticoagulant recently

Which of the following patients is at highest risk of developing an epidural haematoma?

A. A 40-year-old man on an intravenous heparin infusion, stopped 4 hours ago
B. A 40-year-old man on treatment-dose subcutaneous low molecular weight heparin (LWMH), last dose 24 hours ago
C. A 40-year-old man on aspirin, taken 6 hours ago
D. A 40-year-old man on clopidogrel, stopped 4 days ago
E. A 40-year-old man on prophylactic-dose subcutaneous LMWH, last dose 12 hours ago

Answer: D

Short explanation
Clopidogrel should ideally be stopped for 7 days before performing a lumbar puncture. Aspirin can continue to be taken, without significantly increasing the risk of complications. The heparin (and LMHW) times listed are all the times after which a lumbar puncture can be safely performed.

Long explanation
The majority of the work around central neuraxial access and drugs that affect coagulation has focussed on regional anaesthesia. However, the process of lumbar puncture is procedurally similar to spinal anaesthesia and may be extrapolated. The Association of Anaesthetists of Great Britain and Ireland (AAGBI), has published guidelines for regional anaesthesia in patients with abnormalities of coagulation, either due to the administration of drugs or that of pathological process. Serious complications of central neuraxial access are rare, and the Royal College of Anaesthetists' third National Audit Project (NAP3) found the incidence of vertebral canal haematoma was less than 1 per 100,000 after neuroaxial blockade.

The degree to which this risk is increased when abnormalities of coagulation are present cannot be quantified. However, intensive care practitioners, who are likely to be performing epidural analgesia and lumbar punctures, must be aware of the risks of such procedures. Risk is not binary, for example, using a platelet count cutoff would suggest that a platelet count one above and one below are safe and unsafe, respectively. Rather, it is a spectrum of risk.

The exception to this is with drugs that effect coagulation. The pharmacokinetics of these can be used to calculate when it will be acceptable to perform the block after a dose has been given. The timings for the some of the common agents are as follows:

Unfractionated heparin, for both prophylaxis and treatment, should be stopped for 4 hours, or until the activated prothrombin time is normal. LMWH should be stopped for 12 hours and 24 hours for prophylactic and treatment doses, respectively. Clopidogrel and prasugrel should both be stopped for 7 days, and patients on warfarin should have an international normalized ratio <1.5. If fondaparinux is being administered, then anti-Xa levels and discussion with a haematologist may be needed. Patients taking other novel oral anticoagulants (e.g. rivaroxaban) should be discussed with a haematologist. NSAIDs, aspirin and dipyridamole do not need any additional precautions.

Reference

Association of Anaesthetists of Great Britain and Ireland, Obstetric Anaesthetists' Association and Regional Anaesthesia UK. Regional anaesthesia and patients with abnormalities of coagulation. *Anaesthesia*. 2013;68:966–972.

H25. You have admitted a patient with an acute exacerbation of chronic obstructive pulmonary disorder (COPD) for a trial of noninvasive ventilation (NIV).

Which of the following approaches, with reference to inspiratory positive airway pressure (IPAP) and expiratory positive airway pressure (EPAP), is the best way to commence this?

A. Commence an IPAP of 10 cmH_2O, and EPAP 5 cmH_2O, and titrate up the IPAP by 5 cmH2O every 10 minutes until a response is achieved.
B. Commence an IPAP of 10 cmH_2O, EPAP 5 cmH_2O, and titrate up the IPAP by 5 cmH_2O, and the EPAP by 2 cmH_2O every 10 minutes until a response is achieved.
C. Commence an IPAP of 14 cmH_2O, EPAP 4 cmH_2O, and titrate up both by 2 cmH_2O every 5 minutes until the IPAP is 20 cmH_2O and the EPAP is 10 cmH_2O.
D. Commence an IPAP of 14 cmH_2O, EPAP 4 cmH_2O, and titrate up both by 2 cmH_2O every 5 minutes until a response is achieved.
E. Commence an IPAP of 14 cmH_2O, EPAP 4 cmH_2O, and titrate up both by 2 cmH_2O every 10 minutes until a response is achieved, or the IPAP is 20 cmH_2O and the EPAP is 10 cmH_2O.

Answer: A

Short explanation

The British Thoracic Society (BTS) guidelines recommend an initial IPAP of 10 cmH_2O and EPAP of 4 to 5 cmH_2O. The IPAP can be gradually increased by 2- to 5-cm increments at a rate of approximately 5 cmH_2O each 10 minutes until a therapeutic response is achieved or patient tolerability has been reached.

Long explanation

NIV is a valuable tool for the management of type 2 respiratory failure in COPD. There is robust evidence that mortality is reduced by up to half in this patient group, with a number needed to treat of 10.

Patients with respiratory acidosis (pH <7.35, $PaCO_2$ >6 kPa) after 1 hour of medical treatment should be offered NIV. The best position to commence NIV is sitting or semi-recumbent. The patient may already have positioned themselves in the position that allows them the best ventilation, and NIV should be commenced without changing this position. Although NIV can be delivered via a number of different types of mask, a full-face mask is the best tolerated and should be chosen first.

Starting with relatively low pressures, such as an IPAP of 10 cmH$_2$O and an EPAP of 4 to 5 cmH$_2$O will generally be well tolerated by the patient and will improve compliance with the intervention. Increasing the pressures too quickly will undermine this. IPAP should therefore be increased by small amounts, to total roughly 5 cmH$_2$O every 10 minutes, stopping if the patient tolerability necessitates, or if hypercarbia is improving. Most patients will require an IPAP of 20 cmH$_2$O, while EPAP should not usually be increased above 4 to 5 cmH$_2$O. Oxygen levels should be tailored to maintain target saturations, usually 88 to 92%.

References

Ram F, Picot J, Lightowler J, Wedzicha JA. Non-invasive positive pressure ventilation for treatment of respiratory failure due to exacerbations of chronic obstructive pulmonary disease. *Cochrane Database Syst Rev*. 2004(1):CD004104.

Royal College of Physicians, British Thoracic Society. *Intensive Care Society Chronic obstructive pulmonary disease: non-invasive ventilation with bi-phasic positive airways pressure in the management of patients with acute type 2 respiratory failure. Concise Guidance to Good Practice* series, No 11. London: Royal College of Physicians, 2008.

H26. You have just inserted a percutaneous tracheostomy into a 50-year-old patient to aid ventilatory weaning.

Which of the following statements has the strongest recommendation from the Intensive Care Society?

A. In mechanically ventilated patients, the inner tube should be changed every 8 hours.
B. Cuff pressure should not exceed 20 cmH$_2$O.
C. Humidification is essential for all patients.
D. Before decannulation, a fibre-optic inspection of the upper airway should be undertaken.
E. Fenestrated tracheostomy tubes should be the first choice tube type in ventilated patients.

Answer: C

Short answer

Inner tube changes should be balanced against the risk of de-recruitment in ventilated patients. Cuff pressure should not exceed 25 cmH$_2$O. Humidification has strong evidence supporting its use for all patients. Clinical assessment of the upper airway before decannulation should usually suffice, and fenestrated tubes are not recommended in mechanically ventilated patients.

Long answer

The Intensive Care Society has published standards for the care of adult patients with a temporary tracheostomy. This covers all aspects of tracheostomy care, from insertion through to weaning, and was produced to coincide with publication of the 2014 NCEPOD study 'On the Right Trach?'

Use of an inner cannula has the major advantage that if the tracheostomy tube were to become blocked with thick secretions, the obstruction can be immediately relieved by removal of the inner cannula, without having to remove the entire tracheostomy tube. To avoid any obstruction to the inner tube, it should be changed at least every 8 hours in nonventilated patients. In ventilated patients, however, the risk of obstruction needs to be weighed against the risk of de-recruitment, from repeated ventilator disconnections.

Cuff pressure in the tracheostomy tube should be just high enough to prevent air leak and to safely protect the airway. The higher the pressure in the cuff, the more likely it is that there will be ischaemic damage to the mucosa of the trachea. This usually has a capillary pressure of roughly 20 cmH$_2$O, although it will be lower in shocked states. The pressure in the cuff should be checked at least every 8 hours and kept below 20–25 cmH$_2$O.

Adequate humidification of respiratory gases is essential to avoid potentially life-threatening blockage of the tracheostomy. The level of humidification required by each patient will vary and can range from heat and moisture exchange devices (e.g. Swedish nose) to heated water baths.

Other aspects of the general care of the tracheostomy tube that are recommended include changing the entire tube at least every 30 days, always deflating the cuff when a speaking valve is attached and dealing immediately with any failure to pass a suction catheter down the tracheostomy. In addition, all staff caring for patients with a tracheostomy must be able to recognize and manage emergency situations, such as blockage or displacement of the tube.

Reference
Bodenham A, Bell A, Bonner S, et al, on behalf of the Council of the Intensive Care Society. *Standards for the care of adult patients with a temporary Tracheostomy; Standards and Guidelines*. Intensive Care Society Standards. 2014.

H27. You are a member of the trauma team in a major trauma centre and are called to the emergency department (ED). A 30-year-old man has just been brought in after his lower limbs were run over by a truck. He has suffered a traumatic amputation of his left leg and an obvious open fracture to his right leg. He has no obvious head, chest, abdominal or pelvic injuries. He has a heart rate of 145, blood pressure of 90/60, a respiratory rate of 40, and is drowsy. The pre-hospital team have placed a tourniquet on his left thigh, and his right leg has been splinted and covered. Two large-bore cannulae have been sited, and he has received a total of 1000,ml of 0.9% saline.

Which of the following initial fluid management strategies would be most appropriate?

A. Warmed Hartmann's solution
B. 0.9% saline solution
C. O negative packed red cells, fresh frozen plasma (FFP) and platelets
D. Human albumin solution 4.5%
E. Cross-matched packed red cells, FFP and platelets

Answer: C

Short explanation
This patient has class IV shock, and the immediate priority is to stop ongoing haemorrhage. Initial fluid resuscitation should be low volume with 'permissive hypotension' until early bleeding control is achieved. Large volumes of crystalloid can lead to coagulopathy and worsen bleeding. Transfusion of packed red cells should be accompanied by FFP and platelets. Cross-matching of blood typically takes 40 to 60 minutes, which is an unacceptable delay. Transfusion of either crystalloid or synthetic colloid in this situation may cause harm.

Long explanation
Military experience of major trauma over the past 15 years and the development of the major trauma centre network in the United Kingdom have led to changes in the way major trauma is treated. It is no longer reasonable to infuse large volumes of

crystalloids into patients with shock from traumatic haemorrhage. Evidence for use of synthetic colloids is equivocal at best. The focus is now on low volume fluid resuscitation with permissive hypotension (in the absence of traumatic brain or spinal injury) and early use of blood and blood products with early bleeding control.

The Task Force for Advanced Bleeding Care in Trauma has published evidence-based guidelines on the subject for the last decade. The most recently published, from 2013, contain many recommendations, including the following:

- Minimizing the time between injury and surgery for bleeding control
- The use of tourniquets for severe bleeding from limb wounds
- The use of the ATLS classification of shock using physiological parameters (heart rate, systolic blood pressure [SBP], pulse pressure, respiratory rate, urine output and conscious level)
- Immediate bleeding control procedures if the source of haemorrhage is obvious (e.g. surgery, pelvic binding, reduction of displaced long-bone fractures) in unstable patients.
- Early use of imaging for blunt abdominal trauma. Ultrasound (focused assessment with sonography in trauma scan) is commonplace, but some trauma centres may have computed tomography (CT) facilities within the emergency department that reduce the transfer time and risk traditionally associated with CT scanning in trauma.
- Early and regular monitoring of coagulation parameters, including the use of point-of-care coagulation testing.

With regard to fluid resuscitation in traumatic haemorrhagic shock, an initial target for SBP should be 80 to 90 mmHg until bleeding is controlled. In patients who have haemorrhagic shock and a traumatic brain or spinal injury, then a higher target is necessary for perfusion (mean arterial pressure = 80 mmHg).

Traditional trauma recommendations were for early, aggressive fluid resuscitation with large volumes of crystalloid, followed by blood products, with an aim to restore normal blood pressure. This is now widely believed to increase the pressure on any initial clot that has formed and also to contribute to trauma coagulopathy. Larger volumes of crystalloid and/or colloid have been found to be associated with a greater likelihood of developing coagulopathy and increased mortality in large retrospective studies.

Where fluid is indicated, early use of blood products is recommended. Red blood cells should be administered early, along with plasma, platelets and clotting factor concentrates. Increasing the amount of plasma transfusion is another feature of modern trauma care. Plasma/red blood cell ratios of 1:2 or 1:1 have been recommended, although the optimal ratio has yet to be elucidated with large scale trials.

References

Egea-Guerrero JJ, Freire-aragón MD, Serrano-lázaro A, Quintana-díaz M. Update in intensive care: trauma and critical care update. Resuscitative goals and new strategies in severe trauma patient resuscitation. *Med Intensiva.* 2014;38(8):502–512.

Spahn DR, Bouillon B, Cerny V, et al. Management of bleeding and coagulopathy following major trauma: an updated European guideline. *Crit Care.* 2013;17(2): R76.

H28. A 64-year-old man with a history of angina and depression is brought in by ambulance after taking an overdose of atenolol. He is spontaneously ventilating, with SpO$_2$ 99% (FiO$_2$ 0.6) and Glasgow Coma Score of 15. Monitoring shows heart rate 30 bpm, blood pressure 85/37 mmHg, and a 12-lead electrocardiogram (ECG) shows complete heart block with a wide QRS complex. He has intravenous (IV) access in situ and has been given 500 µg of atropine with no response.

What is the most appropriate next step?

A. Commence transcutaneous pacing
B. Administer 600 µg glycopyrrolate
C. Arrange transvenous pacing
D. Administer glucagon IV
E. Administer adrenaline 2–10 µg/min IV

Answer: D

Short explanation

Although all the answers feature in the Resuscitation Council (UK) algorithm for managing an adult bradycardia, this patient has taken an overdose of beta-blocker, and therefore the specific antidote (glucagon) should be given. Transcutaneous pacing is painful in an alert patient and other pharmacological agents are unlikely to be effective. Transvenous pacing may ultimately be necessary but not before treating the overdose.

Long explanation

The Resuscitation Council (UK) publishes algorithms for the management of cardiac arrest and peri-arrest scenarios, including a section on arrhythmias. Although bradycardia (heart rate <60 bpm) can sometimes be physiological, in the presence of adverse signs it requires emergency treatment.

Initial management of a bradycardia includes a structured ABC approach, administration of oxygen, attaining good-calibre IV access, full monitoring and a 12-lead ECG. In the presence of adverse features (shock, syncope, heart failure or myocardial ischaemia), 500 µg of atropine IV should be given. If the patient remains bradycardic and symptomatic, then further doses of atropine can be given (up to 3 mg), transcutaneous pacing can be initiated, or second-line drugs can be used (e.g. isoprenaline 5 µg/min IV, adrenaline 2–10 µg/min IV).

The recommendation is that transcutaneous pacing should be initiated if atropine has been unsuccessful, but the exception to this is in the presence of specific drugs that should be treated with antidotes. Examples of these include treatment of beta-blocker or calcium channel overdose with glucagon, and treatment of digoxin toxicity with digoxin-specific antibody fragments. If these are unsuccessful, then transcutaneous pacing will be required, which may require sedation and analgesia.

If there are no adverse features present or there has been a sufficient response to the initial dose of atropine, then there may still be a risk of asystole. This is particularly likely if there are ventricular pauses of more than 3 seconds, any recent asystole, complete heart block or second-degree heart block (Mobitz type). If these are present then full monitoring should continue and transvenous pacing considered.

Reference

Resuscitation Council (UK). Chapter 8: Peri-arrest arrhythmias. *Advanced Life Support*. Sixth Edition. 2011. https://www.resus.org.uk/pages/periarst.pdf (accessed May 2015).

H29. You are asked to review a 27-year-old man in the emergency department with burns.

Which of the following features is LEAST likely to warrant referral to the regional burns unit?

A. A circumferential burn to the left forearm that is non-blanching and 6% of total body surface area

B. A burn to the back that is 7% total body surface area (BSA) and is non-blanching.

C. A full-thickness burn to the soles of both feet

D. A chemical burn to the right thigh

E. Partial-thickness burns of 12% total body surface area.

Answer: D

Short explanation

The National Burn Care Referral Guidance (2012) clarifies which burns should be referred. They state that all non-blanching burns, significant burns to special areas (e.g. hands, feet) and burns covering 10% of more of the total BSA should be referred to a specialist burn service. Chemical or electrical burns require discussion not necessarily referral.

Long explanation

Specialized Burn Services are stratified into three levels.

Burn centres provide the highest level of inpatient care available, caring for the most complex injuries in a dedicated ward with immediate theatre availability and specialist critical care services. Burn units, an intermediate level of inpatient care, provide a dedicated ward that can cope with all but the most complex injuries. Burn facilities are the basic level of inpatient care, for noncomplex injuries. This covers burn care without a dedicated burn ward and may be a standard plastic surgical ward.

The details of what type of injury should be referred where is as follows:

- Burn Centres – burns with a affected BSA of more than 25% if associated with an inhalation injury, or more than 40% if not, should be referred to a burns centre. Any case with 25% or more BSA affected should be discussed with this centre.
- Burn Unit – any burn with a BSA of 10% or more should be referred to a burn unit, up to the preceding thresholds. They should also receive any burn from 5% BSA if it is non-blanching, any burns that are circumferential, or burns to areas such face, feet, hands, perineum or genitalia which are felt to be significant.
- Burn Facilities – any burn of more than 2% and less than 10% BSA, or any full thickness burn, should be referred to these facilities, even in the presence of inhalation injury, unless any of the criteria above are met. Discussion should be had for any chemical or electrical burn or for any nonsignificant burns to the areas listed here.

Reference

National network for burn care (NNBC). National Burn Care Referral Guidance; version 1. Approved February 2012. http://www.webarchive.org.uk/wayback/archive/20130325152202/http://www.specialisedservices.nhs.uk/library/35/National_Burn_Care_Referral_Guidance.pdf (accessed May 2015).

H30. A 45-year-old 63-kg female is ventilated in the intensive care unit for community-acquired pneumonia and requires noradrenaline to maintain a mean arterial pressure ≥65 mmHg. On day 3, her urine output declines, with 150 ml urine in 6 hours. Her creatinine has risen to 148 µmol/l from a baseline value of 84 µmol/l on admission.

Which of the following is the LEAST important measure to implement in this patient's management?

A. Check and adjust drug doses
B. Active management of blood glucose to avoid hyperglycaemia
C. Renal tract ultrasound
D. Avoid all nephrotoxic drugs and contrast media where possible
E. Optimize patient haemodynamics using cardiac output monitoring

Answer: A

Short explanation

This patient has stage 1 acute kidney injury (AKI) according to the Kidney Disease Improving Global Outcomes (KDIGO) guidelines. They recommend all of the above in the management of patients with acute kidney injury. However, they recommend once stage 2 AKI is reached drug doses should be reviewed and adjusted as required.

Long explanation

The KDIGO guideline has defined the severity of acute kidney injury (AKI) according to three stages:

Stage 1: creatinine 1.5 to 1.9 times baseline OR ≥26.5 µmol/l increase
Urine output <0.5 ml/kg/hr for 6 to 12 hours
Stage 2: creatinine 2.0 to 2.9 times baseline
Urine output <0.5 ml/kg/hr for ≥12 hours
Stage 3: creatinine ≥3.0 times baseline OR ≥353.6 µmol/l
Commencement of renal replacement therapy
Decrease in estimated glomerular filtration rate <35 ml/min/1.73 m^2 (in patients <18 years)
Urine output<0.3 ml/kg/hr for ≥24 hours OR anuria for ≥12 hours

The KDIGO guidelines also recommend management strategies and interventions for patients with AKI according to its severity.

- The following should be initiated for all patients identified at being at high risk of developing AKI:
  - Avoidance of nephrotoxic agents where possible. If imaging with contrast is planned consider whether other imaging modalities may be suitable alternatives or reduce the dose of contrast.
  - Ensure adequate volume status (fluid resuscitation with isotonic crystalloids rather than colloids) and maintenance of appropriate perfusion pressure (with vasopressors as required).
  - The use of functional haemodynamic monitoring should be considered
- Maintain normoglycaemia particularly avoiding hyperglycaemia.
- Once stage 1 disease is reached, a noninvasive diagnostic workup should be instigated. Invasive tests should also be considered to identify the cause of the acute kidney injury.
- Once stage 2 disease is present additional recommendations include the following:
  - All drugs should be reviewed and appropriate dose adjustments made.

- o The need for renal replacement therapy and ICU admission for critical care should be considered.
- Once stage 3 disease is present, the guidelines recommend the avoidance of subclavian lines if possible. This is due to the increased risk of stenosis developing with these lines which may affect the provision of subsequent permanent access if required.

Further recommendations include the following:

- Avoid protein restriction to delay the initiation of renal replacement therapy.
- Diuretics should only be used to manage fluid overload.
- Enteral nutrition is preferred to parenteral nutrition.
- The use of fenoldopam, low-dose dopamine, atrial natriuretic peptide should not be used to prevent or treat AKI.

Reference

Kidney Disease: Improving Global Outcomes. KDIGO 2012 clinical practice guideline for the evaluation and management of chronic kidney disease. *Kidney Int.* 2013;3(Suppl 1):1–150.

Index

Printed in the United States
by Baker & Taylor Publisher Services